Lecture Notes in Computer Science 15729

The series Lecture Notes in Computer Science (LNCS), including its subseries Lecture Notes in Artificial Intelligence (LNAI) and Lecture Notes in Bioinformatics (LNBI), has established itself as a medium for the publication of new developments in computer science and information technology research, teaching, and education.

LNCS enjoys close cooperation with the computer science R & D community, the series counts many renowned academics among its volume editors and paper authors, and collaborates with prestigious societies. Its mission is to serve this international community by providing an invaluable service, mainly focused on the publication of conference and workshop proceedings and postproceedings. LNCS commenced publication in 1973.

Guilhem Gamard · Julien Leroy
Editors

Combinatorics on Words

15th International Conference, WORDS 2025
Nancy, France, June 30 – July 4, 2025
Proceedings

Springer

Editors
Guilhem Gamard
Université de Lorraine
Nancy, France

Julien Leroy
Université de Liège
Liège, Belgium

ISSN 0302-9743 ISSN 1611-3349 (electronic)
Lecture Notes in Computer Science
ISBN 978-3-031-97547-9 ISBN 978-3-031-97548-6 (eBook)
https://doi.org/10.1007/978-3-031-97548-6

This Springer imprint is published by the registered company Springer Nature Switzerland AG
The registered company address is: Gewerbestrasse 11, 6330 Cham, Switzerland

Preface

This volume of Lecture Notes in Computer Science contains the proceedings of the 15th International Conference WORDS, which was organized by the LORIA laboratory at University of Lorraine and took place from June 30 to July 4, 2025 in Nancy, France.

WORDS is the main conference series devoted to the mathematical theory of words. In particular, the combinatorial, algebraic, and algorithmic aspects of words are emphasized. Motivations may also come from other domains such as theoretical computer science, bioinformatics, digital geometry, symbolic dynamics, numeration systems, text processing, number theory, etc.

The conference WORDS takes place every two years. The first conference of the series was held in Rouen, France in 1997. Since then, the locations of WORDS conferences have been: Rouen, France (1999 and 2021), Palermo, Italy (2001), Turku, Finland (2003 and 2013), Montréal, Canada (2005 and 2017), Marseille, France (2007), Salerno, Italy (2009), Prague, Czech Republic (2011), Kiel, Germany (2015), Loughborough, UK (2019), and Umeå, Sweden (2023).

For the seventh time in the history of WORDS, a refereed proceedings volume was published in the *Lecture Notes in Computer Science* series of Springer. There were 31 submissions from 15 countries, and each of them was single-blind reviewed by two or three referees. (Only 4 submissions had two referees.) The selection process was undertaken by the Program Committee with the help of generous reviewers. From these submissions, 20 papers were selected to be published and presented at WORDS.

In addition to the contributed presentations, WORDS featured five invited talks by the following speakers:

- Daniel Gabrić (University of Guelph, Canada)
  Combinatorial problems on closed and privileged words
- France Gheeraert (Université de Picardie Jules Vernes, France)
  The link between return words and extensions of factors
- Idrissa Kaboré (Université Polytechnique de Bobo-Dioulasso, Burkina Faso)
  On the abelian complexity of infinite words
- Martin Lustig (Aix-Marseille Université, France)
  S-adic expansions, invariant measures and substitutional subshifts
- Markus Whiteland (Loughborough University, UK)
  Glimpses into Recent Developments in Subword Combinatorics and Binomial Coefficients

We would like to take this opportunity to thank all the invited speakers and all the authors for their contributions. We are also grateful to all Program Committee members and the additional reviewers—all listed below—for their hard work that led to the selection of papers published in this volume. Additionally, thanks to Mathilde Bouvel, Damien Jamet, Thomas Stoll, and Benjamin Testart for their comments during the preparation of these proceedings.

This edition of WORDS was financially supported by the following entities, to which we express our gratitude: the IECL and LORIA laboratories and the IAEM doctoral school, all from the University of Lorraine—which contributed directly as well; the Inria institute, the ANR project *ArithRand*, the ERC Synergy grant *DynAMiCs*, the *Métropole du Grand Nancy*, and Google.

Finally we would like to warmly thank the local organizers of this event: Marie Baron, Julien Bernat, Mathilde Bouvel, Damien Jamet, Thomas Stoll, Pierre-Adrien Tahay, and Benjamin Testart, who ensured that this conference was a success! Our gratitude also goes to Nathalie Bethus, Nathalie Fritz, and Paola Schneider for their remarkable work in administrative support.

May 2025

Guilhem Gamard
Julien Leroy

Organization

Program Committee Chairs

Guilhem Gamard	University of Lorraine, France
Julien Leroy	University of Liège, Belgium

Steering Committee

Golnaz Badkobeh	City St. George's, University of London, UK
Valérie Berthé	University Paris Cité, France
Julien Cassaigne	Aix-Marseille University, France
Émilie Charlier	University of Liège, Belgium
Ľubomíra Dvořáková	Czech Technical University, Czech Republic
Gabriele Fici	University of Palermo, Italy
Anna Frid	Aix-Marseille University, France
Robert Mercas	Loughborough University, UK
Jean Néraud	University of Rouen-Normandie, France
Jarkko Peltomäki	University of Turku, Finland
Dominique Perrin	Gustave Eiffel University, France
Narad Rampersad	University of Winnipeg, Canada
Daniel Reidenbach	Keele University, UK
Christophe Reutenauer	Université du Québec à Montréal, Canada
Jeffrey O. Shallit	University of Waterloo, Canada
Mikhail Volkov	Ural Federal University, Russia

Program Committee

Mélodie Andrieu	University of the Littoral Opal Coast, France
Nicolas Bédaride	Aix-Marseille University, France
Ľubomíra Dvořáková	Czech Technical University, Czech Republic
Gabriele Fici	University of Palermo, Italy
Jarosław Grytczuk	Warsaw University of Technology, Poland
Svetlana Puzynina	St Petersburg State University, Russia
Narad Rampersad	University of Winnipeg, Canada

Aleksi Saarela	University of Turku, Finland
Arseny Shur	Bar-Ilan University, Israel
Lama Tarsissi	Sorbonne University Abu Dhabi, United Arab Emirates

Additional Reviewers

Laurent Bulteau
Émilie Charlier
James Currie
Alessandro De Luca
Francesco Dolce
Daniel Gabrić
Léo Gayral
France Gheeraert
Herman Goulet-Ouellet
Štěpán Holub
Damien Jamet
Karel Klouda Kosolobov
Savinien Kreczman
Mélodie Lapointe
Anuran Maity
Victor Marsault
Zuzana Masáková
Edita Pelantová
Jarkko Peltomäki
Pierre Popoli
Christophe Reutenauer
Michel Rigo
Eric Rowland
Josef Rukavicka
Andrew Ryzhikhov
Ville Salo
Paul Sarnighausen-Cahn
Lukas Spiegelhofer
Pierre Stas
Wolfgang Steiner
Manon Stipulanti
Pierre-Adrien Tahay
Léo Vivion
Laurent Vuillon
Wiktor Zuba

Abstract of Invited Talks

Combinatorial Problems on Closed and Privileged Words

Daniel Gabrić

School of Computer Science, University of Guelph, Guelph, Ontario, Canada
dgabric@uoguelph.ca

Abstract. A word w has a *border* u if u is a non-empty proper prefix and suffix of w. A word w is said to be *closed* if w is of length at most 1 or if w has a border that occurs exactly twice in w. A word w is said to be *privileged* if w is of length at most 1 or if w has a privileged border that occurs exactly twice in w. In this talk, we discuss recent combinatorial results on closed and privileged words, including their enumeration and generation. We also highlight key open problems and directions for future research.

The Link Between Return Words and Extensions of Factors

France Gheeraert

Laboratoire Amiénois de Mathématique Fondamentale et Appliquée, Université de Picardie Jules Verne, Amiens, France
france.gheeraert@u-picardie.fr

Abstract. Extensions of factors, defined as the letters that can surround factors, were introduced and studied for their connection with factor complexity. On the other hand, return words, which are the words separating two occurrences of the same factor, originate in the symbolic dynamics approach of words and are mostly known for their link with S-adic representations. Throughout the years, many results showed that information on the number or structure of one of the two could be used to study the other. In this talk, I will present a survey of some of these results.

On the Abelian Complexity of Infinite Words

Idrissa Kaboré

Université Polytechnique de Bobo-Dioulasso, Burkina Faso
ikaborei@yahoo.fr

Abstract. The abelian complexity is a combinatorial tool intervening in the study of infinite words. It was formally introduced in 2009 by Richomme et al. Since then it has been intensively studied. In this talk we give a survey on the developments in the subject with a focus on general results (joint work with Cassaigne in 2016) concerning the links between abelian complexity and other notions on infinite words. We conclude with a few questions.

S-adic Expansions, Invariant Measures and Substitutional Subshifts

Martin Lustig

Institut de Mathmatiques de Marseille, Aix-Marseille Université, Marseille, France
MartinLustig@gmx.de

Abstract. In the first part of my talk I will present joint work with N. Bédaride and A. Hilion on (generalized) S-adic developments of a subshift X, and a related very convenient method ("vector towers") to describe invariant measures on X. In particular I will focus on our construction of d distinct ergodic probability measures for certain minimal subshifts X_d, for any given alphabet rank $r(X_{\mathrm{d}}) = d \geq 1$, and the ensuing construction of a fairly large family of minimal subshifts with topological entropy 0 and infinitely many distinct ergodic probability measures.

In a second part I will discuss the relation between substitutions on one hand and morphisms of free groups on the other, and describe a new invariant for primitive substitutional subshifts X which is expected to characterize X up to recognizable morphisms (work in progress).

Glimpses into Recent Developments in Subword Combinatorics and Binomial Coefficients

Markus A. Whiteland

Department of Computer Science, Loughborough University, Loughborough, UK
m.a.whiteland@lboro.ac.uk

Abstract. Subsequences of strings play a crucial role in combinatorics on words, string algorithmics, formal language theory, and various applied fields such as bioinformatics. They provide rich algebraic structures through concepts such as Simon's congruence, the binomial equivalence relation, and M-equivalence.

The main focus of this talk is on binomial coefficients of words: for two words u and v, the quantity $\binom{u}{v}$ denotes the number of times v appears as a scattered subword in u. These quantities appear, for example, in (generalised) Parikh matrices. Recently, a q-analogue of binomial coefficients of words has been introduced. This analogue defines a polynomial in q such that setting $q = 1$ recovers the classical binomial coefficient of words. This polynomial formulation enables a finer analysis of subword counting and I will discuss their combinatorial interpretations, as well as their connection to Eğecioğlu's q-deformations of Parikh matrices.

Additionally, binomial complexities of infinite words have been the subject of extensive recent research. In this talk, I will review recent results concerning k-binomial complexity functions of certain families of infinite words. If time permits, I will discuss some ongoing work regarding the automaticity of these functions in the context of fixed points of Parikh-collinear morphisms. These results shed new light on the combinatorial behavior of infinite words and their structural constraints.

Overall, this talk aims to provide a glimpse of recent developments of binomial coefficients of words and associated complexity functions. It is based on joint work with M. Golafshan, A. Renard, M. Rigo, M. Stipulanti, and N. Wingate.

Keywords: Binomial coefficients of words · Parikh matrices · q-deformations · binomial complexity functions

Contents

And Now There Are Four: Another Brick in the Wall of the Optimal Upper Bound on the MP-Ratio

Kristina Ago(✉) and Bojan Bašić

Department of Mathematics and Informatics, University of Novi Sad, Trg Dositeja Obradovića 4, 21000 Novi Sad, Serbia
kristina.ago@dmi.uns.ac.rs

Abstract. The so-called *MP-ratio* is a kind of measure of how "packed with palindromes" a given word is. The lower bound on the MP-ratio for the set of all n-ary words is (trivially) 1, while the best possible upper bound is an open problem in the general case. It is solved for $n = 2$ (where the optimal upper bound is 4) and for $n = 3$ (where the optimal upper bound is 6). Also, it is known that in the n-ary case the optimal bound is between $2n$ and the order of the growth $n2^{\frac{n}{2}}$. In this article we solve this problem for quaternary words, for which we show that the best possible upper bound on the MP-ratio equals 8. We believe that this is the last case in which the result is $2n$, that is, we believe that for $n \geqslant 5$ there are words whose MP-ratio is strictly larger than $2n$.

Keywords: MP-ratio · Palindrome · Subword

1 Introduction

Various measures of the degree of "palindromicity" of words have been introduced and studied in the literature. Recall, e.g., the so-called *palindromic defect*, which stemmed from the result of Droubay, Justin and Pirillo [11], who showed that the number of different palindromic factors of a given word is bounded from above by its length increased by 1. Although this result is two-decade old, the research on this topic is still very lively, as can be seen, e.g., by the recent articles [5,8–10,12,17,21–23]. Another, less mature measure is the so-called *palindromic length,* introduced by Frid, Puzynina and Zamboni [15] as the minimal number of palindromes into which a given word can be decomposed; see [4,7,13,14,18–20,24] for some recent results on this topic. (It is interesting that there are also works, such as the mentioned [4] and [19], in which those two approaches are combined in certain ways.)

One more such measure, which is the main topic of the present article, is the so-called *MP-ratio.* It is based on palindromic (scattered) subwords of a given word (in contrast to the two previously mentioned measures, which are based on palindromic factors). It is a rational number greater than or equal to 1, such that,

G. Gamard and J. Leroy (Eds.): WORDS 2025, LNCS 15729, pp. 1–12, 2025.
https://doi.org/10.1007/978-3-031-97548-6_1

the larger it is, the more the given word is "packed with palindromes" (therefore, words whose MP-ratio is 1 are an extremal case, considered not packed at all, and they are dubbed *minimal-palindromic*). As originally introduced by Holub and Saari [16], it was defined only for binary words, and although it is obvious what is the most natural generalization of the MP-ratio to larger alphabets, the problem was to show whether the notion is well-defined in the general case. This could be the main reason why the MP-ratio attracted considerably less attention in the literature (and another possible reason is that its behavior can be quite hard to grasp; see e.g. [6]). But this is maybe unjustly so, as it turns out that it is connected to some more often researched topics, e.g. abelian (un)borderedness, and furthermore, any binary word can be reconstructed, up to reversal, from the set of its palindromic subwords (see [16], as well as the introductory parts of [1–3], for further reading). For that reason, the present authors decided to undertake the task of demonstrating that the MP-ratio is well-defined for any arity. This was first shown in [1] for the arity 3, and then finally in [2] for all larger arities.

However, one thing is still covered with darkness. Namely, for the arity 3, it is known that the best possible upper bound on the MP-ratio equals 6 (and for the arity 2, it has been found already in the original article: in that case the optimal bound equals 4). On the other hand, for $n > 3$, the only thing that is known about the best possible upper bound on the MP-ratio of n-ary words is that it lies somewhere between $2n$ and the order of the growth $n2^{\frac{n}{2}}$. In the present article we shed a little bit more light on this question, and show that the best possible upper bound on the MP-ratio of quaternary words equals 8. Although one would be tempted, based on the cases $n = 2, 3, 4$, to conjecture that, in general, the optimal bound equals $2n$, our research gives us a feeling that exactly the opposite is the case: we believe that the case $n = 4$ is the last case in which the bound equals $2n$, that is, that for $n \geqslant 5$ the optimal bound is strictly larger than $2n$ (and, in fact, that it indeed grows exponentially). We, however, so far do not have any guess on exactly what that bound should be, nor any formal arguments to support our belief, but let us try to explain our intuition behind this. Namely, in each of the cases $n = 2, 3, 4$, constructions are designed so that each letter is arranged on the left and on the right of w in a very asymmetric way: for example, for $n = 4$, the letters 0 and 1 appear only on one side, while the letters 2 and 3 are arranged so that the "smaller pile" is immediately next to w, while the "larger pile" is on the opposite side. Our experiments suggest that such symmetry breakings are indeed pivotal for constructing (relatively) short MP-extensions, but with 5 (or more) letters the fifth letter cannot be suitably arranged as it will always contribute to the existence of a too long subpalindrome.

Let us present the structure of this article. In Sect. 2 we recall the most important definitions and fix the notation. After that, we present a survey of the most important milestones in this research strand. Finally, we move on to prove the main result of this article. Due to the space limitations, not all the proofs could be included. We decided to cut out proofs of some technical lemmas, and to include just proofs of the main propositions (excepting the last one, whose proof also had to be left out) that, taken together, complete the puzzle that shows our main result.

2 Preliminaries

A *word* is a sequence of symbols taken from a finite nonempty set Σ, which is called the *alphabet* and its elements are called *letters*. Through this article we consider only finite words. For the set of all finite words over Σ we write Σ^*. In the case $|\Sigma| = 2$ we speak about *binary* words, in the case $|\Sigma| = 3$ we speak about *ternary* words, in the case $|\Sigma| = 4$ we speak about *quaternary* words, or simply 4-ary words and, generally, in the case $|\Sigma| = n$ we speak about n-ary words. If $w = a_1a_2 \ldots a_n$ with $a_1, a_2, \ldots, a_n \in \Sigma$, we say that the *length* of w is n, and write $|w| = n$. For a letter a and a word w, we write $|w|_a$ for the number of occurrences of a in w. The *concatenation* (or *product*) of words u and v, $u = a_1a_2 \ldots a_n$ and $v = b_1b_2 \ldots b_m$, is the word $a_1a_2 \ldots a_nb_1b_2 \ldots b_m$, denoted by uv. For a word w and a positive integer k we write w^k for the word $\underbrace{ww\ldots w}_{k}$.

We write $w[i]$ for the ith letter of the word w, and for any pair (i, j) of integers such that $1 \leqslant i \leqslant j \leqslant |w|$ we write $w[i, j]$ for the factor of w that begins at the ith position and ends at the jth position in w. By convention, this operation has precedence over concatenation; in other words, $uv[i]$ (and similarly $uv[i, j]$) will always denote $u(v[i])$, not $(uv)[i]$. In the case $i > j$, as well as $i > |w|$ or $j < 1$, we define $w[i, j] = \varepsilon$ (in particular, only some borderline cases could occur, like $w[i + 1, i] = \varepsilon$).

A word $u \in \Sigma^*$ is called a *factor* (respectively *prefix*, *suffix*) of a word $w \in \Sigma^*$ if there exist words $x, y \in \Sigma^*$ such that $w = xuy$ (respectively $w = uy$, $w = xu$). A word $u \in \Sigma^*$ is a *subword* of $w \in \Sigma^*$ if there exist words $x_1, x_2, \ldots, x_n, x_{n+1} \in \Sigma^*$ and $y_1, y_2, \ldots, y_n \in \Sigma^*$ such that $u = y_1y_2 \ldots y_n$ and $w = x_1y_1x_2y_2 \ldots x_ny_nx_{n+1}$ (or, equivalently, u is a subword of w if u is a subsequence of w). The set of all factors (respectively prefixes, suffixes, subwords) of a word w is denoted by $\mathrm{Fact}(w)$ (respectively $\mathrm{Pref}(w)$, $\mathrm{Suff}(w)$, $\mathrm{Subw}(w)$).

We define the map $\widetilde{\cdot} : \Sigma^* \to \Sigma^*$, called *reversal*, as follows: if $w = a_1a_2 \ldots a_n$, where $a_1, a_2, \ldots, a_n \in \Sigma$, then $\widetilde{w} = a_na_{n-1} \ldots a_1$. A word w is a *palindrome* (or *palindromic*) if $w = \widetilde{w}$. A palindromic subword of a given word will be called a *subpalindrome*.

3 Known Results About the MP-Ratio

It is not hard to see that every word w over an n-ary alphabet contains a subpalindrome of length at least $\lceil \frac{|w|}{n} \rceil$ (the subword that consists of all occurrences of the most frequent letter). An n-ary word is called *minimal-palindromic* if it does not contain palindromic subwords of length greater than $\lceil \frac{|w|}{n} \rceil$. The *MP-ratio* of a given n-ary word w is defined as the quotient $\frac{|rws|}{|w|}$, where r and s are words such that the word rws is minimal-palindromic and the value $|r| + |s|$ is minimal. The pair (r, s) such that the word rws is minimal-palindromic is called an *MP-extension,* and if additionally the value $|r| + |s|$ is minimal, then (r, s) is called an *SMP-extension* of w. We notice that it is not obvious whether an MP-extension (and thus also an SMP-extension) always exists and, therefore, whether the MP-ratio is well-defined at all. But this is indeed established in the literature, and we now make a recap of the main results on this topic.

The first result comes from Holub and Saari [16].

Theorem 1. *The MP-ratio of any binary word is at most* 4.

It means that every binary word has an MP-extension (r, s) such that $|rws| \leqslant 4|w|$ (note that this resolves the question of well-definedness in the binary case). In particular, for a word $w \in \{0,1\}^*$, they proved that the following word is minimal-palindromic:

$$0^{|w|+|w|_1}\ w\ 1^{|w|+|w|_0}.$$

They further proved that this upper bound is the best possible in the binary case, in the sense that, for any positive η, there exists a binary word whose MP-ratio is larger than $4 - \eta$.

There was no further progress on the topic for almost a decade. Then the next step was made in [1], where the ternary case is researched and the following theorem is proved.

Theorem 2. *The MP-ratio is well-defined in the ternary case and it is bounded from above by* 6.

It was also shown that the bound 6 is the best possible. In order to sketch the construction from that theorem, we need to introduce two auxiliary functions. Namely, for $\Sigma = \{0, 1, 2\}$, $w \in \Sigma^*$ and $a, b \in \Sigma$, let

$$g(w,a,b) = \max\left\{2\big|w[i,|w|]\big|_a - \big|w[i,|w|]\big|_b : i = 1,2,\ldots,|w|+1\right\}, \tag{G}$$

and

$$g'(w,a,b) = \max\left(\left\{2\big|w[i,|w|]\big|_a - \big|w[i,|w|]\big|_b : i = 1,2,\ldots,j(a,w)\right\} \cup \{0\}\right),$$

where $j(a, w)$ denotes the position of the last occurrence of a in w (that is, $w[j(a, w)] = a$ and $w[k] \neq a$ whenever $k > j(a, w)$), and $j(a, w) = 0$ if a does not occur in w.

The following property of the function g was proved in the cited article, and as we shall also need it in the present article, we state it here.

Lemma 1. *Let* $w \in \Sigma^*$, *and let* $|w|_b \geqslant |w|_a$. *We have:*

$$g(w,a,b) + g(\widetilde{w},a,b) \leqslant 3|w|_a.$$

(Note: the statement in [1] is a little different from the statement given here, as in [1] the statement includes the stronger assumption that b is a prevalent letter in w. But in the proof only the weaker assumption $|w|_b \geqslant |w|_a$ is used.)

The construction that settled the problem of well-definedness of the MP-ratio in the ternary case is as follows. Given a word $w \in \Sigma^*$, where $|w|_0 \leqslant |w|_1 \leqslant |w|_2$ can be assumed without loss of generality, it can be proved that at least one among the following two words is minimal-palindromic.

$$0^{2|w|-|w|_0} 2^{2|w|-|w|_2-g'(w,0,2)} \; w \; 2^{g'(w,0,2)} 1^{2|w|-|w|_1};$$

$$1^{2|w|-|w|_1} 2^{g'(\widetilde{w},0,2)} \; w \; 2^{2|w|-|w|_2-g'(\widetilde{w},0,2)} 0^{2|w|-|w|_0}.$$

Finally, the result that shows that everything makes sense for words of any arity came one year later [2].

Theorem 3. *For any positive integer n, $n \geqslant 2$, each n-ary word has an MP-extension.*

Concerning the upper bound in the general case, we have the following theorem from the same article.

Theorem 4. *For any positive η, there exists a word from $\{0, 1, \ldots, n-1\}^*$ whose MP-ratio is larger than $2n - \eta$.*

To summarize: for n-ary words, now we know that the best possible upper bound on the MP-ratio for $n = 2$ equals 4, for $n = 3$ equals 6, and for $n \geqslant 4$ lies somewhere between $2n$ and $\Theta(n2^{\frac{n}{2}})$ (the latter expression follows from the construction used to prove Theorem 3); it remains an open problem to narrow (or even better, close) this gap.

In this article we make another step and solve this problem for quaternary words, for which we show that the best possible upper bound on the MP-ratio equals 8.

4 Main Section

The goal of this section is to prove the following result.

Theorem 5. *The MP-ratio of any 4-ary word is at most 8.*

Note that, together with the results summarized in the previous section, we have that the constant 8 is the best possible.

The proof of Theorem 5 will follow by a series of propositions, divided into a few subsections. Let $\Sigma = \{0, 1, 2, 3\}$, and let $w \in \Sigma^*$. We assume, without loss of generality, $|w|_0 \leqslant |w|_1 \leqslant |w|_2 \leqslant |w|_3$. We now define two extensions of w, and the plan is to show that at least one of them is an MP-extension. Those two extensions are:

$$0^{2|w|-|w|_0} 3^{2|w|-|w|_3-g(w,0,3)} 2^{g(\widetilde{w},1,2)} \; w \; 3^{g(w,0,3)} 2^{2|w|-|w|_2-g(\widetilde{w},1,2)} 1^{2|w|-|w|_1};$$

$$1^{2|w|-|w|_1} 2^{2|w|-|w|_2-g(w,1,2)} 3^{g(\widetilde{w},0,3)} \; w \; 2^{g(w,1,2)} 3^{2|w|-|w|_3-g(\widetilde{w},0,3)} 0^{2|w|-|w|_0};$$

let us call them $f_1(w)$ and $f_2(w)$, respectively. (The function g that appears in $f_1(w)$ and $f_2(w)$ is the same g defined by (G).) As we have $|f_1(w)| = |f_2(w)| = 8|w|$, our task is to prove that at least one of the words $f_1(w)$ and $f_2(w)$ does not have a subpalindrome whose length exceeds $2|w|$. This will follow by Propositions 1–5.

4.1 Palindromes $0p0$ and $1p1$

Here we prove that neither of the words $f_1(w)$ nor $f_2(w)$ contains long palindromes of the form $0p0$ or $1p1$.

Proposition 1. *The length of an arbitrary subpalindrome of the form* $0p0$ *in each of the words* $f_1(w)$ *and* $f_2(w)$ *is less than or equal to* $2|w|$.

Proof. Let a palindrome of the form $0p0$ be a subpalindrome of $f_1(w)$. Then

$$0p0 \in \mathrm{Subw}(0^{2|w|-|w|_0}3^{2|w|-|w|_3-g(w,0,3)}2^{g(\widetilde{w},1,2)}\ w).$$

For a palindrome q and a word $x_1x_2\ldots x_m$, where $x_j \in \Sigma^*$ and q is a subword of $x_1x_2\ldots x_m$, we say that q is *bisected by* x_j if and only if $q\big[1, \lceil \frac{|q|}{2} \rceil\big] \in \mathrm{Subw}(x_1x_2\ldots x_j)$ and $q\big[\lfloor \frac{|q|}{2} \rfloor + 1, |q|\big] \in \mathrm{Subw}(x_jx_{j+1}\ldots x_m)$. For the purpose of this proof, we fix $x_1 = 0^{2|w|-|w|_0}$, $x_2 = 3^{2|w|-|w|_3-g(w,0,3)}$, $x_3 = 2^{g(\widetilde{w},1,2)}$, and $x_4 = w$.

If $0p0$ is bisected by $0^{2|w|-|w|_0}$, the only letter that appears in $0p0$ is 0, thus we have

$$|0p0| \leqslant |0^{2|w|-|w|_0}w|_0 = 2|w| - |w|_0 + |w|_0 = 2|w|.$$

If $0p0$ is bisected by $3^{2|w|-|w|_3-g(w,0,3)}$, then there exists a position i in the word w such that, among the letters at the positions $1, 2, \ldots, i-1$, respectively $i, i+1, \ldots, |w|$, only the letters 3, respectively the letters 0, can participate in the palindrome $0p0$. Hence, there can be at most $\big|w[i, |w|]\big|_0$ zeros at the end of $0p0$, and therefore also at the beginning. Further, this means that in the middle there are at most $2|w| - |w|_3 - g(w,0,3) + \big(|w|_3 - \big|w[i,|w|]\big|_3\big)$ letters 3. Altogether, we conclude

$$\begin{aligned} |0p0| &\leqslant 2|w| - |w|_3 - g(w,0,3) + |w|_3 - \big|w[i,|w|]\big|_3 + 2\big|w[i,|w|]\big|_0 \\ &\leqslant 2|w| - |w|_3 - g(w,0,3) + |w|_3 + g(w,0,3) \\ &= 2|w|, \end{aligned}$$

where we used that the expression $2\big|w[i,|w|]\big|_0 - \big|w[i,|w|]\big|_3$ is bounded from above by $g(w,0,3)$ (by the definition of g).

If $0p0$ is bisected by $2^{g(\widetilde{w},1,2)}$, then $0p0$ is of the form $0^{k_0}3^{k_3}2^{k_2}3^{k_3}0^{k_0}$, where $2^{k_2}3^{k_3}0^{k_0} \in \mathrm{Subw}(2^{g(\widetilde{w},1,2)}w)$. This gives $k_0 \leqslant |w|_0$, $k_3 \leqslant |w|_3$, and $k_2 \leqslant g(\widetilde{w},1,2) + |w|_2$. Now we can calculate

$$\begin{aligned} |0p0| &= 2k_0 + 2k_3 + k_2 \\ &\leqslant |w|_0 + 2|w|_3 + g(\widetilde{w},1,2) + |w|_2 \\ &\leqslant 2|w|_0 + 2|w|_3 + 2|w|_1 + |w|_2 \\ &\leqslant 2|w|, \end{aligned}$$

which was to be proved.

And finally, if $0p0$ is bisected by w, then we immediately have $|0p0| \leqslant 2|w|$.

The proof is analogous if a palindrome $0p0$ is a subpalindrome of $f_2(w)$, since $f_2(w) = \widetilde{f_1(\widetilde{w})}$. This proves the proposition. □

Proposition 2. *The length of an arbitrary subpalindrome of the form* $1p1$ *in each of the words* $f_1(w)$ *and* $f_2(w)$ *is less than or equal to* $2|w|$.

Proof. Analogous to the previous one (where 0s and 1s have switched roles, as well as 2s and 3s, and as well as $f_1(w)$ and $f_2(w)$). □

4.2 Palindromes $3p3$

We here treat palindromes of the form $3p3$. We do *not* prove that *neither* $f_1(w)$ nor $f_2(w)$ contains long palindromes of that form (which is false in general), but we prove that this holds for at least one of $f_1(w)$ or $f_2(w)$. We first introduce the following definition.

Definition 1. *Let* p *be a subpalindrome of a word* z.

- *A* descriptor *of* p *(with respect to* z*) is the set* P *with* $P = \{P_1, P_2, \ldots, P_{|p|}\}$, $P_1 < P_2 < \cdots < P_{|p|}$ *and* $p = z[P_1]z[P_2]\ldots z[P_{|p|}]$. *If* a *is a letter, then* $|P|_a$ *denotes the value* $|\{P_i \in P : z[P_i] = a\}|$.
- *The mapping* $\sigma_P : P \to P$, *defined by* $\sigma_P : P_s \mapsto P_{|p|-s+1}$ *is called* mirroring with respect to P.

Note that a descriptor of a subpalindrome might not be uniquely determined, but whenever we use this notion in the article, a descriptor will be chosen in the beginning and then we work with that chosen descriptor. Also note that σ_P is bijective, that σ_P^2 is an identical mapping, and that σ_P is decreasing (that is, whenever $i < j$, then $\sigma_P(i) > \sigma_P(j)$).

We now show a preliminary lemma.

Lemma 2. *Let* $w \in \{1,2,3\}^*$, *let* p *be a subpalindrome of* $2^{g(\widetilde{w},1,2)}w$, *and let* q *be a subpalindrome of* $w2^{g(w,1,2)}$. *Assume that neither* p *nor* q *consists only of the letters* 2. *If* $|p| + |q| > 2|w|$, *then* $|w|_2 > |w|_3$.

Proof. Omitted.

We want to add, while we were working on the previous lemma, the following result also emerged. We find its statement quite elegant (and similar "in spirit" to Theorem 6.1 from [1]) and thus would like to present it here. Who knows, maybe it could be of an independent interest in some further research (though more probably not, but we cannot resist presenting it anyway).

Corollary 1. *Let* $w \in \{1,2\}^*$ *and let* m_1, m_2 *be nonnegative integers such that* $m_1 \leqslant g(\widetilde{w},1,2)$ *and* $m_2 \leqslant g(w,1,2)$. *Let* p *and* q *be subpalindromes of* $2^{m_1}w$ *and* $w2^{m_2}$, *respectively. Then*

$$|p| + |q| \leqslant 2|w| + \frac{m_1 + m_2}{2}.$$

Proof. Omitted.

And now we are ready for the point of this subsection.

Proposition 3. *At least one among the words* $f_1(w)$ *and* $f_2(w)$ *does not contain a subpalindrome of the form* $3p3$ *longer than* $2|w|$.

Proof. Suppose the contrary: in both the words $f_1(w)$ and $f_2(w)$ a longest subpalindrome of the form $3p3$ is longer than $2|w|$. The corresponding subpalindrome of $f_1(w)$ is of the form $3^{l+g(w,0,3)}p_w3^{l+g(w,0,3)}$, where $p_w \in \mathrm{Subw}(2^{g(\widetilde{w},1,2)}w)$, and $l = |v|_3$ for some $v \in \mathrm{Suff}(w)$. Similarly, the corresponding subpalindrome of $f_2(w)$ is of the form $3^{l'+g(\widetilde{w},0,3)}q_w3^{l'+g(\widetilde{w},0,3)}$, where $q_w \in \mathrm{Subw}(w2^{g(w,1,2)})$, and $l' = |t|_3$ for some $t \in \mathrm{Pref}(w)$. Altogether, we have

$$|p_w| + 2|v|_3 + 2g(w,0,3) + |q_w| + 2|t|_3 + 2g(\widetilde{w},0,3) > 4|w|,$$

which gives the following

$$|p_w| + 2|v|_3 + |q_w| + 2|t|_3 > 4|w| - 2g(w,0,3) - 2g(\widetilde{w},0,3) \geqslant 4|w| - 6|w|_0, \quad (*)$$

where we used Lemma 1 in the second inequality.

Now, if the last letter of p_w is earlier than the first letter of q_w, we have:

$$\begin{aligned}
&|p_w| + 2|v|_3 + |q_w| + 2|t|_3 \\
&\quad \leqslant |w|_0 + |w|_1 + |w|_2 + g(w,1,2) + g(\widetilde{w},1,2) + |p_w|_3 + 2|v|_3 + |q_w|_3 + 2|t|_3 \\
&\quad \leqslant |w|_0 + |w|_1 + |w|_2 + 3|w|_1 + 4|w|_3 \\
&\quad \leqslant 4|w| - 6|w|_0,
\end{aligned}$$

where in the first step we used that $|p_w|_i + |q_w|_i \leqslant |w|_i$ for $i = 0, 1$ and $|p_w|_2 + |q_w|_2 \leqslant |w|_2 + g(w,1,2) + g(\widetilde{w},1,2)$, while in the second step we used again Lemma 1 and the inequalities $|p_w|_3 + 2|v|_3 \leqslant 2|w|_3$ and $|q_w|_3 + 2|t|_3 \leqslant 2|w|_3$. This gives a contradiction with $(*)$.

Thus, we may assume that the first letter of q_w is no later than the last letter of p_w. Let us first deal with the case when one of p_w or q_w consists only of the letters 2; without loss of generality, let that be p_w. Then $|p_w| \leqslant g(\widetilde{w},1,2) + |w|_2$ and $|q_w| \leqslant |w| + g(w,1,2)$. Together with Lemma 1 and $|t|_3 + |v|_3 \leqslant |w|_3$, we obtain:

$$|p_w| + 2|v|_3 + |q_w| + 2|t|_3 \leqslant 3|w|_1 + |w|_2 + |w| + 2|w|_3 \leqslant 4|w| - 6|w|_0,$$

which contradicts $(*)$. Finally, we assume that neither p_w nor q_w consists only of the letters 2. We shall show that then $|p_w| + 2|v|_3 + |q_w| + 2|t|_3 \leqslant 2|w| + 2|w|_3$, and as $2|w| + 2|w|_3 \leqslant 4|w| - 6|w|_0$, again a contradiction with $(*)$ will follow. We have $|t|_3 + |v|_3 \leqslant |w|_3$, thus we are left to prove that $|p_w| + |q_w| \leqslant 2|w|$. Define the word w', $|w'| = |w|$, in the following way:

$$w'[i] = \begin{cases} 3, & \text{if } w[i] = 0; \\ w[i], & \text{otherwise.} \end{cases}$$

As we can see, $|w'|_2 \leqslant |w'|_3$ (since $|w|_2 \leqslant |w|_3$). The equalities $g(w', 1, 2) = g(w, 1, 2)$ and $g(\widetilde{w'}, 1, 2) = g(\widetilde{w}, 1, 2)$ are also true, since the introduced modification does not affect $g(\cdot, 1, 2)$. Similarly, let $p_{w'}$ and $q_{w'}$ be the words obtained from p_w and q_w, respectively, by replacing all 0s by 3s. Then $p_{w'}$ and $q_{w'}$ are subpalindromes of the words $2^{g(\widetilde{w'},1,2)}w$ and $w2^{g(w',1,2)}$. If we suppose the contrary: $|p_w| + |q_w| > 2|w|$, we get $|p_{w'}| + |q_{w'}| = |p_w| + |q_w| > 2|w| = 2|w'|$, but then Lemma 2 gives $|w'|_2 > |w'|_3$, which is impossible. Therefore, $|p_w| + |q_w| \leqslant 2|w|$, which gives the announced contradiction. □

4.3 Palindromes $2p2$

The point of this section is analogous to the previous one (the proofs are, however, not analogous, sadly). We first need a technical definition.

Definition 2. *Let p and q be subpalindromes of some word z, and let P and Q be their descriptors. We say that Q is* offset to the right of P *if and only if for each s, s' such that $s \in P$, $s' \in Q$, $s' \leqslant s$, we have $\sigma_Q(s') > \sigma_P(s)$.*

The preparatory part of this section is the following lemma.

Lemma 3. *Let $w \in \{0, 1, 2, 3\}^*$ where $|w|_0 \leqslant |w|_1 \leqslant |w|_2 \leqslant |w|_3$, and let m_1, m_2 be nonnegative integers such that $m_1 \leqslant 3^{g(\widetilde{w},0,3)}$ and $m_2 \leqslant 3^{g(w,0,3)}$. Let p and q be subpalindromes of $3^{m_1}w$ and $w3^{m_2}$, respectively. Let P and Q be descriptors of p and q (with respect to $3^{m_1}w3^{m_2}$), where $\max P \leqslant |3^{m_1}w|$ and the smallest m_1 elements of P are $1, 2, \ldots, m_1$, and $\min Q \geqslant m_1 + 1$ and the largest m_2 elements of Q are $|3^{m_1}w| + 1, |3^{m_1}w| + 2, \ldots, |3^{m_1}w3^{m_2}|$. Suppose that Q is offset to the right of P. Then, if $u = (3^{m_1}w3^{m_2})[Q_1, P_{|p|}]$ if $Q_1 \leqslant P_{|p|}$, and $u = \varepsilon$ otherwise, we have*

$$|p| + |q| \leqslant 4|w|_3 + 2|u|_2 + 4|w|_0.$$

Proof. Omitted.

And the point of this subsection is the following proposition.

Proposition 4. *At least one among the words $f_1(w)$ and $f_2(w)$ does not contain a subpalindrome of the form $2p2$ longer than $2|w|$.*

Proof. Suppose the contrary: in both the words $f_1(w)$ and $f_2(w)$ a longest subpalindrome of the form $2p2$ is longer than $2|w|$. The corresponding subpalindrome of $f_1(w)$ is of the form $2^{|t|_2+g(\widetilde{w},1,2)}q_w2^{|t|_2+g(\widetilde{w},1,2)}$, where $q_w \in \mathrm{Subw}(w3^{g(w,0,3)})$ and $t \in \mathrm{Pref}(w)$. We may assume, without loss of generality, that t is of the maximal possible length. Similarly, the corresponding subpalindrome of $f_2(w)$ is of the form $2^{|v|_2+g(w,1,2)}p_w2^{|v|_2+g(w,1,2)}$, where $p_w \in \mathrm{Subw}(3^{g(\widetilde{w},0,3)}w)$ and $v \in \mathrm{Suff}(w)$ (and again $|v|$ is maximal). Owing to this, we have:

$$|p_w| + 2|v|_2 + 2g(w, 1, 2) + |q_w| + 2|t|_2 + 2g(\widetilde{w}, 1, 2) > 4|w|,$$

which gives

$$|p_w| + 2|v|_2 + |q_w| + 2|t|_2 > 4|w| - 2g(w,1,2) - 2g(\widetilde{w},1,2) \geqslant 4|w| - 6|w|_1, \quad (**)$$

where we used Lemma 1 in the second inequality.

Let P and Q be descriptors of p_w and q_w with respect to $3^{g(\widetilde{w},0,3)}w3^{g(w,0,3)}$, where $\max P \leqslant |3^{g(\widetilde{w},0,3)}w|$ and $\min Q \geqslant g(\widetilde{w},0,3)+1$, and furthermore, $t = w[1, Q_1 - g(\widetilde{w},0,3) - 1]$ and $v = w[P_{|p_w|} - g(\widetilde{w},0,3) + 1, |w|]$.

If $P_{|p_w|} < Q_1$, we have:

$$\begin{aligned}
&|p_w| + 2|v|_2 + |q_w| + 2|t|_2 \\
&\quad \leqslant |w|_0 + |w|_1 + |p_w|_2 + 2|v|_2 + |q_w|_2 + 2|t|_2 + |w|_3 + g(w,0,3) + g(\widetilde{w},0,3) \\
&\quad \leqslant |w|_0 + |w|_1 + 4|w|_2 + |w|_3 + 3|w|_0 \\
&\quad \leqslant 4|w| - 6|w|_1,
\end{aligned}$$

where in the first step we used that $|p_w|_a + |q_w|_a \leqslant |w|_a$ for $a = 0,1$ and $|p_w|_3 + |q_w|_3 \leqslant |w|_3 + g(w,0,3) + g(\widetilde{w},0,3)$, while in the second step we used again Lemma 1 and the inequalities $|p_w|_2 + 2|v|_2 \leqslant 2|w|_2$ and $|q_w|_2 + 2|t|_2 \leqslant 2|w|_2$. This gives a contradiction with (**).

Thus, we may assume $Q_1 \leqslant P_{|p_w|}$. Let $u = w[Q_1 - g(\widetilde{w},0,3), P_{|p_w|} - g(\widetilde{w},0,3)]$.

Assume first that Q is *not* offset to the right of P. Then there exist $s \in P$, $s' \in Q$, $s' \leqslant s$ such that $\sigma_Q(s') \leqslant \sigma_P(s)$. We may assume $s' \leqslant \sigma_Q(s')$ (otherwise we just swap roles of s' and $\sigma_Q(s')$). Let:

$$\begin{aligned}
x &= w[Q_1 - g(\widetilde{w},0,3), s' - g(\widetilde{w},0,3) - 1]; \\
r &= w[s' - g(\widetilde{w},0,3), \sigma_P(s) - g(\widetilde{w},0,3)]; \\
y &= w[\sigma_P(s) - g(\widetilde{w},0,3) + 1, P_{|p_w|} - g(\widetilde{w},0,3)].
\end{aligned}$$

We have $u = xry$ and $w = txryv$. As σ_P maps members of P less than s' to values greater than $\sigma_P(s)$ (since $s' \leqslant s$), we have $|p_w| \leqslant 2|y| + |r|$. Analogously, $|q_w| \leqslant 2|x| + |r|$. Now we can calculate:

$$|p_w| + 2|v|_2 + |q_w| + 2|t|_2 \leqslant 2|y| + 2|x| + 2|r| + 2|v|_2 + 2|t|_2 \leqslant 2|w|,$$

contradicting (**) (since $3|w|_1 \leqslant |w|_1 + |w|_2 + |w|_3 \leqslant |w|$).

Finally, assume that Q is offset to the right of P. Then by (**) we obtain:

$$\begin{aligned}
|p_w| + |q_w| &> 4|w| - 6|w|_1 - 2|v|_2 - 2|t|_2 \\
&= 4|w|_3 + 2|w|_2 + 2|u|_2 - 2|w|_1 + 4|w|_0 \\
&\geqslant 4|w|_3 + 2|u|_2 + 4|w|_0.
\end{aligned}$$

However, this is an immediate contradiction with Lemma 3, which completes the proof. □

4.4 The "Diagonal" Case

The last proposition that will, combined with the previous ones, lead to the proof of Theorem 5 is the point of this subsection.

Proposition 5. *One of the following is true: the word $f_1(w)$ does not contain a subpalindrome of the form* $2p2$ *longer than* $2|w|$*, or the word $f_2(w)$ does not contain a subpalindrome of the form* $3p3$ *longer than* $2|w|$.

Analogously, the word $f_1(w)$ does not contain a subpalindrome of the form $3p3$ *longer than* $2|w|$*, or the word $f_2(w)$ does not contain a subpalindrome of the form* $2p2$ *longer than* $2|w|$.

Proof. Omitted.

4.5 Putting Everything Together

We are finally ready to sew up the proof of Theorem 5.

Proof (Theorem 5). We have $|f_1(w)| = |f_2(w)| = 8|w|$. We claim that at least one of the words $f_1(w)$ and $f_2(w)$ does not have a subpalindrome whose length exceeds $2|w|$.

Propositions 1 and 2 give that neither $f_1(w)$ nor $f_2(w)$ contains such a subpalindrome of the form $0p0$ or $1p1$. Consider subpalindromes of the form $3p3$. Proposition 3 gives that at least one of the words $f_1(w)$ or $f_2(w)$ does not contain such a subpalindrome longer than $2|w|$; without loss of generality, let that be $f_1(w)$. Now, considering subpalindromes of the form $2p2$, Proposition 4 gives that at least one of the words $f_1(w)$ or $f_2(w)$ does not contain such a subpalindrome longer than $2|w|$; if this is again $f_1(w)$, then, altogether, $f_1(w)$ does not contain any subpalindrome longer than $2|w|$, which means that $f_1(w)$ is an MP-extension. So, assume that $f_1(w)$ contains a subpalindrome of the form $2p2$ longer than $2|w|$, while $f_2(w)$ does not contain such one. But then Proposition 5 (the first part) implies that $f_2(w)$ does not contain a subpalindrome of the form $3p3$ longer than $2|w|$; altogether, $f_2(w)$ does not contain any subpalindrome longer than $2|w|$, which means that $f_2(w)$ is an MP-extension, and the proof is thus finished. □

Acknowledgments. The authors were supported by the Ministry of Science, Technological Development and Innovation of the Republic of Serbia (grants no. 451-03-137/2025-03/200125 and 451-03-136/2025-03/200125). The first author was also supported by the Délvidékért Kiss Alapítvány (the *Délvidékért otthon* program).

References

1. Ago, K., Bašić, B.: On highly palindromic words: the ternary case. Discrete Appl. Math. **284**, 434–443 (2020)
2. Ago, K., Bašić, B.: On highly palindromic words: the n-ary case. Discrete Appl. Math. **304**, 98–109 (2021)

3. Ago, K., Bašić, B.: On a theorem concerning partially overlapping subpalindromes of a binary word. Adv. Appl. Math. **134** (2022). Article No. 102302, 21 pp
4. Ambrož, P., Kadlec, O., Masáková, Z., Pelantová, E.: Palindromic length of words and morphisms in class $\mathcal{P}$. Theoret. Comput. Sci. **780**, 74–83 (2019)
5. Baranwal, A.R., Shallit, J.: Repetitions in infinite Palindrome-rich words. In: Mercaş, R., Reidenbach, D. (eds.) WORDS 2019. LNCS, vol. 11682, pp. 93–105. Springer, Cham (2019). https://doi.org/10.1007/978-3-030-28796-2_7
6. Bašić, B.: Counter-intuitive answers to some questions concerning minimal-palindromic extensions of binary words. Discrete Appl. Math. **160**, 181–186 (2012)
7. Bulgakova, D., Frid, A., Scanvic, J.: Prefix palindromic length of the Sierpinski word. In: Diekert, V., Volkov, M. (eds.) DLT 2022. LNCS, vol. 13257, pp. 78–89. Springer, Cham (2022)
8. Currie, J. D., Mol, L., Rampersad, N.: The repetition threshold for binary rich words. Discrete Math. Theor. Comput. Sci. **22** (2020). Article No. 6, 16 pp
9. Currie, J. D., Mol, L., Rampersad, N.: The lexicographically least binary rich word achieving the repetition threshold. Electron. J. Combin. **31** (2024). Article No. 4.77, 8 pp
10. Dolce, F., Pelantová, E.: On morphisms preserving palindromic richness. Fund. Inform. **185**, 1–25 (2022)
11. Droubay, X., Justin, J., Pirillo, G.: Episturmian words and some constructions of de Luca and Rauzy. Theoret. Comput. Sci. **255**, 539–553 (2001)
12. Dvořáková, L., Klouda, K., Pelantová, E.: The asymptotic repetition threshold of sequences rich in palindromes. European J. Combin. **126** (2025). Article No. 104124, 20 pp
13. Frid, A.E.: Sturmian numeration systems and decompositions to Palindromes. European J. Combin. **71**, 202–212 (2018)
14. Frid, A.E., Laborde, E., Peltomäki, J.: On prefix palindromic length of automatic words. Theoret. Comput. Sci. **891**, 13–23 (2021)
15. Frid, A.E., Puzynina, S., Zamboni, L.Q.: On palindromic factorization of words. Adv. in Appl. Math. **50**, 737–748 (2013)
16. Holub, Š, Saari, K.: On highly palindromic words. Discrete Appl. Math. **157**, 953–959 (2009)
17. Mahalingam, K., Maity, A., Pandoh, P.: Rich words in the block reversal of a word. Discrete Appl. Math. **334**, 127–138 (2023)
18. Rukavicka, J.: Palindromic length of words with many periodic palindromes. In: Jirásková, G., Pighizzini, G. (eds.) DCFS 2020. LNCS, vol. 12442, pp. 167–179. Springer, Cham (2020). https://doi.org/10.1007/978-3-030-62536-8_14
19. Rukavicka, J.: Palindromic factorization of rich words. Discrete Appl. Math. **316**, 95–102 (2022)
20. Rukavicka, J.: Palindromic length and reduction of powers. Theoret. Comput. Sci. **930**, 106–115 (2022)
21. Rukavicka, J.: Rich words containing two given factors. In: Mercaş, R., Reidenbach, D. (eds.) WORDS 2019. LNCS, vol. 11682, pp. 286–298. Springer, Cham (2019). https://doi.org/10.1007/978-3-030-28796-2_23
22. Rukavicka, J.: Upper bound for palindromic and factor complexity of rich words. RAIRO Theor. Inform. Appl. **55** (2021). Article No. 1, 15 pp
23. Rukavicka, J.: A unique extension of rich words. Theoret. Comput. Sci. **896**, 53–64 (2021)
24. Saarela, A.: Palindromic length in free monoids and free groups. In: Brlek, S., Dolce, F., Reutenauer, C., Vandomme, É. (eds.) WORDS 2017. LNCS, vol. 10432, pp. 203–213. Springer, Cham (2017). https://doi.org/10.1007/978-3-319-66396-8_19

A Characterization of Algebraic Multivariate Power Series with Sparse Support

Seda Albayrak(✉)

Simon Fraser University, Burnaby, BC V5A 1S6, Canada
seda_albayrak@sfu.ca

Abstract. Christol's theorem [7] characterizes algebraic formal power series over finite fields in terms of automatic sequences, establishing a fundamental link between algebraicity and computability. A refinement of this result by Bell and the author [2] provides an algebraic characterization of formal power series whose support is a sparse automatic set, showing that sparseness can be characterized in terms of certain key transformations such as the Frobenius map, multiplicative scaling, and power transformations. In this paper, we extend these results to the multivariate setting, building on Salon's [16] generalization of Christol's theorem. By using a characterization of sparse regular languages, we look at algebraic multivariate formal power series whose supports are sparse subsets of $\mathbb{N}^d$ and characterize them in terms of above-mentioned transformations, providing a structural understanding of the interplay between algebraicity and sparseness in multiple variables.

Keywords: Automatic sequences · Christol's theorem · sparse regular languages · algebraic power series · multivariate series · formal power series

1 Introduction

Let K be a field. Consider the formal power series $F(t) = \sum_{n\geq 0} f(n)t^n \in K[[t]]$. Recall that $F(t)$ is said to be algebraic over $K(t)$ if there exists a non-zero polynomial $P(x,t)$ over $K[t]$ such that $P(F(t),t) = 0$. Now, let p be a prime number and q be a power of p. Christol's theorem, in [7,8], states that a power series $F(t) = \sum_{n\geq 0} f(n)t^n \in \mathbb{F}_q[[t]]$ is algebraic over $\mathbb{F}_q(t)$ if and only if the sequence $f(n)$ is p-automatic. That is, $f(n)$ is generated by a finite-state automaton from the base-p representation of n. This is an important result as it establishes a direct link between algebraicity and computational properties of power series coefficients.

One can define automatic sets using automatic sequences.

Definition 1. *Suppose $k \geq 2$ is an integer. We say that $S \subseteq \mathbb{N}$ is a* k-automatic set *if the characteristic function of S, given by $\chi_S : \mathbb{N} \to \{0,1\}$, defines a k-automatic sequence.*

G. Gamard and J. Leroy (Eds.): WORDS 2025, LNCS 15729, pp. 13–23, 2025.
https://doi.org/10.1007/978-3-031-97548-6_2

We recall that the counting function of a regular language $\mathcal{L}$ is given by $f_{\mathcal{L}}(n) = \#\{w \in \mathcal{L}\colon \text{length}(w) \leq n\}$ and that a regular language is sparse if $f_{\mathcal{L}}(n) = O(n^c)$ for some integer $c \geq 0$. (See [18]). Translating this to sets, we call a k-automatic subset S of $\mathbb{N}$ sparse if $\pi_S(x) := |\{n \leq x\colon n \in S\}| = O((\log x)^c)$ for some integer $c \geq 1$. Otherwise, we say that it is non-sparse.

There are many works building upon Christol's result (See [1,2,13,14,16,17]). In [2], Bell and the author considered sparse automatic sets and proved a refinement of Christol's theorem for formal power series whose support is a sparse automatic set. This characterization involves being the smallest non-trivial subalgebra that is closed under certain natural sparse-preserving operators; namely, closure under Frobenius map, under multiplicative scaling and power transformation. Another work building upon Christol's theorem is its extension to multivariate case by Salon [16,17]. For an integer $d \geq 1$, a formal power series $F(t_1, \ldots, t_d) = \sum_{(n_1,\ldots,n_d)\in\mathbb{N}^d} f(n_1 \ldots n_d) t_1^{n_1} \cdots t_d^{n_d}$, where $f(n_1 \ldots n_d)$ lie in a field K, is said to be algebraic over $K(t_1, \ldots, t_d)$ whenever there exists polynomials $P_0, \ldots, P_r \in K[t_1, \ldots, t_d]$, not all zero, such that

$$\sum_{i=0}^{r} P_i(t_1, \ldots, t_d) F^i(t_1, \ldots, t_d) = 0.$$

Theorem 1 (Salon). *Let p be a prime number, Δ be a finite non-empty set of symbols, and $f(n_1, \ldots, n_d) : \mathbb{N}^d \to \Delta$ be a sequence. We have that $f(n_1, \ldots, n_d)$ is p-automatic if and only if there exists an integer $k \geq 1$ and an injection $i : \Delta \to \mathbb{F}_q$, where $q = p^k$, such that the formal power series $F(t_1, \ldots, t_d) = \sum_{(n_1,\ldots,n_d)\in\mathbb{N}^d} i(f(n_1 \ldots n_d)) t_1^{n_1} \cdots t_d^{n_d}$ is algebraic over $\mathbb{F}_q(t_1, \ldots, t_d)$.*

In light of this correspondence between algebraicity and automaticity holding in multivariate case, our aim is to prove a multivariate version of the refinement of Christol's theorem provided in [2], giving a characterization of algebraic multivariate formal power series with sparse support. For this, we need a description of sparse sets in the d-dimensional setting.

For a multivariate formal power series

$$F(t_1, t_2, \ldots, t_d) = \sum_{i_1, i_2, \ldots i_d \geq 0} f(i_1, \ldots, i_d) t_1^{i_1} \cdots t_d^{i_d} \in \bar{\mathbb{F}}_p[[t_1, \ldots, t_d]],$$

where $d \geq 1$ is an integer, the support of F is the set

$$S = \{(i_1, i_2, \ldots i_d) \in \mathbb{N}^d : f(i_1, \ldots, i_d) \neq (0, \ldots, 0)\} \subseteq \mathbb{N}^d.$$

We can naturally extend the notion of a sparse automatic set to subsets of $\mathbb{N}^d$ with $d \geq 1$ as follows. For a k-automatic subset $S \subseteq \mathbb{N}^d$, define

$$\pi_S(x) = |\{(n_1, \ldots, n_d) \in S\colon n_1 + n_2 + \cdots + n_d \leq x\}|.$$

Then $S \subseteq \mathbb{N}^d$ is sparse if $\pi_S(x) = O((\log x)^c)$ for some integer $c \geq 1$.

Our main theorem is the following.

Theorem 2. *Let p be a prime number, $d \geq 1$ be a natural number, A_d be the ring of multivariate formal power series that are algebraic over $\bar{\mathbb{F}}_p(t_1, t_2, \ldots, t_d)$, and B_d be the subring of A_d consisting of algebraic power series with sparse support. Then B_d is the smallest non-trivial $\bar{\mathbb{F}}_p$-subalgebra of A_d that possesses the following closure properties:*

(P1) If $F(t_1, t_2, \ldots, t_d) \in B_d$ and $F(0, 0, \ldots, 0) = 0$, then

$$F(t_1, t_2, \ldots, t_d) + F(t_1^p, t_2^p, \ldots, t_d^p) + F(t_1^{p^2}, t_2^{p^2}, \ldots, t_d^{p^2}) + \cdots \in B_d;$$

(P2) If $F(t_1, t_2, \ldots, t_d) \in B_d$ and $(\alpha_1, \alpha_2, \ldots, \alpha_d) \in \bar{\mathbb{F}}_p^d$, then

$$F(\alpha_1 t_1, \alpha_2 t_2, \ldots, \alpha_d t_d) \in B_d;$$

(P3) If $F(t_1, t_2, \ldots, t_d) \in B_d$ and

$$t_1^{b_1} t_2^{b_2} \cdots t_d^{b_d} F(t_1^{c_1}, t_2^{c_2}, \ldots, t_d^{c_d}) \in A_d$$

with $(c_1, \ldots, c_d) \in \mathbb{Q}_{>0}^d$ and $(b_1, b_2, \ldots, b_d) \in \mathbb{Q}^d$, then

$$t_1^{b_1} t_2^{b_2} \cdots t_d^{b_d} F(t_1^{c_1}, t_2^{c_2}, \ldots, t_d^{c_d}) \in B_d.$$

Remark 1. Note that we intentionally allow the coefficients in (P3) to be in $\mathbb{Q}$. For example, for $d = 1$, consider the sparse series $S(x) = \sum_{n \geq 0} x^{s_n}$ where $s_n = \frac{k^n}{k-1} - \frac{1}{k-1}$. We have $S(x) = x^{-1/(k-1)} G(x^{1/(k-1)})$ where $G(x) = \sum_{n \geq 0} x^{k^n}$.

The outline of this paper is as follows. In the introduction, we present our main theorem and the context it sits in as well as setting some notation. In Sect. 2, we introduce sparse automatic sets, providing a background on sparse regular languages and their relevance to automatic sequences. We present key characterizations of sparse automatic sets, discussing their structural properties and growth behavior. Additionally, we establish important closure properties, proving that sparse automatic sets remain sparse under dilation, translation, union, and set addition. In Sect. 3, we prove our main theorem, which provides an algebraic characterization of multivariate formal power series with sparse automatic support. We define the notion of Artin–Schreier closure in this setting and show that the collection of sparse algebraic power series forms the smallest non-trivial subalgebra that satisfies this closure property. The proof follows by extending known techniques from the univariate case to higher dimensions, leveraging structural results on sparse automatic sets in multidimensions which are established in Sect. 2.

1.1 Notation

Throughout this paper, for an alphabet Σ, Σ^* denotes the set of all finite words over Σ. If $\Sigma = \{u\}$, we write u^* instead of Σ^*. For an integer $k \geq 2$, we set $\Sigma_k = \{0, \ldots, k-1\}$. Then, for an integer $d \geq 1$ the map

$$[\cdot]_k : \left(\Sigma_k^d\right)^* \to \mathbb{N}^d$$

denotes base-k expansion. That is, the map $[\cdot]_k$ takes a d-tuple of words and outputs the d-tuple of natural numbers formed by taking the base-k expansions of these words. Similarly, we have a natural corresponding map

$$(\cdot)_k : \mathbb{N}^d \to \left(\Sigma_k^d\right)^*$$

that takes a d-tuple of natural numbers and gives the base-k representations of them, padding zeros if needed. For example, take $k = 2$, we have $(1, 11, 21)_2 = (00001, 01011, 10101)$.

2 Sparse Automatic Sets

In this section, we give a brief summary of sparse regular languages and sparse automatic sets. Sparse regular languages and related concepts have been extensively studied (see [10] and the references therein). Sparse automatic sets also arise in many different contexts in number theory and algebraic dynamics [5,9,13–15].

Given a finite alphabet Σ and a language $\mathcal{L} \subseteq \Sigma^*$ over Σ, we have an associated counting function $f_{\mathcal{L}}(n) := |\{w \in \mathcal{L} \colon \text{length}(w) \leq n\}|$. A regular language $\mathcal{L}$ is *sparse* if $f_{\mathcal{L}}(n) = \mathrm{O}(n^d)$ for some nonnegative integer d. There is a precise characterization of sparse regular languages, which has been obtained by several authors [6,10–12,18,19]:

Proposition 1. *Let $\mathcal{L}$ be a regular language. The following are equivalent:*

1. *$\mathcal{L}$ is sparse.*
2. *$\mathcal{L}$ is a finite union of languages of the form $v_1 w_1^* v_2 w_2^* \dots v_s w_s^* v_{s+1}$, where $s \geq 0$, the v_i are possibly empty words, and the w_i are non-empty words over the alphabet $\{0, 1, \dots, k-1\}$.*

A k-automatic subset $S \subseteq \mathbb{N}^d$ is then said to be *sparse* if the sublanguage of $(\Sigma_k^d)^*$ corresponding to base-k expansions of elements of S is a sparse regular language. Translating Proposition 1 into the framework of automatic sets, we see that a k-automatic subset $S \subseteq \mathbb{N}^d$ is sparse if

$$\pi_S(x) = |\{(n_1, \dots, n_d) \in S \colon n_1 + n_2 + \dots + n_d \leq x\}| = \mathrm{O}((\log x)^c) \quad (1)$$

for some $c \geq 1$. We note that if S is not sparse, then there is some $\alpha > 0$ such that $\pi_S(x) > x^\alpha$ for x large ([10, §2.3]), and so there is a natural gap separating sparse and non-sparse automatic subsets of $\mathbb{N}^d$.

By using the characterization in Proposition 1, we get a useful description for sparse sets. We get that every sparse language is a finite union of disjoint simple sparse languages and if one translates this into sparse k-automatic sets, we see that a sparse k-automatic set S can be written as a disjoint union

$$S = S_1 \sqcup S_2 \sqcup \dots \sqcup S_m \quad (2)$$

for some integer $m \geq 1$, where each S_i is a set of natural numbers of the form

$$\{[v_1 w_1^{n_1} v_2 w_2^{n_2} \dots v_s w_s^{n_s} v_{s+1}]_k \colon n_1, n_2, \dots, n_s \geq 0\}. \tag{3}$$

We call a set of this form a *simple sparse* k-automatic set. Ginsburg and Spanier [11] showed the following (see also [2, Lemma 3.4]) about simple sparse sets:

Proposition 2. *Let $k \geq 2$ be a natural number, let s be a nonnegative integer. If S is a non-empty set of the form*

$$\{[v_1 w_1^{n_1} v_2 w_2^{n_2} \dots v_s w_s^{n_s} v_{s+1}]_k \colon n_1, n_2, \dots, n_s \geq 0\}$$

where $v_0, v_1, \dots, v_s$ possibly empty and $w_0, \dots, w_s$ are non-empty words in Σ_k^*, then there exist $c_0, \dots, c_s \in \mathbb{Q}$ such that $(k^\ell - 1)c_i \in \mathbb{Z}$ for some $\ell \geq 0$, $c_0 + c_1 + \dots + c_s \in \mathbb{Z}_{\geq 0}$ and positive integers $\delta_1, \dots, \delta_s$ such that

$$S = \left\{c_0 + c_1 k^{\delta_s n_s} + c_2 k^{\delta_s n_s + \delta_{s-1} n_{s-1}} + \dots + c_s k^{\delta_s n_s + \dots + \delta_1 n_1} \colon n_1, \dots, n_s \geq 0\right\}. \tag{4}$$

Moreover, $n \geq c_0$ for all $n \in S$ and $c_0 \in S$ if and only if $s = 0$.

We note that, for each i, the δ_i correspond to the length of the word w_i which is non-empty by construction and this is why $\delta_i \geq 1$.

A direct consequence of Proposition 2 is the following.

Lemma 1. *Let $d \geq 1$, $k \geq 2$ be natural numbers, and let $S \subseteq \mathbb{N}^d$ be a non-empty simple sparse k-automatic set. Then there exist $s \geq 0$, for each $i = 1, \dots, d$, $c_{j,i} \in \mathbb{Q}$ such that $(k^\ell - 1)c_{j,i} \in \mathbb{Z}$ for some $\ell \geq 0$, $\sum_{j=0}^s c_{j,i} \in \mathbb{Z}_{\geq 0}$ and positive integers $\delta_1, \dots, \delta_s$ such that*

$$S = \left\{ \left(\sum_{j=0}^{s} c_{j,1} p^{\sum_{m=s-j+1}^{s} \delta_m n_m}, \dots, \sum_{j=0}^{s} c_{j,d} p^{\sum_{m=s-j+1}^{s} \delta_m n_m} \right) \colon n_1, \dots, n_s \geq 0 \right\}. \tag{5}$$

Moreover,

1. *for all $(x_1, \dots, x_d) \in S$ we have $c_{0,i} \leq x_i$ for each $i = 1, \dots, d$, and*
2. *$(c_{0,1}, \dots, c_{0,d}) \in S$ if and only if $s = 0$.*

Next, we prove some properties of sparse automatic sets. In particular, we look at how sparse sets behave under certain operations. By considering the growth rate of the corresponding sparse regular languages, we can see that sparseness of subsets of $\mathbb{N}^d$ is closed under dilation and translation.

Lemma 2. *Let $k \geq 2$ a natural number and let $S \subseteq \mathbb{N}^d$ be a sparse k-automatic set. Let $c_i \in \mathbb{Q}$ with $c_i \geq 0$ and $b_i \in \mathbb{Q}$ for $i = 1, \dots, d$. Then*

(a) If $(c_1,\ldots,c_d)\cdot S\subseteq\mathbb{N}^d$*, then it is sparse;*
(b) If $(b_1,\ldots,b_d)+S\subseteq\mathbb{N}^d$*, then it is sparse.*

We also have the following two observations.

Lemma 3. *Let* p *be prime,* $k\geq 2$ *a natural number and let* $S\subseteq\mathbb{N}^d$ *be a sparse* p*-automatic set. Then* $U:=\bigcup_{n\geq 0}p^n\cdot S$ *is also sparse* p*-automatic.*

Proof. Let $\mathcal{L}:=\{(n_1,\ldots,n_d)_p\colon(n_1,\ldots,n_d)\in S\}$ be the sparse regular language over Σ_p^d corresponding to S. Then U corresponds to the regular language $\mathcal{L}\cdot\{(0,\ldots,0)\}^*$, where $(0,\ldots,0)\in\Sigma_p^d$ which is sparse by the characterization of sparse regular languages.

Lemma 4. *Let* $k\geq 2$ *and* $d\geq 1$ *be natural numbers. Let* $A,B\subseteq\mathbb{N}^d$ *be sparse* k*-automatic sets. Then* $A\cup B$ *and* $A+B$ *are also sparse and* k*-automatic.*

Proof. We already know that k-automatic sets are closed under union and set addition (see [3, Theorem 5.6.3], for example). Now, since sparse sets can be decomposed as a finite union of simple sparse sets, and since a finite union of simple sparse sets is sparse, we see $A\cup B$ is sparse. Since $\pi_{A+B}(x)\leq\pi_A(x)\pi_B(x)$ and since A and B are sparse, we get the result.

3 Proof of the Main Theorem

We will prove the main theorem by first showing that the closure properties are indeed satisfied by the collection of algebraic multivariate formal power series with sparse support, and then that it is the smallest such. Therefore, we first make the following definition.

Definition 2. *Let* $B_d\subseteq C_d$ *be subalgebras of the ring of multivariate formal power series* $\bar{\mathbb{F}}_p[[t_1,\ldots,t_d]]$*. We say that* B_d *is* Artin–Schreier closed *in* C_d *if the following hold:*

(P1) if $F(t_1,\ldots,t_d)\in B_d$ *and* $F(0,\ldots,0)=0$*, then* $F(t_1,\ldots,t_d)+F(t_1^p,\ldots,t_d^p)+F(t_1^{p^2},\ldots,t_d^{p^2})+\cdots\in B_d$*;*
(P2) if $F(t_1,\ldots,t_d)\in B_d$ *and* $(\alpha_1,\alpha_2,\ldots,\alpha_d)\in\bar{\mathbb{F}}_p^d$*, then* $F(\alpha_1t_1,\ldots,\alpha_dt_d)\in B_d$*;*
(P3) if $F(t_1,\ldots,t_d)\in B_d$ *and* $t_1^{b_1}\cdots t_d^{b_d}F(t_1^{c_1},\ldots,t_d^{c_d})\in C_d$ *with* $(c_1,\ldots,c_d)\in\mathbb{Q}_{>0}^d$ *and* $(b_1,\ldots,b_d)\in\mathbb{Q}$*, then* $t_1^{b_1}\cdots t_d^{b_d}F(t_1^{c_1},\ldots,t_d^{c_d})\in B_d$*.*

Proposition 3. *Let* p *be prime. Then the collection of sparse algebraic power series in* $\bar{\mathbb{F}}_p[[t_1,t_2,\ldots,t_d]]$ *forms a subalgebra of the ring of algebraic power series; moreover, this subalgebra is Artin–Schreier closed in* $\bar{\mathbb{F}}_p[[t_1,\ldots,t_d]]$*.*

Proof. Fix $d \geq 1$. Let A_d denote the collection of sparse algebraic power series in $\bar{\mathbb{F}}_p[[t_1, \ldots, t_d]]$. Then to show that A_d is a subalgebra, it is sufficient to show that it is closed under sum and multiplication. Let $F(t_1, \ldots, t_d), G(t_1, \ldots, t_d) \in A_d$ and let S_F and S_G denote the supports of F and G respectively. Then the support of $(F+G)(t_1, \ldots, t_d)$ is contained in $S_F \cup S_G$ and by Lemma 4, $S_F \cup S_G$ is sparse and so $F+G$ has sparse support. The support of $(FG)(t_1, \ldots, t_d)$ is contained in $S_F + S_G$, where $S_F + S_G$ is the collection of d-tuples of natural numbers that can be expressed in the form $(a_1, \ldots, a_d) + (b_1, \ldots, b_d)$ with $(a_1, \ldots, a_d) \in S_F$ and $(b_1, \ldots, b_d) \in S_G$. Then $\pi_{S_F+S_G}(x) \leq \pi_{S_F}(x)\pi_{S_G}(x)$ and so again by Lemma 4 since $S_F + S_G$ is sparse, we get that $(FG)(t_1, \ldots, t_d)$ has sparse support.

Now we show the properties $(P1)$–$(P3)$. Let $F(t_1, \ldots, t_d) \in A_d$. First, we show $(P1)$ holds. Suppose $F(0, \ldots, 0) = 0$. Let S be the support of F. Since $F(t_1, \ldots, t_d) \in A_d$, F has sparse support, so S is a sparse set. Denote by T the support of $G(t_1, \ldots, t_d) = F(t_1, \ldots, t_d) + F(t_1^p, \ldots, t_d^p) + F(t_1^{p^2}, \ldots, t_d^{p^2}) + \cdots$. Then T is contained in $U := \cup_{n \geq 0} p^n \cdot S$, which is sparse by Lemma 3. So $G(t_1, \ldots, t_d) \in A_d$.

For $(P2)$, suppose $(\alpha_1, \alpha_2, \ldots, \alpha_d) \in \bar{\mathbb{F}}_p^d$. The support of $F(\alpha_1 t_1, \ldots, \alpha_d t_d)$ is contained in S. Hence $F(\alpha_1 t_1, \ldots, \alpha_d t_d) \in A_d$.

For $(P3)$, suppose $(c_1, \ldots, c_d) \in \mathbb{Q}_{>0}^d$ and $(b_1, \ldots, b_d) \in \mathbb{Q}$. Then the support of

$$t_1^{b_1} \cdots t_d^{b_d} F(t_1^{c_1}, \ldots, t_d^{c_d}) \in \bar{\mathbb{F}}_p[[t_1, t_2, \ldots, t_d]]$$

is contained in

$$(b_1, \ldots, b_d) + (c_1, \ldots, c_d) \cdot S$$

Then, by Lemma 2, $t_1^{b_1} \cdots t_d^{b_d} F(t_1^{c_1}, \ldots, t_d^{c_d}) \in A_d$.

Next, we have a lemma concerning algebraic multivariate formal power series whose support set is a simple sparse set.

Lemma 5. *Let p be a prime and let C_d be the smallest nontrivial $\bar{\mathbb{F}}_p$-subalgebra of $\bar{\mathbb{F}}_p[[t_1, t_2, \ldots, t_d]]$ that is Artin–Schreier closed in the ring $\bar{\mathbb{F}}_p[[t_1, t_2, \ldots, t_d]]$. Then, if $S \subseteq \mathbb{N}^d$ is a non-empty simple sparse set, then*

$$G(t_1, \ldots, t_d) := \sum_{(n_1, \ldots n_d) \in S} t_1^{n_1} \cdots t_d^{n_d} \quad \textit{is in } C_d.$$

Proof. We follow the proof of Lemma 4.4 [2] with adjustments to the multivariate case. By Lemma 1, there exist $s \geq 0$, for each $i = 1, \ldots, d$, $c_{j,i} \in \mathbb{Q}$ such that $(p^\ell - 1)c_{j,i} \in \mathbb{Z}$ for some $\ell \geq 0$, $\sum_{j=0}^s c_{j,i} \in \mathbb{Z}_{\geq 0}$ and positive integers $\delta_1, \ldots, \delta_s$ such that

$$S = \left\{ \left(\sum_{j=0}^{s} c_{j,1} p^{\sum_{m=s-j+1}^{s} \delta_m n_m}, \ldots, \sum_{j=0}^{s} c_{j,d} p^{\sum_{m=s-j+1}^{s} \delta_m n_m} \right) : n_1, \ldots, n_s \geq 0 \right\}. \quad (6)$$

Moreover, we have for all $(x_1,\dots,x_d) \in S$, $c_{0,i} \leq x_i$ for each $i=1,\dots,d$, and $(c_{0,1},\dots,c_{0,d}) \in S$ if and only if $s=0$. For any $d \geq 1$, we prove $G(t_1,\dots,t_d) \in C_d$ by induction on s. If $s=0$, then $G(t_1,\dots,t_d) = t_1^{c_{0,1}} \cdots t_d^{c_{0,d}}$ and so $G(t_1,\dots,t_d) \in C_d$. Suppose the result holds up to $s-1$ with $s>1$, we prove it for s.

For each $i=1,\dots,d$, let N_i be a positive integer coprime to p such that $N_i c_{j,i} \in \mathbb{Z}$ for $j=0,\dots,s$.

Let

$$T = \left\{ \left(N_1 \sum_{j=0}^{s-1} c_{j+1,1} p^{\sum_{m=s-j}^{s-1} \delta_m n_m}, \dots, N_d \sum_{j=0}^{s-1} c_{j+1,d} p^{\sum_{m=s-j}^{s-1} \delta_m n_m} \right) : n_1,\dots,n_{s-1} \geq 0 \right\}. \tag{7}$$

Then $T \subseteq \mathbb{Z}^d$ and since $s>1$, for all $(x_1,\dots x_d) \in S$ we have $x_i > c_{0,i}$ for each $i=1,\dots,d$ and so T is a sparse subset of $\mathbb{N}^d \setminus \{(0,\dots,0)\}$. We let $H(t_1,\dots,t_d) = \sum_{(n_1,\dots,n_d)\in T} t_1^{n_1}\cdots t_d^{n_d}$. Then $H(t_1,\dots,t_d) \in \bar{\mathbb{F}}_p[[t_1,\dots,t_d]]$ and by the induction hypothesis $H(t_1,\dots,t_d) \in C_d$. We also have $H(0,\dots,0)=0$.

We have

$$G(t_1^{N_1},\dots,t_d^{N_d}) = \sum_{(n_1,\dots,n_d)\in(N_1,\dots,N_d)\cdot S} t_1^{n_1}\cdots t_d^{n_d} \tag{8}$$

$$= t_1^{N_1 c_{0,1}} \cdots t_d^{N_d c_{0,d}} \left(\sum_{j\geq 0} H\left(t_1^{p^{j\delta_s}},\dots,t_d^{p^{j\delta_s}}\right)\right). \tag{9}$$

By the proof of Lemma 4.3(i) in [2], which is proved by using Combinatorial Nullstellensatz (see [4]), we have that for any integer $\delta \geq 1$, there exist $a_i \in \bar{\mathbb{F}}_p$, $i=1,\dots,c$ such that we have

$$\sum_{i=1}^{\delta} b_i a_i^{p^j} = \begin{cases} 1 & \text{if } j \equiv 0 \pmod{\delta}, \\ 0 & \text{otherwise.} \end{cases} \tag{10}$$

where $b_i \in \bar{\mathbb{F}}_p$. Let

$$H_i(t_1,\dots,t_d) = \sum_{n\geq 0} a_i^{p^n} H(t_1,\dots,t_d)^{p^n}.$$

Then, since $H(t_1,\dots,t_d) \in C_d$ by assumption, $a_i \in \bar{\mathbb{F}}_p$, and C_d is Artin–Schreier closed, we get $a_i H(t_1,\dots,t_d) \in C_d$. Then, again since C_d is Artin–Schreier closed, we get $H_i(t_1,\dots,t_d) \in C_d$ by $(P1)$. Observe that, taking $\delta=\delta_s$, we get

$$\sum_{i=1}^{\delta_s} b_i H_i(t_1,\dots,t_d) = H(t_1,\dots,t_d) + H(t_1^{p^{\delta_s}},\dots,t_d^{p^{\delta_s}}) + H(t_1^{p^{2\delta_s}},\dots,t_d^{p^{2\delta}}) + \cdots \in C_d$$

by Eq. 10. Therefore,

$$\sum_{j\geq 0} H\left(t_1^{p^{j\delta_s}},\ldots,t_d^{p^{j\delta_s}}\right)\in C_d.$$

Since $G(t_1^{N_1},\ldots,t_d^{N_d})\in \bar{\mathbb{F}}_p[[t_1,\ldots,t_d]]$, i.e. a formal power series, and C_d is Artin–Schreier closed in $\bar{\mathbb{F}}_p[[t_1,\ldots,t_d]]$, by Eq. 9, we get $G(t_1^{N_1},\ldots,t_d^{N_d})\in C_d$. Since $G(t_1,\ldots,t_d)\in \bar{\mathbb{F}}_p[[t_1,\ldots,t_d]]$ and again since C_d is Artin–Schreier closed in $\bar{\mathbb{F}}_p[[t_1,\ldots,t_d]]$, we have $G(t_1,\ldots,t_d)\in C_d$.

We are now ready to prove the main theorem. Our proof relies on the steps used in [2] for the univariate case.

Proof (Proof of Theorem 2). Let B_d be the ring of algebraic multivariate power series in d variables with sparse support and let C_d be the smallest nontrivial $\bar{\mathbb{F}}_p$-subalgebra of $\bar{\mathbb{F}}_p[[t_1,t_2,\ldots,t_d]]$ that is Artin–Schreier closed in the ring $\bar{\mathbb{F}}_p[[t_1,t_2,\ldots,t_d]]$. The statement of Theorem 2 is equivalent to $C_d = B_d$. By Proposition 3, we have $C_d \subseteq B_d$. We want to show $B_d \subseteq C_d$.

Let

$$G(t_1,\ldots,t_d) = \sum_{(n_1,\ldots,n_d)\in\mathbb{N}^d} g(n_1,\ldots,n_d)t_1^{n_1}\cdots t_d^{n_d}\in \bar{\mathbb{F}}_p[[t_1,\ldots,t_d]]$$

be a power series that is algebraic over $\bar{\mathbb{F}}_p[[t_1,\ldots,t_d]]$ with a sparse support set $S\subseteq\mathbb{N}^d$. Then G is in $\mathbb{F}_q[[t_1,\ldots,t_d]]$ for some q that is a power of p. For each nonzero $\alpha\in\mathbb{F}_q$, we define $S_\alpha := \{(n_1,\ldots,n_d)\in\mathbb{N}^d\colon g(n_1,\ldots,n_d)=\alpha\}\subseteq\mathbb{N}^d$ so that $S = \bigcup_{\alpha\neq 0,\alpha\in\mathbb{F}_q} S_\alpha$. For each $0\neq\alpha\in\mathbb{F}_q$, since S is a sparse set, we have that S_α is also sparse. We have

$$G(t_1,\ldots,t_d) = \sum_{\alpha\neq 0,\alpha\in\mathbb{F}_q}\alpha\left(\sum_{(n_1,\ldots,n_d)\in S_\alpha} t_1^{n_1}\cdots t_d^{n_d}\right).$$

Now since S_α is sparse, it can be written as a finite disjoint union of simple sparse sets by Equation (2). So we have

$$\bigsqcup_{i=1}^{m_\alpha} S_{\alpha,i}$$

for some integer $m_\alpha\geq 1$, where each $S_{\alpha,i}$ is a simple sparse set. For nonzero $\alpha\in\mathbb{F}_q$ and $i=1,\ldots,m_\alpha$, we define

$$G_{S_{\alpha,i}}(t_1,\ldots,t_d) := \sum_{(n_1,\ldots,n_d)\in S_{\alpha,i}} t_1^{n_1}\cdots t_d^{n_d}.$$

Then we have

$$G(t_1,\ldots,t_d) = \sum_{\alpha\neq 0,\alpha\in\mathbb{F}_q} \alpha\left(\sum_{i=1}^{m_\alpha} G_{S_{\alpha,i}}(t_1,\ldots,t_d)\right).$$

Now by Lemma 5, each $G_{S_{\alpha,i}}(t_1,\ldots,t_d)$ is in C_d and since C_d is Artin–Schreier closed, we get $G(t_1,\ldots,t_d) \in C_d$.

References

1. Adamczewski, B., Bostan, A., Caruso, X.: A sharper multivariate Christol's theorem with applications to diagonals and Hadamard products. arXiv preprint arXiv:2306.02640 (2023)
2. Albayrak, S., Bell, J.P.: A refinement of Christol's theorem for algebraic power series. Math. Z. **300**(3), 2265–2288 (2022)
3. Allouche, J.-P., Shallit, J.: Automatic Sequences. Theory, Applications, Generalizations. Cambridge University Press, Cambridge (2003)
4. Alon, N.: Combinatorial nullstellensatz. Comb. Probab. Comput. **8**(1–2), 7–29 (1999)
5. Bell, J.P., Ghioca, D., Moosa, R.: Effective isotrivial Mordell-Lang in positive characteristic. arXiv preprint arXiv:2010.08579 (2020)
6. Bell, J.P., Moosa, R.: F-sets and finite automata. Journal de Théorie des Nombres de Bordeaux **31**(1), 101–130 (2019)
7. Christol, G.: Ensembles presque périodiques k-reconnaissables. Theoret. Comput. Sci. **9**, 141–145 (1979)
8. Christol, G., Kamae, T., Mendès France, M., Rauzy, G.: Suites algébriques, automates et substitutions. Bull. Soc. Math. France **108**, 401–419 (1980)
9. Derksen, H.: A Skolem-Mahler-Lech theorem in positive characteristic and finite automata. Invent. Math. **168**(1), 175–224 (2007)
10. Gawrychowski, P., Krieger, D., Rampersad, N., Shallit, J.: Finding the growth rate of a regular or context-free language in polynomial time. Int. J. Found. Comput. Sci. **21**(4), 597–618 (2010)
11. Ginsburg, S., Spanier, E.: Bounded regular sets. Proc. Am. Math. Soc. **17**, 1043–1049 (1966)
12. Ibarra, O.H., Ravikumar, B.: On sparseness, ambiguity, and other decision problems for acceptors and transducers. In: STACS 86 (Orsay. 1986). LNCS, vol. 210, pp. 171–179. Springer, Berlin (1986)
13. Kedlaya, K.S.: Finite automata and algebraic extensions of function fields. Journal de Théorie des Nombres de Bordeaux **18**(2), 379–420 (2006)
14. Kedlaya, K.S.: On the algebraicity of generalized power series. Beiträge Algebra Geom. **58**(3), 499–527 (2017)
15. Moosa, R., Scanlon, T.: F-structures and integral points on Semiabelian varieties over finite fields. Am. J. Math. **126**(3), 473–522 (2004)
16. Salon, O.: Suites automatiques à multi-indices et algébricité. Comptes Rendus de l'Académie des Sciences Paris, Série I Mathématique **305**(12), 501–504 (1987)
17. Salon, O.: Suites automatiques à multi-indices (with an appendix by Shallit, J.). Séminaire de Théorie des Nombres Bordeaux **4**, 1–27 (1986–1987)

18. Szilard, A., Yu, S., Zhang, K., Shallit, J.: Characterizing regular languages with polynomial densities. In: Havel, I.M., Koubek, V. (eds.) MFCS 1992. LNCS, vol. 629, pp. 494–503. Springer, Heidelberg (1992). https://doi.org/10.1007/3-540-55808-X_48
19. Trofimov, V.I.: Growth functions of some classes of languages. Kibernetika **17**(6), 9–12, 149 (1981). (in Russian). English translation in Cybernetics **17**, 727–731 (1981)

Avoiding Abelian and Additive Powers in Rich Words

Jonathan Andrade[1] and Lucas Mol[2](✉)

[1] Department of Mathematics and Statistics, University of Victoria, Victoria, BC, Canada
jonathanandrade@uvic.ca
[2] Department of Mathematics and Statistics, Thompson Rivers University, Kamloops, BC, Canada
lmol@tru.ca

Abstract. This paper concerns the avoidability of abelian and additive powers in infinite rich words. In particular, we construct an infinite additive 5-power-free rich word over $\{0, 1\}$ and an infinite additive 4-power-free rich word over $\{0, 1, 2\}$. The alphabet sizes are as small as possible in both cases, even for abelian powers.

Keywords: Power avoidance · Rich words · Abelian powers · Additive powers

1 Introduction

This paper concerns the avoidability of certain types of repetitions in words. We consider words over alphabets where it makes sense to sum the letters. Over such an alphabet, the *sum* of a word w is simply the sum of the letters that make up w. We focus on alphabets that are subsets of the integers with the usual sum.

We consider repetitions of three different forms. For an integer $k \geq 2$, a word w that can be written in the form $w = w_1 w_2 \cdots w_k$, where

- w_i is equal to w_1 for all $i \in \{2, \ldots, k\}$ is called an *ordinary k-power*;
- w_i is an anagram of w_1 (i.e., w_i can be obtained from w_1 by permuting the order of the letters) for all $i \in \{2, \ldots, k\}$ is called an *abelian k-power*;
- w_i has the same length and the same sum as w_1 for all $i \in \{2, \ldots, k\}$ is called an *additive k-power*.

We often say *square* instead of 2-power, and *cube* instead of 3-power.

A word is *ordinary* (resp. *abelian*, *additive*) *k-power-free* if it contains no ordinary (resp. abelian, additive) k-power as a factor. Note that every ordinary k-power is an abelian k-power, and that every abelian k-power is an additive k-power. It follows that every additive k-power-free word is abelian k-power-free, and that every abelian k-power-free word is ordinary k-power-free.

G. Gamard and J. Leroy (Eds.): WORDS 2025, LNCS 15729, pp. 24–36, 2025.
https://doi.org/10.1007/978-3-031-97548-6_3

Our motivating problem is the following. Given a finite alphabet $X \subseteq \mathbb{Z}$, is there an infinite word over X that is ordinary (resp. abelian, additive) k-power-free? If so, then we say that ordinary (resp. abelian, additive) k-powers are *avoidable* over X. If $|X| < 2$, then it is easy to see that there is no such infinite word, so we may assume that $|X| \geq 2$.

Thue [25] solved this problem completely for ordinary k-powers by showing that there is an infinite ordinary cube-free binary word, and an infinite ordinary square-free ternary word. (It is easy to show that there are only finitely many ordinary square-free words over a binary alphabet.) In fact, Thue showed that there is a binary word that contains no *overlap*, i.e., a word of the form $axaxa$, where a is a single letter and x is a (possibly empty) word. Thue's results are considered the origin of the area of *combinatorics on words.*

Erdős [12] was first to ask about the avoidability of abelian k-powers, and this variant of the problem has since been solved completely. Dekking [9] constructed an infinite abelian 4-power-free binary word, and an infinite abelian cube-free ternary word. Keränen [14] constructed an infinite abelian square-free word over a four-letter alphabet. In all three cases, the alphabet size is as small as possible.

Finally, we summarize the progress on the problem for additive k-powers. First note that over an integer alphabet of size two, words of the same length have the same sum if and only if they are anagrams of one another. Thus, over an integer alphabet of size 2, the notion of additive k-power is the same as the notion of abelian k-power. Therefore, it follows from Dekking's result mentioned above that there is an infinite additive 4-power-free word over every subset of $\mathbb{Z}$ of size at least two. Cassaigne, Currie, Schaeffer, and Shallit [5] were the first to construct an infinite additive cube-free word over a finite subset of $\mathbb{Z}$ (namely $\{0, 1, 3, 4\}$). Lietard and Rosenfeld [15] extended this result by showing that there is an infinite additive cube-free word over *every* subset of $\mathbb{Z}$ of size 4, except possibly those subsets equivalent to $\{0, 1, 2, 3\}$, i.e., subsets that are arithmetic progressions of length 4. It is still unknown whether there is an infinite additive cube-free word over $\{0, 1, 2, 3\}$. Rao [20] constructed infinite additive cube-free words over several integer alphabets of size 3. It is still unknown whether there exists an infinite additive square-free word over a finite subset of $\mathbb{Z}$. However, Rao and Rosenfeld [21] constructed an infinite additive square-free word over a finite subset of $\mathbb{Z}^2$.

In this paper, we consider the avoidability of powers in *rich* words. A word of length n is called *rich* if it contains $n+1$ distinct palindromes as factors, and an infinite word is called *rich* if all of its finite factors are rich. Since their (implicit) introduction by Droubay, Justin, and Pirillo [11], the language of rich words has been well-studied. It is known that the set of infinite rich words contains several highly structured classes of words, including Sturmian words, episturmian words, and complementary symmetric Rote words (see [4, 11]).

There are several papers concerning the avoidability of ordinary powers in rich words. Pelantová and Starosta [19] proved that every infinite rich word, over *any* finite alphabet, contains an ordinary square. (In fact, they proved that every infinite rich word contains infinitely many distinct overlaps as factors.)

On the other hand, it is known that there are infinite ordinary cube-free binary rich words; in fact, the results of Baranwal and Shallit [3] and Currie, Mol, and Rampersad [6] describe exactly which *fractional* ordinary powers (between squares and cubes) are avoidable in binary rich words.

In this paper, we establish two results about the avoidability of abelian and additive powers in rich words.

Theorem 1. There is an infinite additive 5-power-free rich word over $\{0, 1\}$.

Theorem 2. There is an infinite additive 4-power-free rich word over $\{0, 1, 2\}$.

These results are best possible in the sense that there are only finitely many abelian 4-power-free rich words over any binary alphabet, and only finitely many abelian cube-free rich words over any ternary alphabet. Using a backtracking algorithm, we determined that the longest abelian 4-power-free binary rich word has length 2411, and that the longest abelian cube-free ternary rich word has length 180. These long backtracking searches were made possible by the use of the data structure `eerTree` of Rubinchik and Shur [22].

The constructions that we use to prove Theorems 1 and 2 are obtained by iterating morphisms. Let $\beta : \{0, 1\}^* \to \{0, 1\}^*$ and $\gamma : \{0, 1, 2\}^* \to \{0, 1, 2\}^*$ be the morphisms defined as follows:

$$\beta(0) = 00001 \qquad \gamma(0) = 2$$
$$\beta(1) = 01101 \qquad \gamma(1) = 101$$
$$\gamma(2) = 10001$$

Let $\mathbf{B} = \beta^\omega(0)$ and $\mathbf{\Gamma} = \gamma^\omega(1)$. We prove Theorem 1 by showing that $\mathbf{B}$ is rich and additive 5-power-free, and we prove Theorem 2 by showing that $\mathbf{\Gamma}$ is rich and additive 4-power-free.

In order to prove the additive power-freeness of $\mathbf{B}$ and $\mathbf{\Gamma}$, we implemented the algorithm of Currie, Mol, Rampersad, and Shallit [7], which is capable of deciding whether words produced by iterating certain morphisms are additive k-power-free. Our implementation of the algorithm can be found at https://github.com/lgmol/Additive-Powers-Decision-Algorithm.

In order to prove that $\mathbf{B}$ and $\mathbf{\Gamma}$ are rich, we take two different approaches. We prove that $\mathbf{B}$ is rich using the automatic theorem-proving software `Walnut`. We prove that $\mathbf{\Gamma}$ is rich more or less directly using a well-known characterization of rich words. We note that the morphism γ^2 can be "conjugated" to a morphism in the class P_{ret}, which has been studied by several authors in the context of rich words [2,10]. While there is a simple criterion for deciding whether a word obtained by iterating a *binary* morphism in class P_{ret} is rich [10], no such simple criterion is known for ternary morphisms. Ultimately, we found it slightly easier to work with γ than the corresponding morphism in class P_{ret}.

The layout of the remainder of the paper is as follows. In Sect. 2, we provide the necessary background. In Sect. 3, we prove Theorem 1. In Sect. 4, we prove Theorem 2. Finally, in Sect. 5, we discuss some open problems related to our work.

2 Background

We provide only a very brief overview of the basic background material on combinatorics on words. See the books of Lothaire [16–18] or the more recent book of Shallit [24] for a more comprehensive treatment.

2.1 Words

An *alphabet* is a nonempty finite set of symbols, which we refer to as *letters*. In this paper, we focus on alphabets that are finite subsets of the integers. Let X be such an alphabet. A *word* over X is a finite or infinite sequence of letters from X. We let X^* denote the set of all finite words over the alphabet X. The *length* of a finite word w, denoted by $|w|$, is the number of letters that make up w. We let ε denote the *empty word*, which is the unique word of length 0. The *sum* of a finite word w, denoted by $\sum w$, is the sum of the letters that make up w.

For a finite word x and a finite or infinite word y, the *concatenation* of x and y, denoted by xy, is the word consisting of all of the letters of x followed by all of the letters of y. Suppose that a finite or infinite word w can be written in the form $w = xyz$, where x, y, and z are possibly empty words. Then the word y is called a *factor* of w, the word x is called a *prefix* of w, and the word z is called a *suffix* of w. If x and z are nonempty, then y is called an *internal factor* of w. If yz is nonempty, then x is called a *proper prefix* of w, and if xy is nonempty, then z is called a *proper suffix* of w.

2.2 Morphisms

For alphabets X and Y, a *morphism* from X^* to Y^* is a function $h : X^* \to Y^*$ that satisfies $h(uv) = h(u)h(v)$ for all words $u, v \in X^*$. We say that h is *strictly growing* if $|h(a)| \geq 2$ for all $a \in X$. For an integer $k \geq 1$, we say that h is *k-uniform* if $|h(a)| = k$ for all $a \in X$.

Let $h : X^* \to X^*$ be a morphism. For all words $x \in X^*$, we define $h^0(x) = x$, and $h^n(x) = h(h^{n-1}(x))$ for all integers $n \geq 1$. For a letter $a \in X$, we say that h is *prolongable* on a if $h(a) = ax$ for some word $x \in X^*$ and $h^n(x) \neq \varepsilon$ for all $n \geq 0$. If h is prolongable on a with $h(a) = ax$, then it is easy to show that for every integer $n \geq 1$, we have

$$h^n(a) = axh(x)h^2(x) \cdots h^{n-1}(x).$$

Thus, each $h^n(a)$ is a prefix of the infinite word

$$h^\omega(a) = axh(x)h^2(x)h^3(x) \cdots .$$

Note that $h^\omega(a)$ is a fixed point of h, i.e., $h(h^\omega(a)) = h^\omega(a)$, where the morphism h is extended to infinite words in the natural way.

2.3 Rich Words

For a finite word $w = w_1 w_2 \cdots w_{n-1} w_n$, where the w_i are letters, the *reversal* of w, denoted $\overline{w}$, is the word obtained by writing the letters of w in the opposite order, i.e.,

$$\overline{w} = w_n w_{n-1} \cdots w_2 w_1.$$

A word w is a *palindrome* if $\overline{w} = w$, i.e., if w reads the same backward as forward. Some examples of palindromes in the English language include `racecar`, `level`, and `kayak`.

Droubay, Justin, and Pirillo [11] were first to observe that every word of length n contains at most $n+1$ distinct palindromes (including the empty word) as factors. This explains the eventual use of the term *rich* for words of length n containing $n+1$ distinct palindromes as factors; these words are "rich" in palindromes.

Let us briefly explain the idea behind Droubay, Justin, and Pirillo's observation. A factor u of a word w is said to be *unioccurrent* in w if u occurs exactly once in w. For example, in the word `011011`, the factor `11011` is unioccurrent, while the factor `11` is not. Observe that every finite word has at most one unioccurrent palindromic suffix (which must necessarily be its longest palindromic suffix). Further, for a finite word w and a letter a, every palindromic factor of wa that is not a factor of w must be a unioccurrent palindromic suffix of wa. So adding a letter to the end of a finite word can add at most one new palindromic factor to the word. Since the empty word contains just one palindromic factor (namely ε itself), it follows by a straightforward induction that every word of length n contains at most $n+1$ distinct palindromic factors.

We will use the following basic results about rich words, which can be proven using simple variations on the idea from the previous paragraph. These results were first proven by Droubay, Justin, and Pirillo [11], but we state them using the updated terminology of Glen, Justin, Widmer, and Zamboni [13].

Theorem 3. A finite word w is rich if and only if every prefix (resp. suffix) of w has a unioccurrent palindromic suffix (resp. prefix).

Lemma 1. If a finite word w is rich, then every factor of w is rich.

Theorem 4. An infinite word $\mathbf{w}$ is rich if and only if every finite prefix of $\mathbf{w}$ has a unioccurrent palindromic suffix.

2.4 A Decision Algorithm for Additive k-Power-Freeness

In order to establish that $\mathbf{B}$ is additive 5-power-free and $\mathbf{\Gamma}$ is additive 4-power-free, we implemented the decision algorithm described by Currie, Mol, Rampersad, and Shallit [7], which applies to fixed points of certain "affine" morphisms. A morphism $f : X^* \to Y^*$ where $X, Y \subseteq \mathbb{Z}$ is called *affine* if there exist $a, b, c, d \in \mathbb{Z}$ such that for all $x \in X$, we have $|f(x)| = a + bx$ and $\sum f(x) = c + dx$. In this case, we define the matrix

$$M_f = \begin{bmatrix} a & b \\ c & d \end{bmatrix}.$$

Theorem 5 (Currie, Mol, Rampersad, and Shallit [7]). Let $f : X^* \to X^*$ be a strictly growing affine morphism such that f is prolongable on a. If M_f is invertible and every eigenvalue λ of M_f satisfies $|\lambda| > 1$, then it is possible to decide whether or not $f^\omega(a)$ is additive k-power-free.

Recall the morphisms β and γ defined in Sect. 1, and let $\delta = \gamma^2$. So we have

$$\begin{array}{lll} \beta(\mathtt{0}) = \mathtt{00001} & \gamma(\mathtt{0}) = \mathtt{2} & \delta(\mathtt{0}) = \mathtt{10001} \\ \beta(\mathtt{1}) = \mathtt{01101} & \gamma(\mathtt{1}) = \mathtt{101} & \delta(\mathtt{1}) = \mathtt{1012101} \\ & \gamma(\mathtt{2}) = \mathtt{10001} & \delta(\mathtt{2}) = \mathtt{101222101} \end{array}$$

and $\mathbf{B} = \beta^\omega(\mathtt{0})$ and $\mathbf{\Gamma} = \gamma^\omega(\mathtt{1}) = \delta^\omega(\mathtt{1})$. First note that β and δ are strictly growing, and that β is prolongable on $\mathtt{0}$, while δ is prolongable on $\mathtt{1}$. Further, for all $x \in \{\mathtt{0}, \mathtt{1}\}$, we have $|\beta(x)| = 5 + 0x$ and $\sum \beta(x) = 1 + 2x$, so β is affine, and we have

$$M_\beta = \begin{bmatrix} 5 & 0 \\ 1 & 2 \end{bmatrix}.$$

Similarly, for the morphism δ, we have $|\delta(x)| = 5 + 2x$ and $\sum \delta(x) = 2 + 4x$ for all $x \in \{\mathtt{0}, \mathtt{1}, \mathtt{2}\}$. Therefore, δ is affine, and we have

$$M_\delta = \begin{bmatrix} 5 & 2 \\ 2 & 4 \end{bmatrix}.$$

Note that M_β is invertible with eigenvalues 2 and 5, and that M_δ is invertible with eigenvalues $\frac{9+\sqrt{17}}{2}$ and $\frac{9-\sqrt{17}}{2}$. Hence, the conditions of Theorem 5 are met, and we can use the decision algorithm of Currie, Mol, Rampersad, and Shallit to prove that $\mathbf{B}$ is additive 5-power-free and $\mathbf{\Gamma}$ is additive 4-power free.

The algorithm makes use of the *template method*, which was originally developed by Currie and Rampersad [8] to decide abelian k-power-freeness. Rao and Rosenfeld [21] established a more general version of the method, and found a way to apply it to both abelian and additive powers. The algorithm which we implement has more restrictive conditions on the morphism and works only for additive powers, but it has the advantage of being easy to implement in general (without worrying about floating point error, for example) and relatively efficient compared to previous template method algorithms.

The essential idea behind the algorithm is that if $f^\omega(a)$ contains a sufficiently long additive k-power, then it must contain some short word that is not "too far" from an additive k-power. We use objects called *templates* to describe structures that are "close to" additive k-powers. The conditions on f ensure that if $f^\omega(a)$ contains a sufficiently long additive k-power, then it contains some short instance of one of only finitely many *ancestor templates*. So to prove that $f^\omega(a)$ is additive k-power-free, it suffices to complete the following three steps.

- **Initial Check:** Check that there is no "short" additive k-power in $f^\omega(a)$.
- **Calculating Ancestors:** Find the set $\mathcal{A}$ of all ancestor templates.
- **Final Check:** Check that there is no "short" word in $f^\omega(a)$ that matches an ancestor template.

The exact length of the "short" factors that need to be checked depends on the integer k, the morphism f, and the set $\mathcal{A}$ of all ancestor templates. For more details concerning the algorithm and its implementation, see the undergraduate Honour's thesis of the first author [1].

Finally, it is interesting to note that while the conditions on the morphism for the algorithm of Rao and Rosenfeld [21] seem less restrictive in general than those of the algorithm we implement, the morphisms γ and $\delta = \gamma^2$ do not satisfy the conditions of Rao and Rosenfeld. In particular, the incidence matrix of the morphism γ, namely

$$\begin{bmatrix} 0 & 1 & 3 \\ 0 & 2 & 2 \\ 1 & 0 & 0 \end{bmatrix},$$

has eigenvalue 1. Rao and Rosenfeld [21, Sect. 6.1] stated that they did not know of such a morphism with an abelian k-power-free fixed point.

3 An Additive 5-Power-Free Binary Rich Word

In this section, we prove Theorem 1.

Proposition 1. The word $\mathbf{B} = \beta^\omega(\mathtt{0})$ is additive 5-power-free.

Proof. We verify that $\mathbf{B}$ is additive 5-power-free by running the algorithm described in Subsect. 2.4. Our implementation of the algorithm can be found at https://github.com/lgmol/Additive-Powers-Decision-Algorithm, and includes this specific run of the algorithm. □

Proposition 2. The word $\mathbf{B}$ is rich.

Proof. Since the morphism β is 5-uniform, the word $\mathbf{B}$ is 5-automatic, and we can use the automatic theorem-proving software `Walnut`, which is capable of deciding statements written in a certain first-order logic about automatic sequences. See the recent book of Shallit [24] for a comprehensive guide to using `Walnut`.

By Theorem 4, it suffices to show that every finite prefix of $\mathbf{B}$ has a unioccurrent palindromic suffix. We express this property in the first-order logic of `Walnut` below. Similar `Walnut` commands were used by Baranwal and Shallit [3] and Schaeffer and Shallit [23] in proving that other infinite words are rich. ïž£

```
morphism b "0->00001 1->01101":
    # defines the morphism beta (using the letter b as a name)


promote B b:
    # defines a DFAO generating the fixed point B of b


def FactorEq "?msd_5 Ak (k<n)=>(B[i+k]=B[j+k])":
    # takes 3 parameters i,j,n and returns true if
    # the length-n factors of B starting at indices i and j are equal
```

```
def Occurs "?msd_5 (m<= n) & ( Ek (k+m<=n) & $FactorEq(i,j+k,m))":
    # takes 4 parameters i,j,m,n and returns true if
    # the length-m factor of B starting at index i
    # occurs in the length-n factor of B starting at index j


def Palindrome "?msd_5 Aj,k ((k<n) & (j+k+1=n)) => (B[i+k]=B[i+j])":
    # takes 2 parameters i,n and returns true if
    # the length-n factor of B starting at index i is a palindrome


def BisRich "?msd_5 An Ej $Palindrome(j,n-j) & ~$Occurs(j,0,n-j,n-1)":
    # returns true if every finite prefix of B has a
    # unioccurrent palindromic suffix, i.e., if B is rich
```

When we run all of these commands, `Walnut` returns `TRUE`. □

Theorem 1 follows immediately from Propositions 1 and 2.

4 An Additive 4-Power-Free Ternary Rich Word

In this section, we prove Theorem 2.

Proposition 3. The word $\mathbf{\Gamma} = \gamma^{\omega}(1)$ is additive 4-power-free.

Proof. We verify that $\mathbf{\Gamma}$ is additive 4-power-free by running the algorithm described in Subsection 2.4 with the morphism $\delta = \gamma^2$. Our implementation of the algorithm can be found at https://github.com/lgmol/Additive-Powers-Decision-Algorithm, and includes this specific run of the algorithm. □

In order to prove that $\mathbf{\Gamma}$ is rich, we start with the following basic lemma.

Lemma 2. If $u \in \{0,1,2\}^*$ is a palindrome, then $\gamma(u)$ is a palindrome.

Proof. We proceed by induction on the length of u. First consider the case that $|u| \leq 1$. The statement holds trivially when $u = \varepsilon$, and it is straightforward to check that $\gamma(0)$, $\gamma(1)$, and $\gamma(2)$ are palindromes. Now suppose for some integer $n \geq 2$ that u has length n, and that for every palindrome $v \in \{0,1,2\}^*$ of length less than n, the word $\gamma(v)$ is a palindrome. Since $|u| \geq 2$, we can write

$$u = awa$$

for some letter $a \in \{0,1,2\}$ and some palindrome $w \in \{0,1,2\}^*$. By the inductive hypothesis, the words $\gamma(w)$ and $\gamma(a)$ are both palindromes. Thus, we have

$$\overline{\gamma(u)} = \overline{\gamma(a)}\ \overline{\gamma(w)}\ \overline{\gamma(a)} = \gamma(a)\gamma(w)\gamma(a) = \gamma(u),$$

and we conclude that $\gamma(u)$ is a palindrome. □

Proposition 4. The word $\mathbf{\Gamma}$ is rich.

Proof. By Theorem 4, it suffices to show that every finite prefix of $\mathbf{\Gamma}$ has a unioccurrent palindromic suffix. We proceed by induction on the length of the prefix. For the base case, we check by computer that every prefix of $\mathbf{\Gamma}$ of length at most 14 has a unioccurrent palindromic suffix.

Now suppose for some integer $n \geq 15$ that every prefix of $\mathbf{\Gamma}$ of length less than n has a unioccurrent palindromic suffix, and let P be the prefix of $\mathbf{\Gamma}$ of length n. Let p be the longest prefix of $\mathbf{\Gamma}$ such that P has $\gamma(p)$ as a prefix, and let $a \in \{\texttt{0},\texttt{1},\texttt{2}\}$ be the unique letter such that $q = pa$ is a prefix of $\mathbf{\Gamma}$. Then we can write

$$P = \gamma(p)r,$$

where r is a proper prefix of $\gamma(a)$. (See Fig. 1.) Note that since $n \geq 15$, we have $|p| \geq 5$.

Since γ is nonerasing and $|\gamma(\texttt{1})| = 3$, we have

$$|q| = |p| + 1 < |\gamma(p)| \leq n.$$

Hence, by the inductive hypothesis, every prefix of q has a unioccurrent palindromic suffix. It follows by Theorem 3 that p and q are rich. Let s be the unioccurrent palindromic suffix of p, and let t be the unioccurrent palindromic suffix of q. Since $|p| \geq 5$ and every letter appears in the prefix of p of length 4, both s and t must have length at least 2.

Case 1: Suppose that $r = \varepsilon$.
We claim that

$$S = \gamma(s)$$

is a unioccurrent palindromic suffix of P. Since s is a palindrome, it follows from Lemma 2 that S is a palindrome. Note that every occurrence of S in P corresponds to an occurrence of s in p. Thus, since s is unioccurrent in p, we conclude that S is unioccurrent in P.

Case 2: Suppose that $r \in \{\texttt{100}, \texttt{1000}\}$.
Since r is only a prefix of the γ-image of the letter 2, we must have $a = \texttt{2}$. Write $t = \texttt{2}t'\texttt{2}$. We claim that

$$S = \overline{r}\gamma(t')r$$

is a unioccurrent palindromic suffix of P. Since t is a palindrome, the word t' is also a palindrome, and it follows easily from Lemma 2 that S is a palindrome. Note that $\overline{r}$ is a suffix of $\gamma(\texttt{2})$, so that S is indeed a suffix of P. Finally, since r (resp. $\overline{r}$) is only a prefix (resp. suffix) of the γ-image of 2, we see that every occurrence of S in P corresponds to an occurrence of $t = \texttt{2}t'\texttt{2}$ in q. Since t is unioccurrent in q, we conclude that S is unioccurrent in P.

Case 3: Suppose that $r \in \{\texttt{1}, \texttt{10}\}$.
Since r is not a prefix of $\gamma(\texttt{0})$, we must have $a \in \{\texttt{1}, \texttt{2}\}$. We have two subcases.

Case 3.1: Suppose that p ends in $\texttt{0}$.
Since 02 is not a factor of $\mathbf{\Gamma}$, we must have $a = \texttt{1}$. Write $t = \texttt{1}t'\texttt{1}$. We claim that

$$S = \overline{r}\gamma(t')r$$

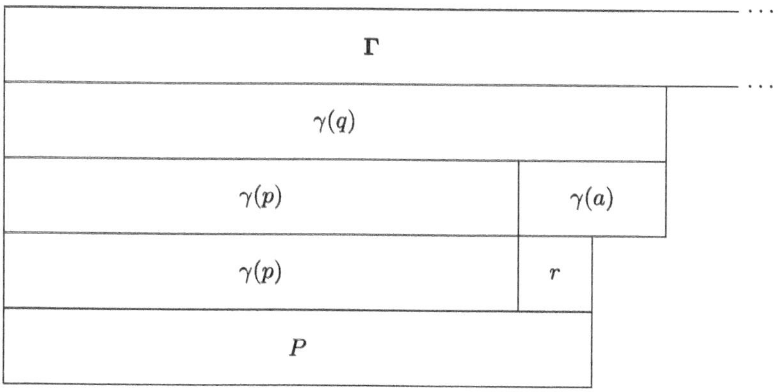

Fig. 1. The prefix P of $\mathbf{\Gamma}$

is a unioccurrent palindromic suffix of P. The proof that S is a palindromic suffix of P can be completed as in Case 2. Since p ends in $\mathtt{0}$ and t' is a palindromic suffix of p, we see that t' begins and ends in $\mathtt{0}$. Since $\mathtt{20}$ and $\mathtt{02}$ are not factors of $\mathbf{\Gamma}$, we see that every occurrence of S in P corresponds to an occurrence of $t = \mathtt{1}t'\mathtt{1}$ in q. Since t is unioccurrent in q, we conclude that S is unioccurrent in P.

Case 3.2: Suppose that p ends in $\mathtt{1}$ or $\mathtt{2}$.
Write $p = \mathtt{1}p'$. Since p is rich, we see by Lemma 1 that p' is also rich. Let s' be the unioccurrent palindromic suffix of p'. (Note that s' only differs from s in the case that $s = p$.) Since p' has prefix $\mathtt{0121}$, we must have $|s'| \geq 2$.

We claim that

$$S = \overline{r}\gamma(s')r$$

is a unioccurrent palindromic suffix of P. Since s' is a palindrome, we see by Lemma 2 that $\gamma(s')$ is a palindrome, and in turn that S is a palindrome.

Next, we show that S is a suffix of P. First note that since s' is a proper suffix of p, there is some letter b such that bs' is a suffix of p, and in turn $bs'a$ is a suffix of q. Suppose towards a contradiction that $b = \mathtt{0}$. Then $bs'a = \mathtt{0}s'a$. Since p ends in $\mathtt{1}$ or $\mathtt{2}$ and s' is a palindromic suffix of p, we see that s' must begin and end in either $\mathtt{1}$ or $\mathtt{2}$. Since $\mathtt{02}$ is not a factor of $\mathbf{\Gamma}$, we see that s' must begin and end in $\mathtt{1}$. Write $s' = \mathtt{1}s''\mathtt{1}$, so that $bs'a = \mathtt{01}s''\mathtt{11}$. Since every occurrence of $\mathtt{01}$ in $\mathbf{\Gamma}$ is followed by $\mathtt{1}$ or $\mathtt{2}$, while every occurrence of $\mathtt{11}$ in $\mathbf{\Gamma}$ is preceded by $\mathtt{0}$, we see that s'' begins in $\mathtt{1}$ or $\mathtt{2}$ and ends in $\mathtt{0}$. But this contradicts the fact that s' is a palindrome. So we have $b \in \{\mathtt{1}, \mathtt{2}\}$, which means that $\overline{r}$ is a suffix of $\gamma(b)$, and it follows that S is a suffix of P.

Finally, since every occurrence of S in P must arise from an occurrence of s' in p', and s' is unioccurrent in p', we conclude that S is unioccurrent in P. $\square$

Theorem 2 follows immediately from Propositions 3 and 4.

5 Open Problems

The first problem that we mention is analogous to the one considered by Lietard and Rosenfeld [15].

Problem 1. Determine the subsets of $\mathbb{Z}$ over which there is an infinite additive 4-power-free rich word.

We have shown that there is an infinite additive 4-power-free rich word over $\{0, 1, 2\}$, and that there is no such word over any subset of $\mathbb{Z}$ of size 2. It is possible that there is an infinite additive 4-power-free rich word over *every* subset of $\mathbb{Z}$ of size 3. In fact, based on computational evidence, the word $\mathbf{\Gamma}_{\mathtt{a},\mathtt{b},\mathtt{c}}$ obtained from $\mathbf{\Gamma}$ by applying the morphism defined by $\mathtt{0} \mapsto \mathtt{a}$, $\mathtt{1} \mapsto \mathtt{b}$, and $\mathtt{2} \mapsto \mathtt{c}$, appears to be additive 4-power-free for many choices of distinct integers $\mathtt{a}$, $\mathtt{b}$, and $\mathtt{c}$. Perhaps ideas similar to those of Lietard and Rosenfeld [15] could be applied here.

It is known that every sufficiently long rich word (over *any* alphabet) contains an ordinary square [19], and hence an additive square. However, a positive answer to the following problem seems plausible to us.

Problem 2. Is there an infinite additive cube-free rich word over some finite subset of $\mathbb{Z}$?

Disclosure of Interests. The authors have no competing interests to declare that are relevant to the content of this article.

Acknowledgments. Research of Lucas Mol is supported by NSERC grant RGPIN-2021-04084. We thank the anonymous reviewers for a careful reading of the paper and helpful suggestions. When searching for a ternary morphism with a rich and additive 4-power-free fixed point, we originally found the morphism δ, and did not notice that it is the square of the morphism γ. We thank Pascal Ochem for alerting us to this fact, which simplified the proof of Proposition 4.

References

1. Andrade, J.: Avoiding additive powers in words. Undergraduate Honour's thesis, Thompson Rivers University (2024). https://tru.arcabc.ca/islandora/object/tru%3A6415
2. Balková, Ľ, Pelantová, E., Starosta, Š: Infinite words with finite defect. Adv. Appl. Math. **47**(3), 562–574 (2011). https://doi.org/10.1016/j.aam.2010.11.006
3. Baranwal, A.R., Shallit, J.: Repetitions in infinite palindrome-rich words. In: WORDS 2019. LNCS, vol. 11682, pp. 93–105. Springer-Verlag (2019). https://doi.org/10.1007/978-3-030-28796-2_7
4. Blondin Massé, A., Brlek, S., Labbé, S., Vuillon, L.: Palindromic complexity of codings of rotations. Theoret. Comput. Sci. **412**(46), 6455–6463 (2011). https://doi.org/10.1016/j.tcs.2011.08.007

5. Cassaigne, J., Currie, J.D., Schaeffer, L., Shallit, J.: Avoiding three consecutive blocks of the same size and same sum. J. ACM **61**(2), 1–17 (2014). https://doi.org/10.1145/2590775
6. Currie, J.D., Mol, L., Rampersad, N.: The repetition threshold for binary rich words. Discrete Math. Theoret. Comput. Sci. **22**(1), 1–16 (2020). Article no. 6. https://doi.org/10.23638/DMTCS-22-1-6
7. Currie, J.D., Mol, L., Rampersad, N., Shallit, J.: Extending Dekking's construction of an infinite binary word avoiding abelian 4-powers. SIAM J. Discrete Math. **38**(4), 2913–2925 (2024). https://doi.org/10.1137/23M1558513
8. Currie, J.D., Rampersad, N.: Fixed points avoiding Abelian k-powers. J. Comb. Theory Ser. A **119**(5), 942–948 (2012). https://doi.org/10.1016/j.jcta.2012.01.006
9. Dekking, F.M.: Strongly non-repetitive sequences and progression-free sets. J. Comb. Theor. Ser. A **27**(2), 181–185 (1979). https://doi.org/10.1016/0097-3165(79)90044-X
10. Dolce, F., Pelantová, E.: On morphisms preserving palindromic richness. Fund. Inform. **185**(1), 1–25 (2022). https://doi.org/10.3233/FI-222102
11. Droubay, X., Justin, J., Pirillo, G.: Episturmian words and some constructions of de Luca and Rauzy. Theoret. Comput. Sci. **255**(1–2), 539–553 (2001). https://doi.org/10.1016/S0304-3975(99)00320-5
12. Erdős, P.: Some unsolved problems. Magyar Tud. Akad. Mat. Kutató Int. Közl. **6**, 221–254 (1961). https://users.renyi.hu/~p_erdos/1961-22.pdf
13. Glen, A., Justin, J., Widmer, S., Zamboni, L.Q.: Palindromic richness. European J. Combin. **30**(2), 510–531 (2009). https://doi.org/10.1016/j.ejc.2008.04.006
14. Keränen, V.: Abelian squares are avoidable on 4 letters. In: ICALP 1992. LNCS, vol. 623, pp. 41–52. Springer (1992). https://doi.org/10.1007/3-540-55719-9_62
15. Lietard, F., Rosenfeld, M.: Avoidability of additive cubes over alphabets of four numbers. In: DLT 2020. LNCS, vol. 12086, pp. 192–206. Springer-Verlag (2020). https://doi.org/10.1007/978-3-030-48516-0_15
16. Lothaire, M.: Combinatorics on Words. Encyclopedia of Mathematics and its Applications, vol. 17. Addison-Wesley (1983). https://doi.org/10.1017/CBO9780511566097
17. Lothaire, M.: Algebraic Combinatorics on Words. Encyclopedia of Mathematics and its Applications, vol. 90. Cambridge University Press (2002). https://doi.org/10.1017/CBO9781107326019
18. Lothaire, M.: Applied Combinatorics on Words Encyclopedia of Mathematics and its Applications, vol. 105. Cambridge University Press (2005). https://doi.org/10.1017/CBO9781107341005
19. Pelantová, E., Starosta, Š: Languages invariant under more symmetries: overlapping factors versus palindromic richness. Discrete Math. **313**(21), 2432–2445 (2013). https://doi.org/10.1016/j.disc.2013.07.002
20. Rao, M.: On some generalizations of abelian power avoidability. Theoret. Comput. Sci. **601**, 39–46 (2015). https://doi.org/10.1016/j.tcs.2015.07.026
21. Rao, M., Rosenfeld, M.: Avoiding two consecutive blocks of same size and same sum over $\mathbb{Z}^2$. SIAM J. Discrete Math. **32**(4), 2381–2397 (2018). https://doi.org/10.1137/17M1149377
22. Rubinchik, M., Shur, A.M.: EERTREE: an efficient data structure for processing palindromes in strings. European J. Combin. **68**, 249–265 (2018). https://doi.org/10.1016/j.ejc.2017.07.021
23. Schaeffer, L., Shallit, J.: Closed, palindromic, rich, privileged, trapezoidal, and balanced words in automatic sequences. Electron. J. Combin. **23**(1), 1–19 (2016). Article no. P1.25. https://doi.org/10.37236/5752

24. Shallit, J.: The Logical Approach To Automatic Sequences: Exploring Combinatorics on Words with `Walnut`. London Mathematical Society Lecture Note Series, vol. 482. Cambridge University Press (2022). https://doi.org/10.1017/9781108775267
25. Thue, A.: Über die gegenseitige Lage gleicher Teile gewisser Zeichenreihen. Norske vid. Selsk. Skr. Mat. Nat. Kl. **1**, 1–67 (1912)

Maximal 2-Dimensional Binary Words of Bounded Degree

Alexandre Blondin Massé[1,3], Alain Goupil[2,3], Raphael L' Heureux[2], and Louis Marin[4(✉)]

[1] Université du Québec à Montréal, Montréal, Canada
[2] Université du Québec à Trois-Rivières, Trois-Rivières, Canada
[3] LACIM, Montréal, Canada
[4] LIGM, Université Gustave Eiffel, Champs-sur-Marne, France
louis.marin@univ-eiffel.fr

Abstract. Let $d \in \{0,1,2,3,4\}$ and W be a 2-dimensional word of dimensions $h \times w$ on the binary alphabet $\{\square, \blacksquare\}$, where $h, w \in \mathbb{Z}_{>0}$. Assume that each occurrence of the letter $\blacksquare$ in W is adjacent to at most d letters $\blacksquare$ and let $|W|_\blacksquare$ be the number of letters $\blacksquare$ in W. We provide an exact formula for the maximum value of $|W|_\blacksquare$ for fixed (h, w). As a byproduct, we deduce an upper bound on the length of maximum snake polyominoes contained in a $h \times w$ rectangle.

1 Introduction

We adapt the terminology for 2-dimensional words from [6,9]. For $a, b \in \mathbb{Z}_{>0}$, let $[\![a,b]\!] = \{n \in \mathbb{Z}_{>0} \mid a \le n \le b\}$. Let A be a finite alphabet and $h, w \in \mathbb{Z}_{>0}$. A *2-dimensional word W of dimension $h \times w$ on A* is a matrix of h rows and w columns with entries in A. The set of all 2-dimensional words of dimensions $h \times w$ on A is denoted by $\mathcal{W}_{h\times w}(A)$. Given $W \in \mathcal{W}_{h\times w}(A)$ and $a \in A$, we denote by $|W|_a$ the number of occurrences of a in W. For $i \in [\![1,h]\!]$ and $j \in [\![1,w]\!]$, the entry a of W at the intersection of the i-th row and j-th column is written $W[i,j]$ and is called *the a-cell of W at (i,j)*. The *horizontal concatenation* of $U \in \mathcal{W}_{h\times w_1}(A)$ and $V \in \mathcal{W}_{h\times w_2}(A)$ is the word $U \oplus V \in \mathcal{W}_{h\times(w_1+w_2)}(A)$ defined by

$$(U \oplus V)[i,j] = \begin{cases} U[i,j], & \text{if } j \le w_1; \\ V[i,j-w_1], & \text{if } j > w_1. \end{cases}$$

The *vertical concatenation* $U \ominus V$ of $U \in \mathcal{W}_{h_1\times w}(A)$ and $V \in \mathcal{W}_{h_2\times w}(A)$, is defined similarly. Given $W \in \mathcal{W}_{h\times w}(A)$ and $m, n \in \mathbb{Q}$ such that $mh, nw \in \mathbb{Z}$, the *(m,n)-th power of W* is the word $W^{m\times n} \in \mathcal{W}_{mh\times nw}(A)$ such that $W^{m\times n}[i,j] = W[(i-1 \bmod h)+1, (j-1 \bmod w)+1]$. The *neighborhood of $W[i,j]$*, denoted by $N_W(i,j)$, is defined as

$$N_W(i,j) = \{(i-1,j),(i+1,j),(i,j-1),(i,j+1)\} \cap ([\![1,h]\!] \times [\![1,w]\!])$$

G. Gamard and J. Leroy (Eds.): WORDS 2025, LNCS 15729, pp. 37–48, 2025.
https://doi.org/10.1007/978-3-031-97548-6_4

The i-th row of W is denoted by $W[i]$ and the factor obtained by selecting all rows between i and i', both included, is denoted by $W[\![i, i']\!]$. For any $a \in A$, the *a-degree* of $W[i, j]$ is given by

$$\deg_{a,W}(i,j) = \begin{cases} \sum_{(i',j') \in N_W(i,j)} \mathbb{I}(W[i',j'] = a) & \text{if } W[i,j] = a; \\ 0, & \text{if } W[i,j] \neq a, ddd \end{cases}$$

where $\mathbb{I}$ is the usual indicator function.

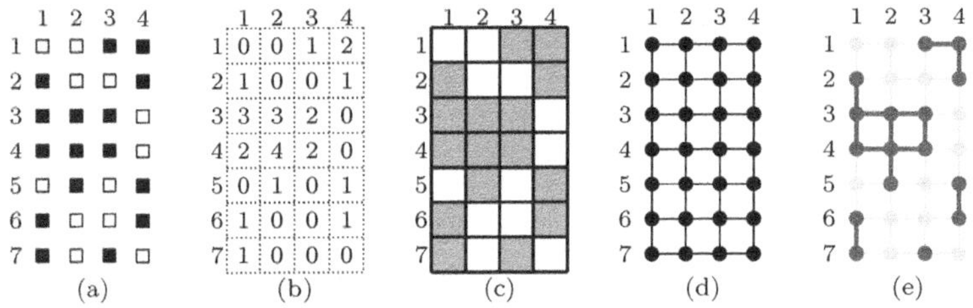

Fig. 1. (a) A 7×4 word W on the alphabet $\{\square, \blacksquare\}$. (b) The $\blacksquare$-degree word of W. (c) The geometric representation of W. (d) The grid graph $G_{7,4}$. (e) The subgraph (in green) of $G_{7,4}$ (in pale gray) induced by W. Row and column indices of W appear in blue.

For example, if W is the word illustrated in Fig. 1(a), then $\deg_{\blacksquare,W}(2,3) = 0$ and $\deg_{\blacksquare,W}(3,3) = 2$. The *$a$-degree word* of W, denoted by $\deg_a(W)$, is the 2-dimensional word of dimensions $h \times w$ on the alphabet $\{0,1,2,3,4\}$ with entry at (i,j) given by $\deg_{a,W}(i,j)$ (Fig. 1(b)). In this paper, all binary words are taken on the alphabet $\{\square, \blacksquare\}$ and for brevity, we write $\mathcal{W}_{h\times w}$ instead of $\mathcal{W}_{h\times w}(\{\square, \blacksquare\})$ and "degree" instead of "$\blacksquare$-degree". $|W|_\blacksquare$ is also called the *area* of W. Given $d \in [\![1,4]\!]$, the set of all 2-dimensional words of dimensions $h \times w$ with $\blacksquare$-cells of degree bounded by d is denoted by $\mathcal{W}^{\leq d}_{h\times w}$:

$$\mathcal{W}^{\leq d}_{h\times w} = \{W \in \mathcal{W}_{h\times w} \mid \deg_\blacksquare(W)[i,j] \leq d, \text{for } (i,j) \in [\![1,h]\!] \times [\![1,w]\!]\}.$$

For any $d \in [\![1,4]\!]$, $h, w \in \mathbb{Z}_{>0}$, let $\max_{\leq d}(h,w) = \max\{|W|_\blacksquare : W \in \mathcal{W}^{\leq d}_{h\times w}\}$, be the maximum number of filled cells of degree at most d in a $h \times w$ 2D word. Clearly, $\max_{\leq d}(h,w)$ is symmetric with respect to (h,w) i.e. $\max_{\leq d}(h,w) = \max_{\leq d}(w,h)$. Moreover, let $G = (V,E)$ be a simple graph and $U \subseteq V$. Then U is called a *dominating set* of G if, for every vertex $v \in V \backslash U$, there exists $u \in U$ such that $\{u,v\} \in E$, i.e. v has at least one neighbor in U. The *domination number of* G, denoted by $\gamma(G)$, is the minimal cardinality of a dominating set of G.

The main result of this document is the following theorem.

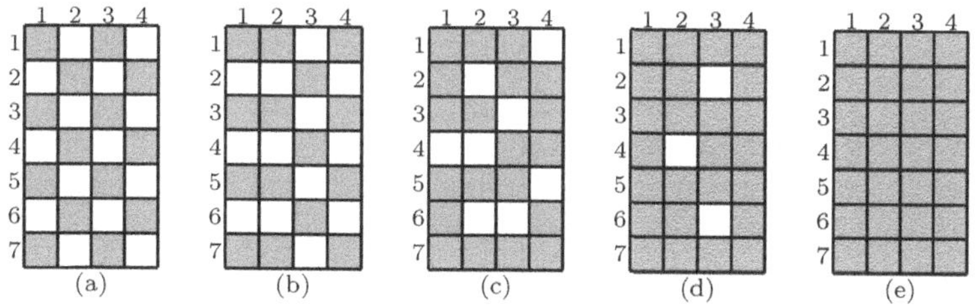

Fig. 2. 7×4 d-full words, for (a) $d = 0$, (b) $d = 1$, (c) $d = 2$, (d) $d = 3$ and (e) $d = 4$.

Theorem 1. *Let $d \in [\![1, 4]\!]$, $h, w \in \mathbb{Z}_{>0}$, $h \geq w$ and $m_{\leq d}(h, w)$ be defined by*

$$m_{\leq 0}(h,w) = \lceil hw/2 \rceil,$$

$$m_{\leq 1}(h,w) = \begin{cases} hw/2, & \text{if } h, w \equiv_2 0; \\ (h-1)w/2 + \lceil 2w/3 \rceil, & \text{if } h \equiv_2 1 and w \equiv_2 0; \\ h(w-1)/2 + \lceil 2h/3 \rceil, & \text{otherwise,} \end{cases}$$

$$m_{\leq 2}(h,w) = \begin{cases} hw, & \text{if } w = 1 or h = w = 2; \\ 3hw/4 + 1/2, & \text{if } h \equiv_2 1, h \geq 3 and w = 2; \\ 3hw/4, & \text{if } h \equiv_2 0, h \geq 4 and w = 2; \\ 2hw/3 + 2, & \text{if } w = 3 or h \equiv_3 w \equiv_3 0; \\ 2hw/3 + 4/3, & \text{if } w \geq 4 and h \equiv_3 w \not\equiv_3 0; \\ 2hw/3 + 1, & \text{if } w \geq 4, w \equiv_3 0 and h \not\equiv_3 w; \\ 2hw/3 + 2/3, & \text{otherwise,} \end{cases}$$

$$m_{\leq 3}(h,w) = \begin{cases} hw, & \text{if } 1 \leq w \leq 2; \\ hw - \gamma(G_{h-2,w-2}), & \text{otherwise,} \end{cases}$$

$$m_{\leq 4}(h,w) = hw.$$

Then $\max_{\leq d}(h,w) = m_{\leq d}(h,w)$.

A 2-dimensional word $W \in \mathcal{W}_{h \times w}^{\leq d}$ is called *d-full* or maximal if $|W|_{\blacksquare} = \max_{\leq d}(h,w)$. The set of d-full words of degree at most d is denoted $\mathcal{MW}_{h \times w}^{\leq d}$. In Fig. 2, examples of d-full words of dimensions 7×4 are illustrated, for $d \in [\![1, 4]\!]$.

Motivation. The study of multidimensional words has generated interest in the combinatorics on words community [3,6,9]. A famous conjecture, presented by Nivat in his invited talk at ICALP [10], stating that 2-dimensional words with low rectangular complexity are periodic, has generated extensive literature [4,5,11,12]. Recently, Mahalingam and Pandoh have considered $2D$ generalizations of palindromes [8]. Theorem 1 can be seen as a result on pattern avoidance [2] as words in $\mathcal{W}_{h \times w}^{\leq d}$ avoiding specific 3×3 rectangles. Also, beyond its intrinsic interest, Theorem 1 has consequences on the study of polyominoes. It is also convenient to interpret binary words in $\mathcal{W}_{h \times w}^{\leq d}$ from geometric and graph-theoretic

points of view.
Let $W \in \mathcal{W}_{h\times w}^{\leq d}$. W can also be seen as a $h \times w$ rectangle in which $\blacksquare$ cells are filled unit cells, and $\square$ cells are empty unit cells (Fig. 1(c)). A *horizontal* (resp. *vertical*) *n-pillar in W* is a factor of the form $\blacksquare^{1\times n}$ (resp. $\blacksquare^{n\times 1}$) in W. An n-pillar is *maximal* when it is not a proper factor of a larger pillar in its row (or column) and it is a solitary pillar when it is maximal and its adjacent cells are empty. For $n \geq 3$, a *north n-bench* is a factor of the form $W = \begin{smallmatrix}\blacksquare\\\blacksquare\end{smallmatrix}\left(\begin{smallmatrix}\square\\\blacksquare\end{smallmatrix}\right)^{1\times(n-2)}\begin{smallmatrix}\blacksquare\\\blacksquare\end{smallmatrix}$. The bottom filled row is called the *seat* of the bench and the two extremal columns are called the *legs* of the bench. *South, east and west n-benches* are defined similarly, with the two legs indicating the cardinal direction. The sequence $(|W[1]|_{\blacksquare}, |W[2]|_{\blacksquare}, \ldots, |W[h]|_{\blacksquare})$ giving the number of filled cells in each row $W[i]$ of a word W is called the *row distribution* of W. The row distribution of the word in Fig. 1 is $(2, 2, 3, 3, 2, 2, 2)$. We denote by W^t the transpose of the matrix W so that $W^t[j]$ is the j-th column of W.

$W \in \mathcal{W}_{h\times w}^{\leq d}$ is also related to graph-theoretical concepts. Recall that the *grid graph $G_{h,w}$ of dimensions $h \times w$* is the simple graph with set of vertices $[\![1, h]\!] \times [\![1, w]\!]$ and set of edges $\{(i, j), (i', j')\}$, where $i, i' \in [\![1, h]\!]$, $j, j' \in [\![1, w]\!]$ and $(i - i')^2 + (j - j')^2 = 1$ (Fig. 1(d)). The *subgraph induced by W in $G_{h,w}$*, denoted by $G[W]$, is the subgraph of $G_{h,w}$ induced by the set of vertices $\{(i, j) \in [\![1, h]\!] \times [\![1, w]\!] : W[i, j] = \blacksquare\}$ (Fig. 1(e)). There exists an extensive literature studying the induced subgraphs of bounded degree, in particular induced subgraphs of planar graphs [1]. Theorem 1 provides an exact solution for the case of grid graphs where the induced subgraphs are linear forests of bounded degree d.

Finally, the study of words in $\mathcal{W}_{h\times w}^{\leq d}$ is useful to the study of polyominoes. A *snake polyomino* (or simply a *snake*) is a polyomino inducing a chain subgraph, while a *snake forest* (also called a *linear forest*) is a set of polyominoes inducing a forest of chains as a subgraph. Therefore, snakes and snake forests are subsets of $\mathcal{W}_{h\times w}^{\leq 2}$ and Theorem 1 provides an upper bound for their maximal size in a $h \times w$ rectangle, which is tight for $w \in [\![1, 4]\!]$. We can also use the construction of 2-full words in the proof of Theorem 1 to establish a lower bound for the maximal size of snakes contained in a $h \times w$ rectangle. Figure 2(c) shows a snake polyomino of maximal size. The remainder of this paper is devoted to the proof of Theorem 1.

2 The Cases $d \in \{0, 1, 3, 4\}$

The case $d = 4$ is immediate. In this section, we study the cases $d \in \{0, 1, 3\}$, keeping the case $d = 2$ for Sect. 3. The reasons behind this division are (1) the cases $d \in \{0, 1\}$ are easy to prove, (2) the case $d = 3$ is proved by establishing a relation with the dominating set problem on grid graphs, which is solved in [7], (3) the case $d = 2$ is involved and requires a thorough combinatorial study.

We start with the case $d = 0$.

Lemma 1. *For any $(h, w) \in [\![1, h]\!] \times [\![1, w]\!]$, $\max_{\leq 0}(h, w) = \lceil hw/2 \rceil$.*

(a) 18 dominoes, 0 monomino (b) 12 dominoes, 1 monomino

Fig. 3. The word $W = \left(\begin{smallmatrix}\blacksquare&\square\\\square&\blacksquare\end{smallmatrix}\right)^{h/2\times w/2}$ superimposed with an arbitrary dominoes/monomino tiling, when the dimensions are (a) 6×6 (b) 5×5.

Proof. Let $W = \left(\begin{smallmatrix}\blacksquare&\square\\\square&\blacksquare\end{smallmatrix}\right)^{h/2\times w/2} \in \mathcal{W}_{h\times w}$ (see Fig. 3). Since each $\blacksquare$-cell of W is surrounded by $\square$-cells, $\deg_{\blacksquare,W}(i,j) = 0$ for all $(i,j) \in [\![1,h]\!] \times [\![1,w]\!]$, which implies that $W \in \mathcal{W}^{\leq 0}_{h\times w}$. Since $|W|_{\blacksquare} = \lceil hw/2\rceil$, we have $\max_{\leq 0}(h,w) \geq \lceil hw/2\rceil$. We prove that $\lceil hw/2\rceil$ is an upper bound with a pigeonhole principle argument. Let $W \in \mathcal{W}^{\leq 0}_{h\times w}$. Then W can be partitioned into $\lfloor hw/2\rfloor$ dominoes and $hw \bmod 2 \in \{0,1\}$ monomino (see Fig. 3). Since at most one cell of each domino can be filled and since a monomino contains at most one filled cell, there are at most $\lfloor hw/2\rfloor + hw \bmod 2 = \lceil hw/2\rceil$ filled cells in W, i.e. $\max_{\leq 0}(h,w) \leq \lceil hw/2\rceil$. Hence, $\max_{\leq 0}(h,w) = \lceil hw/2\rceil$.

Due to lack of space, we only provide a sketch of the proof for the case $d = 1$.

Lemma 2. *Theorem 1 holds for* $d = 1$.

Proof (sketched). This is proved using an argument similar to the one presented in the proof of Lemma 1, by partitioning the rectangle with 2×2 and 1×3 tiles. The detailed proof is available in the Arxiv version.

For the case $d = 3$, we need additional definitions. The *boundary* $\partial(W)$ of W is the set $\partial(W) = \{(i,j) \in [\![1,h]\!] \times [\![1,w]\!] \mid i \in \{1,h\} or j \in \{1,w\}\}$. We say that $(i,j) \in \partial(W)$ is *in a corner* of W if $i \in \{1,h\}$ and $j \in \{1,w\}$. Otherwise, (i,j) is *on the side* of W. Since an expression for the domination number of any grid graph was recently proved in [7], we take advantage of that expression to reduce the problem.

Lemma 3. *Theorem 1 holds for* $d = 3$.

Proof If $1 \leq h \leq 2$ or $1 \leq w \leq 2$, then it suffices to define $W[i,j] = \blacksquare$ for all $(i,j) \in [\![1,h]\!] \times [\![1,w]\!]$, which makes W obviously 3-full. Now assume that $h,w \geq 3$.

First, we show that there exists a 3-full word $W \in \mathcal{W}_{h\times w}$ such that $W[i,j] = \blacksquare$ for all $(i,j) \in \partial(W)$, i.e. the cells on the boundary of W are filled. Arguing by contradiction, assume that no such word exists and let W be a 3-full word with an empty cell on the left side on W, i.e. there exists $i \in [\![h]\!]$ such that $W[i,1] = \square$. Clearly, the $\square$-cell cannot be in a corner of W, i.e. $i \neq 1,h$: If it is the case, then the word W' obtained from W by replacing the entry at $(i,1)$ by $\blacksquare$ would satisfy $W' \in \mathcal{W}^{\leq 3}_{h\times w}$ and $|W'|_{\blacksquare} = |W|_{\blacksquare} + 1$, contradicting the assumption that W

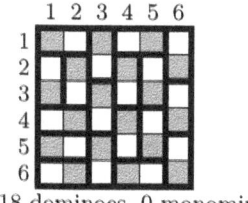

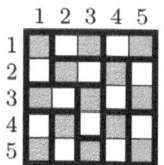

is 3-full. Hence, $i \neq 1, h$. Moreover, the cell at the right of $(i,1)$ must be filled, i.e. $W[i,2] = \blacksquare$: If it is not the case, then the word W'' obtained from W by replacing the entry at $(i,1)$ by $\blacksquare$ would satisfy $W' \in \mathcal{W}^{\leq 3}_{h\times w}$ and $|W'|_\blacksquare = |W|_\blacksquare + 1$, also contradicting the assumption that W is 3-full. Finally, observe that the word W''' obtained from W by replacing the entry at $(i,1)$ by $\blacksquare$ and the entry at $(i,2)$ by $\square$ satisfies $W' \in \mathcal{W}^{\leq 3}_{h\times w}$, $|W'|_\blacksquare = |W|_\blacksquare$ and has one empty cell less than W on its boundary, contradicting the assumption that the boundary of W contains a minimum number of empty cells. Hence, there exists a 3-full word $W \in \mathcal{W}_{h\times w}$ with its boundary contains only filled cells.

To conclude, let $G_{h-2,w-2}$ be the grid subgraph with set of vertices $[\![2, h-1]\!] \times [\![2, w-1]\!]$ and let $U = \{(i,j) \in [\![1,h]\!] \times [\![1,w]\!] \mid W[i,j] = \square\}$. We observe that $W \in \mathcal{W}^{\leq 3}_{h\times w}(\{\square, \blacksquare\})$ only if U is a dominating set of $G_{h-2,w-2}$. Therefore, W is 3-full if and only if U is a minimum dominating set of $G_{h-2,w-2}$. □

3 The Case $d = 2$

This section is devoted to the proof of the case $d = 2$. As a first step, we introduce a simple but useful concept that facilitates the discussion:

Definition 1 (Excess of a word). *Let $W \in \mathcal{W}_{h\times w}$. The* excess of W, *denoted by $e(W)$, is defined by $e(W) = |W|_\blacksquare - 2hw/3$. The maximal excess that can be realized by a word of dimensions $h\times w$ of degree d less or equal to 2 is $e_{max}(h,w) = max\{e(W) : W \in \mathcal{W}^{\leq 2}_{h\times w}\}$.*

In other words, the excess of W is the surplus, in number of filled cells, that W has when 2/3 of the area of its bounding rectangle is filled, and the maximal excess that a word can have if it is contained in a $h\times w$ rectangle R is $e_{max}(h,w)$. If R is the bounding rectangle of W, we sometimes write $e(R)$ instead of $e(W)$.

An immediate property of the excess function is that it is additive with respect to horizontal and vertical concatenation:

Proposition 1. *Let $W_1 \in \mathcal{W}_{h\times w_1}, W_2 \in \mathcal{W}_{h\times w_2}$ be two words with respective sizes $h \times w_1$ and $h \times w_2$. Similarly let $W_3 \in \mathcal{W}_{h_1\times w}, W_4 \in \mathcal{W}_{h_2\times w}$ be two words of respective sizes $h_1 \times w$ and $h_2 \times w$. Then $e(W_1 \oplus W_2) = e(W_1) + e(W_2)$ and $e(W_3 \ominus W_4) = e(W_3) + e(W_4)$.*

Using Definition 1, the case $d = 2$ of Theorem 1 can be reformulated:

Theorem 2. *Let $h, w \in \mathbb{Z}_{>0}$, with $h \geq w$. Then*

$$e_{max}(h,w) = \begin{cases} hw/3, & \textit{if } w = 1 or h = w = 2; \\ hw/12, & \textit{if } h \equiv_2 0, h \geq 4 and w = 2; \\ hw/12 + 1/2 & \textit{if } h \equiv_2 1, h \geq 3 and w = 2; \\ 2 & \textit{if } w = 3 or h \equiv_3 w \equiv_3 0; \\ 4/3 & \textit{if } w \geq 4 and h \equiv_3 w \not\equiv_3 0; \\ 1 & \textit{if } w \geq 4, hw \equiv_3 0 and h \not\equiv_3 w; \\ 2/3 & \textit{otherwise.} \end{cases} \quad (1)$$

This equivalent form is particularly convenient in comparison with the definition of $\max_{\leq d}(h, w)$, since the excess becomes bounded whenever $h, w \geq 3$. Finally, for $h, w \in \mathbb{Z}_{>0}$, let

$$\hat{\mathbf{e}}_{\mathbf{max}}(h, w) = m_{\leq 2}(h, w) - 2/3(w \cdot h). \tag{2}$$

Theorem 2 claims that $\hat{\mathbf{e}}_{\mathbf{max}}(h, w) = e_{max}(h, w)$. The remainder of the paper is devoted to its proof. We begin with the cases where the width w is small.

Lemma 4 (Base cases). *Theorem 2 holds for* $(h, w) \in (\mathbb{Z}_{>0} \times [\![1, 6]\!]) \cup (7, 7)$.

Proof. It is immediate that if $w = 1$, then a h-pillar is both 2-full and element of $\mathcal{W}_{h\times 1}^{\leq 2}$ which yields $e_{max} = h - 2h/3 = h/3$. Due to space restriction, the proof of the other base cases are presented in the Arxiv version, except for the case $w = 4$ which is detailed in Subsect. 3.1.

3.1 The Subcase $w = 4$

Notice that

$$\hat{\mathbf{e}}_{\mathbf{max}}(h, 4) = \begin{cases} 4/3, & \text{if } h \equiv_3 1; \\ 1, & \text{if } h \neq 3 and h \equiv_3 0; \\ 2/3, & \text{if } h \equiv_3 2. \end{cases} \tag{3}$$

Proposition 2. *Let $h \geq 4$ be an integer. Then $e_{max}(h, 4) = \hat{\mathbf{e}}_{\mathbf{max}}(h, 4)$.*

The proof of Proposition 2 is combinatorial and requires the examination of several cases. We first introduce some lemmas.

Lemma 5. *For any integer $h \geq 4$, $e_{max}(h, 4) \geq \hat{\mathbf{e}}_{\mathbf{max}}(h, 4)$. Moreover, if $h > 4$, there exists a snake $S \in \mathcal{W}_{h\times 4}^{\leq 2}$ such that $e(S) = \hat{\mathbf{e}}_{\mathbf{max}}(h, 4)$.*

Proof. Let

$$A = \begin{smallmatrix}\blacksquare\blacksquare\blacksquare\blacksquare\\ \blacksquare\square\square\blacksquare\\ \square\blacksquare\blacksquare\blacksquare\\ \blacksquare\blacksquare\square\square\end{smallmatrix},\ B = \begin{smallmatrix}\blacksquare\blacksquare\blacksquare\blacksquare\\ \blacksquare\square\square\blacksquare\\ \blacksquare\square\blacksquare\blacksquare\\ \square\blacksquare\blacksquare\square\end{smallmatrix},\ C = \begin{smallmatrix}\blacksquare\square\square\blacksquare\\ \blacksquare\blacksquare\blacksquare\blacksquare\end{smallmatrix},\ D = \begin{smallmatrix}\blacksquare\blacksquare\square\blacksquare\\ \blacksquare\square\square\blacksquare\\ \blacksquare\blacksquare\blacksquare\blacksquare\end{smallmatrix},\ \tilde{D} = \begin{smallmatrix}\blacksquare\square\blacksquare\blacksquare\\ \blacksquare\square\square\blacksquare\\ \blacksquare\blacksquare\blacksquare\blacksquare\end{smallmatrix},\ E = \begin{smallmatrix}\blacksquare\blacksquare\blacksquare\blacksquare\\ \blacksquare\square\square\blacksquare\end{smallmatrix},$$

$$U = \begin{smallmatrix}\blacksquare\square\blacksquare\blacksquare\\ \blacksquare\square\blacksquare\square\\ \blacksquare\square\blacksquare\blacksquare\end{smallmatrix},\ V = \begin{smallmatrix}\blacksquare\blacksquare\square\blacksquare\\ \square\blacksquare\square\blacksquare\\ \blacksquare\blacksquare\square\blacksquare\end{smallmatrix},\ X = \begin{smallmatrix}\blacksquare\blacksquare\square\blacksquare\\ \blacksquare\square\blacksquare\blacksquare\\ \blacksquare\square\blacksquare\square\end{smallmatrix},\ Y = \begin{smallmatrix}\blacksquare\square\blacksquare\blacksquare\\ \blacksquare\blacksquare\square\blacksquare\\ \square\blacksquare\square\blacksquare\end{smallmatrix},\ W_4 = \begin{smallmatrix}\blacksquare\blacksquare\blacksquare\blacksquare\\ \blacksquare\square\square\blacksquare\\ \blacksquare\square\square\blacksquare\\ \blacksquare\blacksquare\blacksquare\blacksquare\end{smallmatrix},$$

and for each integer $h \geq 5$, let W_h be defined by

$$W_h = \begin{cases} A \ominus (U \ominus V)^{(h-6)/6\times 1} \ominus C, & \text{if } h \equiv_6 0; \\ B \ominus (X \ominus Y)^{(h-7)/6\times 1} \ominus D, & \text{if } h \equiv_6 1; \\ E \ominus (U \ominus V)^{(h-2)/6\times 1}, & \text{if } h \equiv_6 2; \\ A \ominus (U \ominus V)^{(h-9)/6\times 1} \ominus U \ominus C, & \text{if } h \equiv_6 3; \\ B \ominus (X \ominus Y)^{(h-10)/6\times 1} \ominus X \ominus \tilde{D}, & \text{if } h \equiv_6 4; \\ E \ominus (U \ominus V)^{(h-5)/6\times 1} \ominus U, & \text{if } h \equiv_6 5. \end{cases}$$

It suffices to observe that, for each integer $h \geq 4$, $W_h \in \mathcal{W}_{h\times d}^{\leq 2}$ and $e(W_h) = \hat{\mathbf{e}}_{\mathbf{max}}(h, 4)$. Moreover, W_h is a snake for $h \neq 4$. $\square$

To prove that $e_{max}(h,4) \le \hat{\mathbf{e}}_{\mathbf{max}}(h,4)$, we proceed by contradiction: We assume that there exists a word $W \in \mathcal{W}^{\le 2}_{h\times 4}$, where h is as small as possible, such that $e(W) > \hat{\mathbf{e}}_{\mathbf{max}}(h,4)$, and show that W cannot exist. Such a word W is called a *minimal counter-example* (MCE).

Lemma 6. *Let $h \ge 1$ be an integer. Then the following statements hold.*

(i) $\hat{\mathbf{e}}_{\mathbf{max}}(h,4) \le 2$;
(ii) If $h \ne 1$ *and* $h \equiv_3 1$, *then* $\hat{\mathbf{e}}_{\mathbf{max}}(h,4) - \hat{\mathbf{e}}_{\mathbf{max}}(h-2,4) = 2/3$;
(iii) If $h \ne 3$ *and* $h \equiv_3 0$, *then* $\hat{\mathbf{e}}_{\mathbf{max}}(h,4) - \hat{\mathbf{e}}_{\mathbf{max}}(h-4,4) = 1/3$;
(iv) If $h \ne 3$, *then* $\hat{\mathbf{e}}_{\mathbf{max}}(h,4) \le 4/3$;
(v) If $h \ge 4$ *and* k *is an integer such that* $1 \le k \le h-1$ *and* $h-k \ne 3$, *then* $\hat{\mathbf{e}}_{\mathbf{max}}(h,4) - \hat{\mathbf{e}}_{\mathbf{max}}(h-k,4) \ge -2/3$.

Proof. Follows from the definition of $\hat{\mathbf{e}}_{\mathbf{max}}(h,4)$. □

Lemma 7. *Let W be a MCE of height $h \ge 4$, with $h \equiv_3 2$, and $U \in \mathcal{W}^{\le 2}_{3\times 4}$ an inner factor of W. Then*

(i) $U \ne$ .
(ii) $|U|_{■} \le 8$.

Proof. (i) Arguing by contradiction, assume the opposite. Then there exists a factor $U' \in \mathcal{W}^{\le 2}_{5\times 4}$ of W, having U as an inner factor. Since all cells in the top and bottom rows of U have degree 2, then $|U'[1]|_{■} = |U'[5]|_{■} = 0$, so that $e(U') = -10/3$. Write $W = P \ominus U' \ominus S$. Then $e(W) = e(P) + e(U') + e(S) \le 2 - 10/3 + 2 = 2/3 = \hat{\mathbf{e}}_{\mathbf{max}}(h,4)$, contradicting $e(W) > \hat{\mathbf{e}}_{\mathbf{max}}(h,4)$.

(ii) We know from (i) that $|U|_{■} \le 9$. Again by contradiction, assume that $|U|_{■} = 9$. Then U belongs to the following set, up to symmetry:

Let U' be a factor of height 4 containing U, such that either $|U'[1]|_{■} \le 1$ or $|U'[4]|_{■} \le 1$. Such a factor exists since U has at least one of its top or bottom row with at least 3 cells of degree 2. Therefore, $e(U') \le -5/3$. Write $W = P \ominus U' \ominus S$, where P has height h'. There are three subcases to consider according to the value of h' mod 3. If $h' \equiv_3 0$, then $e(W) = e(P) + e(U') + e(S) \le 1 - 5/3 + 4/3 = 2/3 = \hat{\mathbf{e}}_{\mathbf{max}}(h,4)$, contradicting $e(W) > \hat{\mathbf{e}}_{\mathbf{max}}(h,4)$. If $h' \equiv_3 1$, then $e(W) = e(P) + e(U') + e(S) \le 4/3 - 5/3 + 1 = 2/3 = \hat{\mathbf{e}}_{\mathbf{max}}(h,4)$, contradicting $e(W) > \hat{\mathbf{e}}_{\mathbf{max}}(h,4)$. Finally, if $h' \equiv_3 2$, then $e(W) = e(P) + e(U') + e(S) \le 2/3 - 5/3 + 2/3 = -1/3$, also contradicting $e(W) > \hat{\mathbf{e}}_{\mathbf{max}}(h,4) = 2/3$. □

Lemma 8. *There does not exist any MCE of height $h \ge 4$.*

Proof. By contradiction, assume that there exists a word W of height $h \geq 4$ that is a MCE. There are three cases to consider, according to the value of $h \bmod 3$.

Case $h \equiv_3 1$. Let $W = P \ominus S$, where P has height $h-2$ and S has height 2. Then $e(S) \leq 2/3$. Therefore $e(P) = e(W) - e(S) > \hat{\mathbf{e}}_{\mathbf{max}}(h,4) - 2/3 = \hat{\mathbf{e}}_{\mathbf{max}}(h-2,4) + 2/3 - 2/3 = \hat{\mathbf{e}}_{\mathbf{max}}(h-2,4)$, contradicting the minimality of W.

Case $h \equiv_3 0$. Let $W = P \ominus S$, where P has height $h-4$ and S has height 4. First, assume that $e(S) \leq 1/3$. Then $e(P) = e(W) - e(S) > \hat{\mathbf{e}}_{\mathbf{max}}(h,4) - 1/3 > \hat{\mathbf{e}}_{\mathbf{max}}(h-4,4) + 1/3 - 1/3 = \hat{\mathbf{e}}_{\mathbf{max}}(h-4,4)$, contradicting the minimality of W. Hence, $e(S) \geq 4/3$. By exhaustive enumeration, this implies that S belongs to the following set, up to symmetry:

$S \in$  .

Write $W = P' \ominus S'$, where P' has height $h-5$ and S' has height 5. Then $|S'[1]|_{\blacksquare} \leq 1$, since S has at least 3 cells of degree 2 on each of its side. Therefore, $e(S') \leq -1/3$. But $e(P') = e(W) - e(S') > \hat{\mathbf{e}}_{\mathbf{max}}(h,4) + 1/3 = \hat{\mathbf{e}}_{\mathbf{max}}(h-5,4) - 1/3 + 1/3 = \hat{\mathbf{e}}_{\mathbf{max}}(h-5,4)$, contradicting the minimality of W.

Case $h \equiv_3 2$. Write $W = P \ominus U_1 \ominus \cdots \ominus U_k \ominus S$, where both P and S have height 4 and U_i has height 3 for $i = 1, 2, \ldots, k$. Using an argument similar to the previous subcase, we have $e(P), e(S) \leq 1/3$. Moreover, by Lemma 7, $e(U_i) = 0$ for $i = 1, 2, \ldots, k$. Hence, $e(W) = 1/3 + 0 + \ldots + 0 + 1/3 = 2/3$, contradicting $e(W) > \hat{\mathbf{e}}_{\mathbf{max}}(h,4)$. □

Proposition 2 follows from Lemmas 5 and 8.

3.2 The General Case $w > 6$

Lemma 9. *Let $W \in \mathcal{MW}^{\leq 2}_{h \times w}$ such that $h, w > 6$. Then*

$$e(W) \geq \hat{\mathbf{e}}_{\mathbf{max}}(h,w).$$

Proof. We prove that for $i, j \leq 2$, $h = 3k_1 + j, w = 3k_2 + i$, there exist words $W \in \mathcal{W}^{\leq 2}_{h \times w}$ such that $e(W) = \hat{\mathbf{e}}_{\mathbf{max}}(h,w)$. In a $h \times w$ rectangle R, start by inserting a Q-shape of excess 1 in the top left corner (purple cells in Fig. 4a). Then concatenate this Q-shape with the Q-shape from which the top left corner cell has been removed k_1 times horizontally and k_2 times vertically. Repeat this insertion in order to fill R except for the bottom i rows and the right j columns. Then fill the remaining bottom i rows with rows $R[3+i'], 1 \leq i' \leq i$ and the right j columns with columns $R^t[3+j'], 1 \leq j' \leq j$. Then fill the bottom right $i \times j$ rectangle with 0, 1 or 3 cells (purple cells) in bottom right of R (Figs. 4a to 4d). This produces a $h \times w$ word $W \in \mathcal{W}^{\leq 2}_{h \times w}$ with $e(W) = \hat{\mathbf{e}}_{\mathbf{max}}(W)$.

The following lemma completes the proof of Theorem 2. It is separated in 6 cases. Due to space restriction, we only provide the complete proofs of the first 5 cases, and leave the sixth, and more technical, cases in the Arxiv version. Before that, we need additional results.

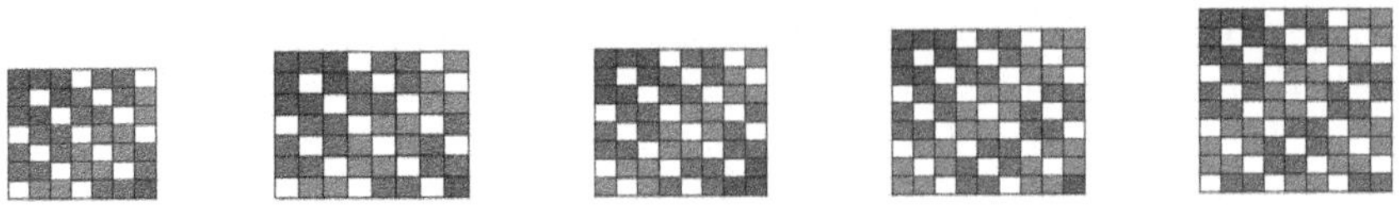

(a) 7×7 (b) 7×8 (c) 8×8 (d) 9×9 (e) 10×9

Fig. 4. Some 2-full words.

Corollary 1. *Let $k \geq 2$ and $W \in \mathcal{W}^{\leq 2}_{(3k+1)\times 5}$ such that $W^t[\![1,4]\!] \in \mathcal{MW}^{\leq 2}_{3k+1\times 4}$. Then $e(W^t[5]) \leq -5/3$.*

Corollary 2. *Let $k \geq 1$ and $W \in \mathcal{W}^{\leq 2}_{(3k+2)\times 6}$ such that $W^t[\![1,5]\!] \in \mathcal{MW}^{\leq 2}_{(3k+2)\times 5}$. Then $e(W^t[6]) \leq -4/3$.*

Corollary 3. *Let $W \in \mathcal{W}^{\leq 2}_{h\times 4}$ such that $W^t[\![1,3]\!] \in \mathcal{MW}^{\leq 2}_{h\times 3}$. Then*

(i) $|W^t[4]|_\square \geq \lfloor (h-2)/3 \rfloor + 3$;
(ii) If $h \equiv_3 1$, then $|W^t[4]|_\square \geq \lfloor (h-2)/3 \rfloor + 4$;
(iii) If $h \equiv_3 2$, then $|W^t[4]|_\square \geq \lfloor (h-2)/3 \rfloor + 5$.

Lemma 10. *Let $W \in \mathcal{MW}^{\leq 2}_{h\times w}$ such that $h, w > 6$. Then*

$$e(W) \leq \hat{\mathbf{e}}_{\mathbf{max}}(h,w).$$

The proof of these results are included in the Arxiv version. Moreover, by combining Lemma 4, Lemma 9 and Lemma 10, we have a proof of Theorem 2 and, by extension, Theorem 1.

Proof. We prove that there exists no $W \in \mathcal{W}^{\leq 2}_{h\times w}$ such that $e(W) > \hat{\mathbf{e}}_{\mathbf{max}}(h,w)$. There are 6 cases up to symmetry and we prove each of them next.

Case 1. Let $W \in \mathcal{W}^{\leq 2}_{3k_1\times 3k_2}$ with $k_1, k_2 \geq 3$. By minimal counterexample, assume that W is minimal such that $e(W) > \hat{\mathbf{e}}_{\mathbf{max}}(h,w)$. We have that $\hat{\mathbf{e}}_{\mathbf{max}}(3k_1, 3k_2) = 2$ so we assume $e(W) = 3$.

Write $W = T_1 \ominus T_2$ where each factor T_1, T_2 has respective height $3k_{1,1} + 1, 3k_{1,2} + 2$ for some $k_{1,1} \geq 1, k_{1,2} \geq 1$. By minimality hypothesis, we have $e(T_i) \leq \hat{\mathbf{e}}_{\mathbf{max}}(3k_{1,1}+1, 3k_2) = \hat{\mathbf{e}}_{\mathbf{max}}(3k_{1,2}+2, 3k_2) = 1$, so that $e(W) = e(T_1) + e(T_2) \leq 1 + 1 = 2$, in contradiction with the hypothesis $e(W) = 3$.

Case 2. Let $W \in \mathcal{W}^{\leq 2}_{(3k_1+2)\times(3k_2+2)}$ with $k_1, k_2 \geq 2$. By minimal counterexample, assume that W is minimal such that $e(W) > \hat{\mathbf{e}}_{\mathbf{max}}(h,w)$. We have that $\hat{\mathbf{e}}_{\mathbf{max}}(3k_1+2, 3k_2+2) = 4/3$ so we assume $e(W) = 7/3$.

Write $W = T_1 \ominus T_2$, where each factor T_1, T_2 has respective height $3k_{1,1} + 1, 3k_{1,2} + 1$ for some $k_{1,1}, k_{1,2} \geq 1$. We have by minimality hypothesis $e(T_i) \leq \hat{\mathbf{e}}_{\mathbf{max}}(3k_{1,1}+1, 3k_2+2) = \hat{\mathbf{e}}_{\mathbf{max}}(3k_{1,2}+1, 3k_2+2) = 2/3$, which implies $e(W) = e(T_1) + e(T_2) \leq 2/3 + 2/3 = 4/3$, in contradiction with the hypothesis $e(W) = 7/3$.

Case 3. Let $W \in \mathcal{W}^{\leq 2}_{(3k_1+1)\times(3k_2+1)}$ with $k_1, k_2 \geq 2$. By minimal counterexample, assume that W is minimal such that $e(W) > \hat{\mathbf{e}}_{\mathbf{max}}(h, w)$. We have that $\hat{\mathbf{e}}_{\mathbf{max}}(3k_1 + 1, 3k_2 + 1) = 4/3$. Therefore, we may assume $e(W) = 7/3$. The case $k_1 = k_2 = 2$ is proved as one of the base cases in Lemma 4 ($\mathcal{W}^{\leq 2}_{7\times 7}$).

Assume $k_1 \geq 3$ and write $W = T_1 \ominus T_2$, where the factors T_1, T_2 have respective heights $3k_{1,1}+2, 3k_{1,2}+2$ for some $k_{1,1}, k_{1,2} \geq 1$. By minimality hypothesis $e(T_i) \leq \hat{\mathbf{e}}_{\mathbf{max}}(3k_{1,1} + 2, 3k_2 + 1) = \hat{\mathbf{e}}_{\mathbf{max}}(3k_{1,2} + 2, 3k_2 + 1) = 2/3$, so that $e(W) = e(T_1) + e(T_2) \leq 2/3 + 2/3 = 4/3$, in contradiction with the hypothesis $e(W) = 7/3$.

Case 4. Let $W \in \mathcal{W}^{\leq 2}_{(3k_1+1)\times 3k_2}$ with $k_1 \geq 2, k_2 \geq 3$. By minimal counterexample, assume that W is minimal such that $e(W) > \hat{\mathbf{e}}_{\mathbf{max}}(h, w)$. We have that $\hat{\mathbf{e}}_{\mathbf{max}}(3k_1 + 1, 3k_2) = 1$ so we assume $e(W) = 2$.

Now write $W = W^t[\![1, 4]\!] \ominus W^t[\![5, 3k_2]\!]$. By minimality hypothesis, we have $e(W^t[\![1, 4]\!]) \leq 4/3$ and $e(W^t[\![5, 3k_2]\!]) \leq 2/3$. Therefore, $e(W^t[\![1, 4]\!]) = 4/3$ and $e(W^t[\![5, 3k_2]\!]) = 2/3$. But, by Corollary 1, $e(W^t[\![1, 4]\!]) = 4/3$ implies $e(W^t[5]) \leq -5/3$. Thus $e(W^t[\![1, 5]\!]) \leq -1/3$, so that $e(W^t[\![6, 3k_2]\!]) \geq 7/3$, in contradiction with the minimality hypothesis.

Case 5. Let $W \in \mathcal{W}^{\leq 2}_{3k_1\times(3k_2+2)}$ with $k_1 \geq 3, k_2 \geq 2$. By minimal counterexample, assume that W is minimal such that $e(W) > \hat{\mathbf{e}}_{\mathbf{max}}(h, w)$. We have that $\hat{\mathbf{e}}_{\mathbf{max}}(3k_1, 3k_2+2) = 1$ so we assume $e(W) = 2$. Now let $W = W[\![1, 5]\!] \ominus W[\![6, 3k_1]\!]$. By minimality hypothesis we have $e(W[\![1, 5]\!]) \leq 4/3$ and $e(W[\![6, 3k_1]\!]) \leq \hat{\mathbf{e}}_{\mathbf{max}}(3(k_1-2)+1, 3k_2+2) = 2/3$. This implies $e(W) = e(W[\![1, 5]\!]) + e(W[\![6, 3k_1]\!]) \leq 4/3 + 2/3 = 2$. Hence, both $W[\![1, 5]\!]$ and $W[\![6, 3k_1]\!]$ are 2-full. Therefore, since $W[\![1, 5]\!]$ is 2-full, by Corollary 2, we deduce that $e(W[6]) \leq -4/3$. Moreover, since $e(W[\![6, 3k_1]\!]) = 2/3$, we must have $e(W[\![7, 3k_1]\!]) \geq 2$. If $k_1 > 3$, then $\hat{\mathbf{e}}_{\mathbf{max}}(3(k_1 - 2), 3k_2 + 2) = 1$, which means $e(W[\![7, 3k_1]\!]) \geq 2 > 1 = \hat{\mathbf{e}}_{\mathbf{max}}(3(k_1 - 2), 3k_2 + 2)$, contradicting the minimality hypothesis. If $k_1 = 3$, then $W[\![7, 3k_1]\!] = W[\![7, 9]\!]$ and must be 2-full with $e(W[\![7, 9]\!]) = 2$ which then forces $e(W[6]) = -4/3$. However, Corollary 3 *iii)* stipulates that, if $W[\![7, 9]\!]$ is 2-full, then $W[6]$ contains at least $\lfloor \frac{(3k_2+2)-2}{3} \rfloor + 5 = k_2 + 5$ empty cells. This means $e(W[6]) \leq -13/3$, which is a contradiction.

Case 6. Let $W \in \mathcal{W}^{\leq 2}_{3k_1+2\times 3k_2+1}$ with $k_1, k_2 \geq 2$. By minimal counterexample, assume that W is minimal such that $e(W) > \hat{\mathbf{e}}_{\mathbf{max}}(h, w)$. We have that $\hat{\mathbf{e}}_{\mathbf{max}}(3k_1 + 2, 3k_2 + 1) = 2/3$ so we assume $e(W) = 5/3$. Partition W as $W[\![1, 4]\!] \ominus W[\![5, 3k_1 + 2]\!]$. We have by minimality of W that

$$e(W[\![1, 4]\!]), e(W[\![5, 3k_1 + 2]\!]) \leq 4/3.$$

There are two cases to consider : either $e(W[\![1, 4]\!]) = 4/3$ and $e(W[\![5, 3k_1 + 2]\!]) = 1/3$ or $e(W[\![1, 4]\!]) = 1/3$ and $e(W[\![5, 3k_1 + 2]\!]) = 4/3$. The proof that these two cases lead to contradictions is given in the Arxiv version.

4 Conclusion

The results presented in this paper are opening to several questions. One of these is in the following.

Conjecture 1. For $h > w \geq 5$ and $h \equiv_3 w$, 2-full words $W \in \mathcal{MW}^{\leq 2}_{h\times w}$ are unique up to symmetries of the plane plus a local symmetry (a flip) on hook snakes.

Thanks to Theorem 1, the length of maximal snakes in a $h \times w$ rectangle can be given upper and lower bounds but the question of finding their exact length in terms of h and w is open. Theorem 1 raises analogous questions for polycubes : is there a maximal ratio analogous to 2/3 for sets of cubic cells of bounded degree in a given rectangular parallelepiped. This question is open to our knowledge. Theorem 1 also lead to the question of enumerating d-full words with respect to width w. More interesting questions rise when languages with more than two letters are used as labels of the unit cells in words W.

References

1. Amini, O., Peleg, D., Pérennes, S., Sau, I., Saurabh, S.: Degree-constrained subgraph problems: hardness and approximation results. In: International Workshop on Approximation and Online Algorithms, pp. 29–42. Springer (2008)
2. Berstel, J.: Combinatorics on Words: Christoffel Words and Repetitions in Words, Volume 27. American Mathematical Soceity (2009)
3. Carpi, A.: Multidimensional unrepetitive configurations. Theoret. Comput. Sci. **56**(2), 233–241 (1988)
4. Cyr, V., Kra, B.: Nonexpansive $\mathbb{Z}^2$-subdynamics and Nivat's conjecture. Trans. Am. Math. Soc. **367**(9), 6487–6537 (2015)
5. Epifanio, C., Koskas, M., Mignosi, F.: On a conjecture on bidimensional words. Theoret. Comput. Sci. **299**(1–3), 123–150 (2003)
6. Giammarresi, D., Restivo, A.: Two-dimensional languages. In: Handbook of Formal Languages: Volume 3 Beyond Words, pp. 215–267. Springer (1997)
7. Gonçalves, D., Pinlou, A., Rao, M., Thomassé, S.: The domination number of grids. SIAM J. Discret. Math. **25**(3), 1443–1453 (2011)
8. Mahalingam, K., Pandoh, P.: HV-palindromes in two-dimensional words. Int. J. Found. Comput. Sci. **33**(05), 389–409 (2022)
9. Morita, K.: Two-dimensional languages. In: Formal Languages and Applications, pp. 427–437. Springer (2004)
10. Nivat, M.: Invited talk at ICALP. In: ICALP 1997, Bologna (1997)
11. Sander, J.W., Tijdeman, R.: The rectangle complexity of functions on two-dimensional lattices. Theoret. Comput. Sci. **270**(1–2), 857–863 (2002)
12. Szabados, M.: Nivat's conjecture holds for sums of two periodic configurations. In: Theory and Practice of Computer Science: 44th International Conference on Current Trends in Theory and Practice of Computer Science, SOFSEM 2018, Krems, Austria, 29 January–2 February 2018, Proceedings 44, pp. 539–551. Springer (2018)

Factorizations and Monoids

Fabio Burderi(✉)

Dipartimento d'Ingegneria, Università degli studi di Palermo, 90128 Palermo, Italy
fabio.burderi@unipa.it

Abstract. In this paper we introduce the notion of generalized factorization for an arbitrary submonoid $M \subseteq A^*$, where A^* is the free monoid generated by an alphabet A, generalizing, in this way, the notion of factorization of A^*. Then we give a characterization of the free product of two submonoids of A^* in terms of unambiguous products of monoids. To do this we make use of the notion of coding partition of a set $X \subseteq A^+$, where A^+ is the free semigroup generated by an alphabet A. Moreover, given a coding partition of a set $X \subseteq A^+$, we will show how to construct a generalized factorization of X^*.

1 Introduction

The notion of coding partition of a set of words was introduced in [4] and deepened in [1] and [3], to study decipherability conditions weaker than Unique Decipherability (UD). In the context of information theory, language theory and combinatorics on words the theory of UD codes plays a central role (see [2]), however, to handle some special problems in information transmission or in the study of natural languages, it is necessary to weaken the condition of unique decipherability (see [5–7]). By the notion of coding partition, codes that are not UD are partitioned in classes so that, roughly speaking, ambiguities remain confined within the classes of the partition and, in this way, we can recover the "unique decipherability" at the level of the classes of the partition. This notion will be used in the second part of the paper. In the first part we will introduce the notions of decomposition and generalized factorization.

Given an alphabet A and a sequence $(X_1, X_2, \ldots, X_n)$ of subsets of A^*, the product $X_1X_2 \cdots X_n$ is defined as the set of words $x_1x_2 \cdots x_n$, obtained concatenating words from the X_i's. The product is said to be unambiguous if any word $w \in X_1X_2 \cdots X_n$ has only one factorization $w = x_1x_2 \cdots x_n$, with $x_i \in X_i$. As we will see it is possible to generalize both the notions of product and unambiguous product, to a family of subsets of A^* indexed by a totally ordered set, provided that the empty word belongs to all but a finite number of sets of the family (in this way we will obtain a generalization of Kleene star). If a set $X \subseteq A^*$ is the unambiguous product of a family $(X_i)_{i \in I}$, we will say that the family $(X_i)_{i \in I}$, is a *decomposition* of the set X. We will give a formulation of the notion of decomposition that makes use of formal series. Moreover if M is a monoid, each X_i is a base (that is X_i is the minimal set of generators of X_i^*), and the family $(X_i^*)_{i \in I}$, is a decomposition of M, then we will say that the family $(X_i)_{i \in I}$, is a *generalized factorization* of M. From an algebraic point of view we will deal with monoids that are submonoids of a free monoid. The

G. Gamard and J. Leroy (Eds.): WORDS 2025, LNCS 15729, pp. 49–60, 2025.
https://doi.org/10.1007/978-3-031-97548-6_5

factorization of free monoids is studied in theory of codes and it is related to notion of circular codes (see [2,8]). Moreover by [2, Theorem 8.2.], it is shown a method to construct all bisections of a free monoid. In this paper we generalize the notion of factorization to any submonoid of a free monoid. The question now arises whether we are able to find factorizations of a submonoid of a free monoid. If the monoid M is not freely indecomposable, the positive answer is given is Sect. 4: we will show how to find generalized bisections and trisections of M. To do this we use the notion of coding partition and, with this tool, we will also be able to characterize the free product of two submonoids of a free monoid.

2 Definitions and Preliminary Results

(We would like to point out that, regrettably, all proofs have been omitted due to page constraints.)

Let A be an alphabet. We let A^* denote the set of finite words over the alphabet A, and A^+ the set of nonempty finite words. Then A^* is a monoid under the concatenation operation of two words, with the empty word, denoted by ε, as the neutral element; A^* is called the *free monoid* over A. Following the notation of [1,4,6] and [3], in this paper, a *code* X over A, is just a nonempty subset $X \subseteq A^+$. Its elements are called *code words*; the submonoid and the subsemigroup of A^* generated by X, are denoted by X^* and X^+ respectively; the elements of X^* are called *messages*. We stress that, usually, the word "code" denote what we will call *uniquely decipherable* codes (see below).

For each submonoid $M \subseteq A^*$, there is a unique minimal set X of generators of M, called the *base* of M:

$$X = (M \setminus \varepsilon) \setminus (M \setminus \varepsilon)^2$$

(see [2]); i.e. if Y is a set of generators of M, then $X \subseteq Y$ and the set X is unique. A set $X \subseteq A^*$ is called a *base* if X is the base of X^*.

A word $x \in A^*$ is a *factor* of a word $w \in A^*$ if there exist $u, v \in A^*$ such that $w = uxv$.

A *factorization* of a word $w \in A^*$, is a finite sequence $(v_1, v_2, \ldots, v_s)$ of words in A^*, such that $s \geq 0$ and $w = v_1 v_2 \cdots v_s$. If $X \subseteq A^+$ is a code, a factorization $(x_1, x_2, \ldots, x_s)$, with $x_1, \ldots, x_s \in X$, into code words of a message $w \in X^*$ is said an *X-factorization* of w.

A code X is said to be *uniquely decipherable* (*UD*) if every message has a unique factorization into code words, i.e., for all $n, m \geq 0$, the equality

$$x_1 x_2 \cdots x_n = y_1 y_2 \cdots y_m,$$

$x_1, \ldots, x_n, y_1, \ldots, y_m \in X$, implies $n = m$ and $x_1 = y_1, \ldots, x_n = y_n$.

Let X be a code and let

$$P = \{X_i \mid i \in I\}$$

be a partition of X: i.e. $\bigcup_{i \in I} X_i = X$ and $X_i \cap X_j = \emptyset$ if and only if $i \neq j$.

A *P-factorization* of a message $w \in X^+$ is a factorization $w = z_1 z_2 \cdots z_t$, where

- for each i, $z_i \in X_k^+$, for some $k \in I$
- if $t > 1$, $z_i \in X_k^+ \Rightarrow z_{i+1} \notin X_k^+ \quad (1 \leq i \leq t-1)$.

The partition P is called a *coding partition* if any element $w \in X^+$ has a *unique* P-factorization, i.e., if

$$w = z_1 z_2 \cdots z_s = u_1 u_2 \cdots u_t,$$

where $z_1 z_2 \cdots z_s$, $u_1 u_2 \cdots u_t$ are P-factorizations of w, then $s = t$ and $z_i = u_i$ for $i = 1, \ldots, s$.
We observe that the trivial partition $P = \{X\}$ is always a coding partition.

Let X be a code. A *relation* on X is a pair of factorizations $(x_1, x_2, \cdots, x_s)$, $(y_1, y_2, \cdots, y_t)$ into code words of the same message $z \in X^+$, hence $z = x_1 x_2 \cdots x_s = y_1 y_2 \cdots y_t$. The relation is said to be nontrivial if the factorizations are distinct. Usually we will use the notation $x_1 x_2 \cdots x_s = y_1 y_2 \cdots y_t$ to refer to the relation; when no confusion arises, sometimes we let z denote both the word z and the relation $x_1 x_2 \cdots x_s = y_1 y_2 \cdots y_t$. We say that the relation $x_1 x_2 \cdots x_s = y_1 y_2 \cdots y_t$ is *prime* if for all $i < s$ and for all $j < t$ one has $x_1 x_2 \cdots x_i \neq y_1 y_2 \cdots y_j$.

All the results of this section come from [4] and [3].

Theorem 1. *Let $P = \{X_i \mid i \in I\}$ be a partition of a code X. The partition P is a coding partition if and only if for every prime relation $x_1 x_2 \cdots x_s = y_1 y_2 \cdots y_t$, the code words $x_i, y_j, 1 \leq i \leq s, 1 \leq j \leq t$, belong to the same class of the partition.* □

We use the following partial order between partitions of a set X: if P_1 and P_2 are two partitions of X then $P_1 \leq P_2$ if the classes of P_1 are unions of classes of P_2 and we say that P_2 is *finer* than P_1.

Corollary 1. *Let P and P' be two partitions of a code X with $P' \leq P$. If P is a coding partition then so is P'.* □

Theorem 2. *The set of coding partitions of a code X is a complete lattice.* □

As a consequence of this theorem we can give the following definition. Given a code X, *the* finest coding partition P of X is called the *characteristic* partition of X and it is denoted by $P(X)$.

A code X is called *ambiguous* if it is not *UD*. It is called *totally ambiguous* (*TA*) if $|X| > 1$ and $P(X)$ is the trivial partition: $P(X) = \{X\}$.

Remark 1. In this way *UD* codes and *TA* codes correspond to the two extremal cases, since a code is *UD* if and only if $P(X) = \{\{x\} \mid x \in X\}$.

Theorem 3. *Let $P = \{X_i \mid i \in I\}$ be a coding partition of a code with $|I| > 1$. Then the sets $\{X_{i_1}^+ X_{i_2}^+ \cdots X_{i_n}^+ \mid n \geq 2,\ i_j \in I,\ i_j \neq i_{j+1}$ for $1 \leq j < n,\ i_n \neq i_1\}$ are UD codes.* □

As we will see, given a partition $P = \{X_i \mid i \in I\}$ of a code $X \subseteq A^+$, the condition that every word $w \in X^+$ admits a unique P-factorization has a natural algebraic interpretation in terms of free product of monoids (see [3]).

Let M be a monoid generated by submonoids $M_\lambda, \lambda \in \Lambda$, and let $m \in M$. An expression of m of the form $m_1 m_2 \cdots m_r$, where $r \geq 0$, $1 \neq m_i \in M_{\lambda_i}$, $\lambda_i \neq \lambda_{i+1}$, is said to be in *reduced form* with respect to the M_λ's. By definition M is the free product of the M_λ's if and only if every element of M has an unique expression in reduced form with respect to the M_λ's, and we write $M = Fr_{\lambda \in \Lambda} M_\lambda$. In the finite case we also write $M = M_{\lambda_1} * \cdots * M_{\lambda_n}$.

A monoid M is said *freely indecomposable* if M cannot be expressed as a free product of nontrivial monoids.

The connection between the notion of coding partition and that of free product of monoids is given by the following theorem and corollary.

Theorem 4. *Let $X \subseteq A^+$ be a code, $P = \{X_i \mid i \in I\}$ a partition of X, $M = X^*$, and $M_i = X_i^*$ with $i \in I$. If P is a coding partition of X then M is the free product of the M_i's. Conversely let M be the free product of a family $(M_i)_{i \in I}$ of nontrivial submonoids, X_i a set of generators of M_i and $X := \bigcup_{i \in I} X_i$. Then $P := \{X_i \mid i \in I\}$ is a coding partition of X.* □

Corollary 2. *A monoid $M \subseteq A^*$ is freely indecomposable if and only if any set of generators of M is either a singleton or a totally ambiguous code.* □

3 Decompositions of a Set

In this section we introduce the notion of decomposition of a set $X \subseteq A^*$, and the notion of generalized factorization (resp. bisection and trisection), of a monoid $M \subseteq A^*$. These notions generalize those given in [2, Chap. 8] for a free submonoid of A^*. We recall that a monoid $M \subseteq A^*$ is a *free submonoid* if, say X its base, X is a UD code.

We will give moreover a formulation of the notion of decomposition using formal series.

Definition 1. *Let $X, X_1, \ldots, X_n$ be nonempty subsets of A^*. We say that the family $(X_1, X_2, \ldots, X_n)$ is a* decomposition *of X if $X = X_1 X_2 \cdots X_n$ and the product is unambiguous.*

So if $(X_1, X_2, \ldots, X_n)$ is a decomposition of X, any word $x \in X$ has only one factorization $x = x_1 x_2 \cdots x_n$ with $x_i \in X_i$, $1 \leq i \leq n$; moreover $\varepsilon \in X$ if and only if $\varepsilon \in X_i, \forall i$.

We want to generalize this definition to a family $(X_i)_{i \in I}$ of nonempty subsets of A^*, indexed by a totally ordered set I, provided $\varepsilon \in X_i$ for all but a finite number of indices $i \in I$. To do this we first define the *product* of the sets X_i, $i \in$

I, as follows. Since $\varepsilon \in X_i$ for all but a finite number of indices $i \in I$, we can consider families of words $(x_i)_{i\in I}$, $x_i \in X_i$, such that $x_i = \varepsilon$ for all but a finite number of indices $i \in I$: we call a such family *admissible*. So each admissible family of words is (uniquely) associated with a word w : the product of the words of the family different from ε, if any, or the empty word otherwise. More precisely, if $(x_i)_{i\in I}$, $x_i \in X_i$, is an admissible family, then there exists a finite set of indices $S := \{i_1, i_2 \ldots, i_n\} \subseteq I$, such that $x_i \neq \varepsilon$ if and only if $i \in S$; then this family is associated with the word $w := x_{i_1}x_{i_2}\cdots x_{i_n}$, with $n \geq 0$, $x_{i_k} \in X_{i_k}, 1 \leq k \leq n$, $i_1 < i_2 < \cdots < i_n$. In this way we can define the set X composed of all such words and we will say that X is the *product* of the sets $(X_i)_{i\in I}$, writing

$$X = \widehat{\prod}_{i\in I} X_i.$$

As in the finite case $\varepsilon \in X$ if and only if $\varepsilon \in X_i$, for all $i \in I$.

Remark 2. We note that the product $\widehat{\prod}_{i\in I} X_i$ is a generalization of Kleene star. Indeed for any set $X \subseteq A^*$, setting $X' := X \cup \{\varepsilon\}$, then $X^* = \widehat{\prod}_{i\in\mathbb{N}} X'$.

Finally we can give the next definition.

Definition 2. *Let $(X_i)_{i\in I}$ be a family of nonempty subsets of A^* and let $X \subseteq A^*$. If any word $x \in X$ is associated with exactly one admissible family $(x_i)_{i\in I}$, $x_i \in X_i$, we say that the family $(X_i)_{i\in I}$ is a* decomposition *of X and we write*

$$X = \prod_{i\in I} X_i.$$

Moreover, if $\prod_{i\in I} X_i = \prod_{i\in I} X_{\sigma(i)}$, for any permutations σ of I, we say that the family $(X_i)_{i\in I}$ is a commutative *decomposition of X.*

We observe that if $\varepsilon \in X$ and $(X_i)_{i\in I}$ is a decomposition of X then $X_i \cap X_j = \varepsilon$, $\forall i, j \in I$, $i \neq j$. In the general case the next proposition holds.

Proposition 1. *If $(X_i)_{i\in I}$ is a decomposition of a set $X \subseteq A^*$, then, for each word $w \in A^+$, the number of indices $i \in I$ such that $w \in X_i$ is finite.* □

We observe that if a set $X \subseteq A^*$ is the product of a family $(X_i)_{i\in I}$ of nonempty subsets of A^*, then we can identify the sets of the family that do not contain the empty word. More precisely there exists a finite set

$$S = \{s_1, s_2, \ldots, s_n\} \subseteq I, \; s_1 < s_2 < \cdots < s_n,$$

such that $k \in S$ if and only if $\varepsilon \notin X_k$. Then, using the set S, we can define the product as follow:

$$\widehat{\prod}_{i\in I} X_i = \{x_{i_1}x_{i_2}\cdots x_{i_m} | \, i_k \in I, x_{i_k} \in X_{i_k}, i_1 < \cdots < i_m, \; S \subseteq \{i_1, \ldots, i_m\}\}.$$

Let us give now a formulation of the notion of decomposition that makes use of formal series.

We recall (see [2]) that for $X \subseteq A^*$, $\underline{X}$ denote the characteristic series of X and, for $X \subseteq A^+$, is defined the series $\underline{X}^*$, with the coefficient

$$(\underline{X}^*, w) = |\{(x_1, \ldots, x_n) : n \geq 0,\, x_i \in X,\, w = x_1 \cdots x_n\}|\,.$$

Then $(\underline{X}^*, w)$ counts the distinct factorizations of w in words in X.

Remark 3. For $X \subseteq A^+$, we stress the typographical difference between $\underline{X}^*$ that is equal to $(\underline{X})^*$ and $\underline{X^*}$ that is equal to $\underline{(X^*)}$.

We start observing that, if $(X_1, X_2, \ldots, X_n)$ is a finite family of subsets of A^*, then, by definition, the family is a decomposition of a set $X \subseteq A^*$, if and only if $\underline{X} = \underline{X_1}\,\underline{X_2} \cdots \underline{X_n}$. Now we study the general case.

Let $(\sigma_i)_{i \in I}$ be a family of formal series over an alphabet A with coefficients in a ring with unit K, indexed by a totally ordered set I. We recall that a family $(\sigma_i)_{i \in I}$ of series is said to be *locally finite*, if for all $w \in A^*$ the set $\{i \in I \,|\, (\sigma_i, w) \neq 0\}$ is finite. In this case it is possible to define the series

$$\sigma = \sum_{i \in I} \sigma_i, \; by \; (\sigma, w) := \sum_{i \in I} (\sigma_i, w).$$

Let the family $(\sigma_i)_{i \in I}$ be locally finite and let $J := \{j_1, j_2, \ldots, j_n\}$, be a finite subset of I, with $j_1 < j_2 < \cdots < j_n$. We set

$$\tau_J := \sigma_{j_1} \sigma_{j_2} \cdots \sigma_{j_n},$$

where $\tau_J = 1$, if $J = \emptyset$. So, for all $w \in A^*$,

$$(\tau_J, w) = \sum_{x_1 x_2 \cdots x_n = w} (\sigma_{j_1}, x_1)(\sigma_{j_2}, x_2) \cdots (\sigma_{j_n}, x_n). \tag{1}$$

Let S be the family of all finite subsets of I. We will prove that the family $(\tau_J)_{J \in S}$, is locally finite. Indeed, for any $x \in A^*$, let I_x be the set of indices $i \in I$ such that $(\sigma_i, x) \neq 0$. Since the family $(\sigma_i)_{i \in I}$ is locally finite then, for any $x \in A^*$, I_x is finite. Hence, for each $w \in A^*$, say $F(w)$ the finite set of factors of w, the set $K := \bigcup_{x \in F(w)} I_x$ is also finite. Consequently, for each $w \in A^*$, if $(\tau_J, w) \neq 0$, then $J \subseteq K$ and so the family $(\tau_J)_{J \in S}$ is locally finite. Now, similarly to [2, Chap. 8], we can define the series

$$\sigma = \prod_{i \in I} (1 + \sigma_i), \qquad \text{by} \qquad \sigma := \sum_{J \in S} \tau_J.$$

We observe that the notation $\sigma = \prod_{i \in I}(1 + \sigma_i)$ is justified because, if $I = \{1, 2, \ldots, n\}$ is finite, then

$$\sum_{J \in S} \tau_J = \prod_{i=1}^{n} (1 + \sigma_i).$$

Thus, given a family $(\sigma_i)_{i\in I}$ of formal series such that $(\sigma_i - 1)_{i\in I}$ is locally finite, set, $\forall i \in I$, $\sigma'_i := \sigma_i - 1$, it is defined the series

$$\sigma := \prod_{i\in I}(1+\sigma'_i) = \prod_{i\in I}\sigma_i. \tag{2}$$

Remark 4. We observe that given a family $(\sigma_i)_{i\in I}$ of formal series, the family $(\sigma_i - 1)_{i\in I}$ is locally finite if and only if the family $(\sigma_i)_{i\in I}$ is locally finite and the constant term $(\sigma_i, 1)$ is equal to 1 for all but a finite number of indices $i \in I$.

Let us now suppose that the family $(\sigma_i - 1)_{i\in I}$ is locally finite and consider a finite set $J = \{i_1, i_2, \ldots, i_n\} \subseteq I$, with $i_1 < i_2 < \cdots < i_n$. Then we can define the following product of $2n+1$ formal series:

$$\prod_{I,J}\sigma_i := (\prod_{i<i_1}\sigma_i)\cdot\sigma_{i_1}\ \cdots\ \sigma_{i_{n-1}}\cdot(\prod_{i_{n-1}<i<i_n}\sigma_i)\cdot\sigma_{i_n}\cdot(\prod_{i_n<i}\sigma_i), \tag{3}$$

where, for the $n+1$ series written in parentheses, we used Definition (2).

At this point we can restate the notion of decomposition of a set $X \subseteq A^*$ using formal series. This is done with the following proposition that comes from Proposition 1 and Formula (2).

Proposition 2. *Let $(X_i)_{i\in I}$ be a decomposition of X (that is $X = \prod_{i\in I} X_i$), and let $J = \{i_1, i_2, \ldots, i_n\} \subseteq I$, $i_1 < i_2 < \cdots < i_n$, such that $i_j \in J$ if and only if $\varepsilon \notin X_{i_j}$. Then the family $(\underline{X_i} - 1)_{i\in I}$ is locally finite and*

$$\underline{X} = \prod_{I,J}\underline{X_i}. \tag{4}$$

Conversely if the family $(\underline{X_i} - 1)_{i\in I}$ is locally finite, then the product $\prod_{i\in I}\underline{X_i}$ is defined and, if Eq. (4) holds for a certain finite set $J \subseteq I$, then the family $(X_i)_{i\in I}$ is a decomposition of X, that is $X = \prod_{i\in I} X_i$. □

Now we want to compare the notion of decomposition with that of factorization (see [8, Chap. 5], [2, Chap. 8]). The notion of factorization is defined only for A^* or for a free monoid generated by a *UD* code. It follows that if $(X_i)_{i\in I}$ is a factorization of A^*, then each X_i is a *UD* code and the semigroups X_i^+ are pairwise disjoint. Conversely, if $(X_i)_{i\in I}$ is a decomposition of a set $X \subseteq A^*$, the sets X_i need not be pairwise disjoint and, even in the case $X = A^*$, the sets $X_i \setminus \varepsilon$ need not be *UD* codes (see next examples and Remark 6).

Moreover, given a factorization of A^*, we can easily express it in terms of decompositions. Indeed, if $(X_i)_{i\in I}$ is a factorization of A^*, set $N_i := X_i^*$, then the family $(N_i)_{i\in I}$ is a decomposition of A^*. Conversely, if $(N_i)_{i\in I}$ is a decomposition of A^* with the N_i's free submonoids, then, say X_i the base of N_i, the family $(X_i)_{i\in I}$ is a factorization of A^*.

This last observation allows as to generalize the notions of factorization, bisection and trisection using the next definition.

Definition 3. *Let $M \subseteq A^*$ a monoid and let $(N_i)_{i \in I}$ be a decomposition of M with the N_i's nontrivial monoids. Set X_i the base of N_i, $i \in I$, we say that the family $(X_i)_{i \in I}$ is a generalized factorization of M. If $|I| = 2$ ($|I| = 3$), we say that the family is a generalized bisection (resp. generalized trisection) of M.*

Remark 5. By definition, if $M \subseteq A^*$ is a monoid and $X_i \subseteq A^+$, $i \in I$, are bases, then the family $(X_i)_{i \in I}$ is a generalized factorization of M if and only if $(X_i^*)_{i \in I}$ is a decomposition of M.

We know (see [2, Proposition 8.2.4]) that if M and N are two submonoids of A^* such that $\underline{A^*} = \underline{M}\ \underline{N}$, then M and N are free and the pair (X, Y) of their bases is a bisection of A^*. So the following proposition holds.

Proposition 3. *Let $M \subseteq A^*$ be a monoid, if $M = A^*$, then the notions of bisection of A^* and generalized bisection of A^* agree.* □

Remark 6. Of course, in principle, the notion of generalized factorization of A^* is more general of that of factorization of A^* but, actually, we do not know whether there are generalized trisections (generalized factorizations) of A^* that are not trisections (factorizations).

Let's give some examples about the notions introduced up to now.

Example 1. Let $\{X_1, X_2\}$ be a partition of a set $X \subseteq A^+$. If $B = \{a, b\}$ is a two-letter alphabet, it is easy to see that $(a^*b,\ a)$ and $(a,\ b^+a^+,\ b)$ are respectively a bisection and a trisection of B^*. Guided from this fact we can easily express X^* as a product of monoids as follows: $X^* = (X_1^* X_2)^* X_1^*$, and $X^* = X_1^* (X_2^+ X_1^+)^* X_2^*$. We note that, in general, the previous families $((X_1^* X_2)^*, X_1^*)$ and $(X_1^*, (X_2^+ X_1^+)^*, X_2^*)$ are not decompositions of X^*.

Remark 7. As we will see later (Proposition 6), the construction used in the previous example may be generalized to partitions of a set $X \subseteq A^+$ composed of more than two classes.

Example 2. Let A be an alphabet and let $X_0 = A \cup \{\varepsilon\}$, $X_1 = A^2 \cup \{\varepsilon\}$, $X_2 = A^4 \cup \{\varepsilon\}$ and, in general, $X_i = A^{2^i} \cup \{\varepsilon\}$, for $i \geq 0$. Then the family $(X_i)_{i \geq 0}$ is a commutative decomposition of A^*. Indeed, let $w \in A^*$, $n = |w|$, and let $(c_k c_{k-1} \cdots c_1 c_0)$ be the binary representation of n, with $c_i \in \{0, 1\}$, for $0 \leq i \leq k$. Then we can uniquely decompose w as the concatenation of words x_i of length 0 or 2^i, according to $c_i = 0$ or $c_i = 1$, for $0 \leq i \leq k$. So taking $x_i = \varepsilon$ for $i > k$, the family $(x_i)_{i \geq 0}$ is the unique admissible family associated with w. The fact that the decomposition is commutative is clear.

Example 3. Let $A = \{a, b\}$, $X_1 = \{\varepsilon,\ a,\ xb,\ xba : x \in A^*\}$ and $X_2 = (a^2)^*$. The reader may easily check that $\underline{A^*} = \underline{X_1}\ \underline{X_2}$. So (X_1, X_2), is an example of a finite decomposition of A^* that is not a factorization; we note moreover that $X_1 \cap A^+$ is not a *UD* code.

Now we want to study some properties of the generalized bisections of a not free submonoid $M \subsetneq A^*$, and we will compare these properties with those of the bisections of A^*.

Let $M \subseteq A^*$ be a monoid and let $(X_i)_{i \in I}$ be a family of subsets of A^+ indexed by a totally ordered set I. If the family is a generalized factorization of M then, set Z the base of M, by Eq. (4) with $J = \emptyset$, we can write

$$\underline{M} = \underline{Z^*} = \prod_{i \in I} \underline{X_i^*}. \tag{5}$$

Conversely if the semigroups X_i^+ are pairwise disjoint, then the family $(\underline{X_i^+})_{i \in I}$ is locally finite so the product $\prod_{i \in I} \underline{X_i^*}$ is defined and, if each X_i is a base, Eq. (5) implies that the family $(X_i)_{i \in I}$ is a generalized factorization of Z^*. We note that $Z \subseteq \bigcup_{i \in I} X_i$. Indeed the set $\bigcup_{i \in I} X_i$ is a set of generators of M and, since Z is the base of M, the assertion is clear. In particular if (X, Y) is a generalized bisection of M then

$$\underline{M} = \underline{X^*}\,\underline{Y^*}\,, \qquad Z \subseteq X \cup Y\,, \qquad X^+ \cap Y^+ = \emptyset\,. \tag{6}$$

Proposition 4. *Let $M, N, Q \subseteq A^*$ be monoids such that $\underline{Q} = \underline{M}\,\underline{N}$ and let $x_1, x_2 \in Q$. If $x_1 x_2 \in M$ then $x_2 \in M$; if $x_1 x_2 \in N$ then $x_1 \in N$.*

As we have recalled if M, N are two submonoids of A^* such that $\underline{A^*} = \underline{M}\,\underline{N}$, then M and N are free and the pair (X, Y) of their bases is a bisection of A^*. Moreover if (S, T) is a partition of A^*, there exists a unique bisection (X, Y) of A^* such that $X \subseteq S$ and $Y \subseteq T$. But if a monoid $M \subseteq A^*$ is not free, then it is no longer true, as the following example shows, that for each partition (S, T) of M it is possible to find a generalized bisection (X, Y) of M such that $X \subseteq S$ and $Y \subseteq T$.

Example 4. Let $Z = \{0, 01, 10\}$, $S = \{0, 01\}$, $T = Z^* \setminus S$ and let, by contradiction, (X, Y) be a generalized bisection of Z^* with $X \subseteq S$ and $Y \subseteq T$. By Formula (6), $X = S$ and, since $0, 10 \in Z$ and $010 \in X^*$, by previous proposition, $10 \in X^*$ and this is a contradiction.

The following example shows a generalized bisection of the monoid Z^* considered before.

Example 5. Let $Z = \{0, 01, 10\}$, $X = \{0, (01)^n 10 \mid n \geq 0\}$ and $Y = \{01\}$. Then (X, Y) is a generalized bisection of Z^*.

If a monoid $M \subseteq A^*$ is free, and (X, Y) is a bisection of M, then we know (see [2, Chap. 8]) that X, Y are *UD* codes and $YX \subseteq X \cup Y$. This is no longer true if M is not free. Indeed, in the previous proposition, the monoid Z^* is not free, both X and Y are still *UD* codes, but $YX \not\subseteq X \cup Y$. In this case, $YX \subseteq X^* \cup Y^*$, but this is not always true as the following example shows.

Example 6. Let $Z = \{0, 01, 10\}$, $X = \{10, 0^n01 \mid n \geq 0\}$, $Y = \{0\}$. It is possible to prove that X, Y is a generalized bisection of Z^*; moreover $010 \in YX$ but $010 \notin X^* \cup Y^*$.

In general, finding a generalized bisection of a non-free monoid M is non-trivial. However, as we will see in the next section, if the monoid M is not freely indecomposable, there is a general construction to find both a generalized bisection and a generalized trisection of M.

4 Free Products, Decompositions and Factorizations

In this section, using coding partitions, we first characterize the free product of two submonoids of A^* in terms of unambiguous product of monoids. Then we show how to construct generalized bisections and trisections of a non-freely indecomposable submonoid $M \subseteq A^*$, starting by a non-trivial coding partition of a set X of generators of M.

We start giving the following characterization of a coding partition of a set $X \subseteq A^+$, composed of two classes.

Theorem 5. *Let $P = \{X_1, X_2\}$ be a partition of a set $X \subseteq A^+$. Then P is a coding partition if and only if*

i) $(X_1^*, (X_2^+X_1^+)^*, X_2^*)$ is a decomposition of X^*
ii) $X_2^+X_1^+$ is an unambiguous product
iii) $X_2^+X_1^+$ is a UD code. □

Corollary 3. *Let $P = \{X_1, X_2\}$ be a coding partition of a set $X \subseteq A^+$. If both X_1 and X_2 are bases, then $(X_1, X_2^+X_1^+, X_2)$ is a generalized trisection of X^*.*

Corollary 4. *Let $X \subseteq A^+$ with $|X| > 1$. If X is not totally ambiguous then we can find a generalized trisection of X^*.*

Remark 8. We note that, in Theorem 5, the first and third conditions together do not imply the second one, as the following example shows; therefore they are not sufficient to guarantee that P is a coding partition of X.

Example 7. Let $X_1 = \{bb, babb\}$, $X_2 = \{aa, aaba\}$ be two subsets of $\{a, b\}^*$. It is possible to prove that, set $X := X_1 \cup X_2$,

- $(X_1^*, (X_2^+X_1^+)^*, X_2^*)$ is a decomposition of X^*
- $X_2^+X_1^+$ is a *UD* code.

But the relation $aaba \cdot bb = aa \cdot babb$ shows that the product $X_2^+X_1^+$ is ambiguous and $P = \{X_1, X_2\}$ is not a coding partition of $X_1 \cup X_2$. Moreover, $(X_1, X_2^+X_1^+, X_2)$ is an example of generalized trisection of X^*.

At this point we can give an algebraic interpretation of previous results. We will see how to have a characterization of the free product of two submonoids of A^*, in terms of unambiguous product of monoids. More precisely if M is the free product of M_1 and M_2, then, by definition, the product M_2M_1 is unambiguous; moreover we will prove that there exists a free monoid N, depending only by M_1 and M_2, such that M is equal to the unambiguous product M_1NM_2; finally we will see that these are characteristic properties of a free product of two monoids.

Theorem 6. *Let M, M_1, M_2 be submonoids of A^* and let X_1, $X_2 \subseteq A^+$ be sets of generators of M_1 and M_2 respectively. Then, set $N := (X_2^+X_1^+)^*$, M is the free product of M_1 and M_2 if and only if*

i) $\underline{M} = \underline{M_1}\underline{N}\underline{M_2}$
ii) M_2M_1 *is an unambiguous product*
iii) N *is a free monoid.*

By definition of free indecomposable monoid, follows the next

Corollary 5. *Let M be a submonoid of A^*. Then M is not freely indecomposable if and only if there exist three monoids M_1, M_2 and N such that:*

i) $\underline{M} = \underline{M_1}\underline{N}\underline{M_2}$
ii) M_2M_1 *is an unambiguous product*
iii) N *is a free monoid.* □

Theorem 7. *If $P = \{X_1, X_2\}$ is a coding partition of a set $X \subseteq A^+$, then $((X_1^*X_2)^*, X_1^*)$ is a decomposition of X^*. Moreover, if X_1 and X_2 are bases, $(X_1^*X_2, X_1)$ is a generalized bisection of X^*.*

Now we can assert that, if $P = \{X_1, X_2\}$ is a coding partition of a set $X \subseteq A^+$, with X_1, X_2 bases, then by Corollary 3 and Theorem 7, we can find both a generalized bisection and a generalized trisection of X^*:

$$(X_1^*X_2,\, X_1), \quad (X_1,\, X_2^+X_1^+,\, X_2).$$

We note the analogy with the following bisection and a trisection of A^* where $A = \{a, b\}$:

$$(a^*b,\, a), \qquad (a,\, b^+a^+,\, b).$$

We will show now a factorization of a free monoid generated by an alphabet A that we will use next in the paper. In the case the alphabet is $A = \{a, b\}$, we will find the bisection $(a, b\,a^*)$ that is the "reverse" of $(a^*b,\, a)$.

Proposition 5. *Consider a totally ordered set I and let $A = \{a_i \mid i \in I\}$ be an alphabet. For each $j \in I$, let $Y_j := \{a_i : i < j,\, i \in I\}$ and let $X_j := a_jY_j^*$. Then $\underline{A^*} = \prod_{j\in I} \underline{X_j}^*$ and the family $(X_j)_{j\in I}$ is a factorization of A^*.*

Starting from any nontrivial partition of a code $X \subseteq A^+$ and using the previous proposition, we are able to find a family $(Z_i)_{i\in I}$ of subsets of X^* such that $X^* = \widehat{\prod}_{i\in I} Z_i$. Indeed the following proposition holds.

Proposition 6. *Let $P = \{X_i \mid i \in I\}$ be a partition of a code $X \subseteq A^+$. Then we can find a family $(Z_j)_{j\in I}$ of subsets of X^* such that $X^* = \widehat{\prod}_{j\in I} Z_j$.*

Finally we will see, using a coding partition of a set $X \subseteq A^+$ and the factorization given in Proposition 5, how to find a generalized factorization of X^*.

Theorem 8. *Let $P = \{X_i \mid i \in I\}$ be a coding partition of a code $X \subseteq A^+$. Let $\forall i \in I$, $Y_i = \bigcup_{j<i} X_j$ and let $Z_i = X_i Y_i^*$. Then the family $(Z_i^*)_{i\in I}$ is a decomposition of X^*. Moreover, if $\forall i \in I$, X_i is a base, then the family $(Z_i)_{i\in I}$ is a generalized factorization of X^*.*

References

1. Béal, M.-P., Burderi, F., Restivo, A.: Coding partitions of regular sets. Int. J. Algebr. Comput. **198**(8), 1011–1023 (2009)
2. Berstel, J., Perrin, D., Reutenauer, C.: Codes and Automata, volume 129 of Encyclopedia of Mathematics and its Applications. Cambridge University Press (2010)
3. Burderi, F.: Full monoids and maximal codes. Int. J. Found. Comput. Sci. **23**(8), 1677–1690 (2012)
4. Burderi, F., Restivo, A.: Coding partitions. Discrete Math. Theoret. Comput. Sci. **9**(2), 227–240 (2007)
5. Gönenç, G.: Unique decipherability of codes with constraints with application to syllabification of Turkish words. In: Proceedings of the International Conference on Computational Linguistics, COLING 1973. Computational And Mathematical Linguistics, vol. 1, pp. 183–194 (1973)
6. Guzmán, F.: Decipherability of codes. J. Pure Appl. Algebra **141**(1), 13–35 (1999)
7. Lempel, A.: On multiset decipherable codes. IEEE T. Inform. Theor. **32**(5), 714–716 (1986)
8. Lothaire, M.: Combinatorics on Words, volume 17 of Encyclopedia of Mathematics and its Applications. Cambridge University Press (1997)

Free Product of Formal Series

Fabio Burderi(✉)

Dipartimento d'Ingegneria, Università degli studi di Palermo, 90128 Palermo, Italy
fabio.burderi@unipa.it

Abstract. In this paper we introduce the notion of free product of formal series. Using this notion, we can characterize the free product of submonoids of A^*, where A^* is the free monoid generated by an alphabet A. Moreover given a set $X \subseteq A^+$, where A^+ is the free semigroup generated by an alphabet A, we can characterize a partition of X by the free product of the formal series associated to the classes of the partition.

1 Introduction

The use of formal series in noncommutative variables cover many areas of mathematics: enumerative combinatorics, formal languages, probability, theory of rational identities in rings, just to name a few. They first appeared in a rigorous theoretical context in the theory of formal languages and finite automata (see Kleene [6], Schützenberger [7,8], and Fliess [5]); for example formal series are used in the factorization of the free monoid A^* generated by an alphabet A (see [1]).

In this paper we define the *free product of formal series*. In [3] it was introduced the notion of *coding partition* of a set $X \subseteq A^+$, relating this notion to that of free product of submonoids of A^*. With the tool of the free product of formal series we will be able to characterize both a partition $P = \{X_i \mid i \in I\}$ and a coding partition, of a set $X \subseteq A^+$, in terms of the formal series $\underline{X_i}$. This, in turn, will give a characterization of the free product of a family of submonoids of A^*.

2 Preliminary Definitions

Let A be an alphabet, i.e. a non-empty set, whose elements are called *letters*. A *word* on the alphabet A is a finite sequence of letters. We let A^* denote the set of words over the alphabet A, and A^+ the set of nonempty words. Then A^* is a monoid under the concatenation operation of two words, with the empty word, denoted by 1, as the neutral element; A^+ is the corresponding semigroup. The length $|w|$ of the word $w = a_1 a_2 \cdots a_n$, with $a_i \in A$, is the number n of letters in w.

A word $x \in A^*$ is a *factor* of a word $w \in A^*$ if there exist $u, v \in A^*$ such that $w = uxv$. A *code* X over A, in this paper, is just a nonempty subset $X \subseteq A^+$. Its elements are called *code words*; the submonoid and the subsemigroup of A^*

G. Gamard and J. Leroy (Eds.): WORDS 2025, LNCS 15729, pp. 61–69, 2025.
https://doi.org/10.1007/978-3-031-97548-6_6

generated by X, are denoted by X^* and X^+ respectively; the elements of X^* are called *messages*. We stress that, usually, the word "code" denote what we will call *uniquely decipherable* codes (see below).

For each submonoid $M \subseteq A^*$, there is a unique minimal set X of generators of M, called the *base* of M:

$$X = (M \setminus 1) \setminus (M \setminus 1)^2$$

(see [1]); i.e. if Y is a set of generators of M, then $X \subseteq Y$ and the set X is unique. A set $X \subseteq A^*$ is called a *base* if X is the base of X^*.

Let $w \in A^*$ be a word. A *factorization* of w is a finite sequence $(v_1, v_2, \dots, v_s)$ of words in A^*, such that $s \geq 0$ and $w = v_1 v_2 \cdots v_s$; the integer $s \geq 0$, is said the *length* of the factorization. If $X \subseteq A^+$ is a code, a factorization $(x_1, x_2, \dots, x_s)$, $x_1, \dots, x_s \in X$, into code words of a message $w \in X^*$, is said an X-*factorization* of w.

Remark 1. If $X \subseteq A^*$ is a base, then the only X-factorization of a word $x \in X$ is the trivial factorization $x = (x)$.

A formal series (or just series) over A with coefficients in a semiring K, is a mapping $\sigma : A^* \to K$. The value of σ on $w \in A^*$ is denoted (σ, w).

The formal series $\sigma + \tau, \sigma\tau$ and $k\sigma$, with $k \in K$, are defined as follows:

$$(\sigma + \tau, w) = (\sigma, w) + (\tau, w),$$

$$(\sigma\,\tau, w) = \sum_{uv=w} (\sigma, u)(\tau, v),$$

$$(k\,\sigma, w) = k(\sigma, w).$$

The neutrals elements for the addition and the multiplication, denoted by 0 and 1 respectively, are defined by

$$(0, w) = 0, \ \forall w \in A^*,$$

$$(1, w) = \begin{cases} 1 & \text{if } w = 1, \\ 0 & \text{otherwise .} \end{cases}$$

A family of series $(\sigma_i)_{i \in I}$, indexed by a totally ordered set I, is said to be *locally finite*, if for all $w \in A^*$, the set $\{i \in I \mid (\sigma_i, w) \neq 0\}$ is finite. In this case a series σ, denoted

$$\sigma = \sum_{i \in I} \sigma_i,$$

can be defined by

$$(\sigma, w) := \sum_{i \in I} (\sigma_i, w).$$

The constant term of σ is the element $(\sigma, 1) \in K$. If σ has zero constant term, then the family $(\sigma^n)_{n\geq 0}$ is locally finite. We denote, as usual, by σ^* and by σ^+ the series

$$\sigma^* = \sum_{n\geq 0} \sigma^n, \qquad \sigma^+ = \sum_{n>0} \sigma^n.$$

We observe that $\sigma^* = \sigma^+ + 1$ and $\sigma\sigma^* = \sigma^*\sigma = \sigma^+$.

In what follows we will take $K = \mathbb{Z}$, and it is easy to see that the series $1-\sigma$ is invertible with $(1-\sigma)^{-1} = \sigma^*$.

For $X \subseteq A^*$, we denote by $\underline{X}$ the *characteristic series* of X defined by

$$(\underline{X}, x) = \begin{cases} 1 & \text{if } x \in X, \\ 0 & \text{otherwise}. \end{cases}$$

For $X, Y \subseteq A^*$ the product XY is defined by

$$XY := \{xy \mid x \in X,\ y \in Y\};$$

it is said to be *unambiguous* if any word $w \in XY$ has only one factorization $w = xy$ with $x \in X$, $y \in Y$.

By definition, if $X, Y \subseteq A^*$, then

$$(\underline{X}\,\underline{Y}, w) = |\{(x, y) \in X \times Y \mid w = xy\}|.$$

So, if $Z = XY$, then $\underline{Z} = \underline{X}\,\underline{Y}$ if and only if the product XY is unambiguous.

Moreover for $X \subseteq A^+$, we have

$$(\underline{X}^*, w) = |\{(x_1, \ldots, x_n) : n \geq 0,\ x_i \in X,\ w = x_1 \cdots x_n\}|,$$

thus $(\underline{X}^*, w)$ counts the distinct factorizations of w in words in X (see [1]).

Remark 2. For $X \subseteq A^+$, we stress the typographical difference between $\underline{X}^*$ that is equal to $(\underline{X})^*$ and $\underline{X^*}$ (which we will use later) that is equal to $\underline{(X^*)}$.

A code X is said to be *uniquely decipherable* (*UD*) if every message has a unique factorization into code words, i.e., for all $n, m \geq 0$, the equality

$$x_1 x_2 \cdots x_n = y_1 y_2 \cdots y_m,$$

$x_1, \ldots, x_n,\ y_1, \ldots, y_m \in X$, implies $n = m$ and $x_1 = y_1, \ldots, x_n = y_n$. Using formal series, X is a *UD* code if and only if $\underline{X^*} = \underline{X}^*$

Let X be a code and let

$$P = \{X_i \mid i \in I\}$$

be a partition of X: i.e. $\bigcup_{i\in I} X_i = X$ and $X_i \cap X_j = \emptyset$ if and only if $i \neq j$.

A *P-factorization* of a message $w \in X^+$ is a factorization $w = z_1 z_2 \cdots z_t$, where

- for each i, $z_i \in X_k^+$, for some $k \in I$
- if $t > 1$, $z_i \in X_k^+ \Rightarrow z_{i+1} \notin X_k^+ \quad (1 \leq i \leq t-1)$.

The partition P is called a *coding partition* if any element $w \in X^+$ has a *unique* P-factorization, i.e., if

$$w = z_1 z_2 \cdots z_s = u_1 u_2 \cdots u_t,$$

where $z_1 z_2 \cdots z_s$, $u_1 u_2 \cdots u_t$ are P-factorizations of w, then $s = t$ and $z_i = u_i$ for $i = 1, \ldots, s$.

We observe that the trivial partition $P = \{X\}$ is always a coding partition.

Let X be a code. A *relation* on X is a pair of factorizations $(x_1, x_2, \cdots, x_s)$, $(y_1, y_2, \cdots, y_t)$ into code words of the same message $z \in X^+$, and hence $z = x_1 x_2 \cdots x_s = y_1 y_2 \cdots y_t$. The relation is said to be nontrivial if the factorizations are distinct. Usually we will use the notation $x_1 x_2 \cdots x_s = y_1 y_2 \cdots y_t$ to refer to the relation; when no confusion arises, sometimes we let z denote both the word z and the relation $x_1 x_2 \cdots x_s = y_1 y_2 \cdots y_t$. We say that the relation $x_1 x_2 \cdots x_s = y_1 y_2 \cdots y_t$ is *prime* if for all $i < s$ and for all $j < t$ one has $x_1 x_2 \cdots x_i \neq y_1 y_2 \cdots y_j$.

All the results of this section comes from [3] and [2].

Theorem 1. *Let $P = \{X_i \mid i \in I\}$ be a partition of a code X. The partition P is a coding partition if and only if for every prime relation $x_1 x_2 \cdots x_s = y_1 y_2 \cdots y_t$, the code words $x_i, y_j, 1 \leq i \leq s, 1 \leq j \leq s$, belong to the same class of the partition.* □

(See [3, Theorem 1].)

We use the following partial order between partitions of a set X: if P_1 and P_2 are two partitions of X then $P_1 \leq P_2$ if the classes of P_1 are unions of classes of P_2 and we say that P_2 is *finer* than P_1.

Corollary 1. *Let P and P' be two partitions of a code X with $P' \leq P$. If P is a coding partition then so is P'.* □

(See [2, Corollary 2].)

Theorem 2. *The set of coding partitions of a code X is a complete lattice.* □

(See [3, Theorem 3].)

As a consequence of this theorem we can give the following definition. Given a code X, *the* finest coding partition P of X is called the *characteristic* partition of X and it is denoted by $P(X)$.

A code X is called *ambiguous* if it is not UD. It is called *totally ambiguous* (TA) if $|X| > 1$ and $P(X)$ is the trivial partition: $P(X) = \{X\}$.

Remark 3. In this way UD codes and TA codes correspond to the two extremal cases, since a code is UD if and only if $P(X) = \{\{x\} \mid x \in X\}$.

Theorem 3. *Let $P = \{X_i \mid i \in I\}$ be a coding partition of a code with $|I| > 1$. Then the sets $X_{i_1}^+ X_{i_2}^+ \cdots X_{i_n}^+$ (such that $n \geq 2$, $i_j \in I$, $i_j \neq i_{j+1}$ for $1 \leq j < n$, $i_n \neq i_1$) are UD codes.* □

(See [2, Theorem 10].)

As we will see, given a partition $P = \{X_i \mid i \in I\}$ of a code $X \subseteq A^+$, the condition that every word $w \in X^+$ admits a unique P-factorization has a natural algebraic interpretation in terms of free product of monoids (see [2]).

Let M be a monoid generated by submonoids $M_\lambda, \lambda \in \Lambda$, and let $m \in M$, so $M = \left(\bigcup_{\lambda \in \Lambda} M_\lambda\right)^*$. An expression of m of the form $m_1 m_2 \cdots m_r$, where $r \geq 0$, $1 \neq m_i \in M_{\lambda_i}$, $\lambda_i \neq \lambda_{i+1}$, is said in *reduced form* with respect to the M_λ's. By definition M is the free product of the M_λ's if and only if every element of M has a unique expression in reduced form with respect to the M_λ's, and we write $M = Fr_{\lambda \in \Lambda} M_\lambda$. In the finite case we also write $M = M_{\lambda_1} * \cdots * M_{\lambda_n}$.
A monoid M is said *freely indecomposable* if M cannot be expressed as a free product of nontrivial monoids.

The connection between the notion of coding partition and that of free product of monoids is given by the following theorem and corollary.

Theorem 4. *Let $X \subseteq A^+$ be a code, $P = \{X_i \mid i \in I\}$ a partition of X, $M = X^*$, and $M_i = X_i^*$ with $i \in I$. If P is a coding partition of X then M is the free product of the M_i's. Conversely let M be the free product of a family $(M_i)_{i \in I}$ of nontrivial submonoids, X_i a set of generators of M_i and $X := \bigcup_{i \in I} X_i$. Then $P := \{X_i \mid i \in I\}$ is a coding partition of X.* □

(See [2, Theorem 13].)

Corollary 2. *A monoid $M \subseteq A^*$ is freely indecomposable if and only if any set of generators of M is either a singleton or a totally ambiguous code.* □

(See [2, Remark 15].)

2.1 Free Product of Formal Series

Now we want to introduce the notion of *free product* of formal series. We will use this notion to characterize a partition $P = \{X_i \mid i \in I\}$ of a set $X \subseteq A^+$ in terms of $\underline{X}^+$ and the free product of the formal series $\underline{X_i}^+$.
Let $(\sigma_i)_{i \in I}$ be a family of formal series over an alphabet A with coefficients in a semiring K. Assume that the family is locally finite and furthermore that $(\sigma_i, 1) = 0$, $\forall i \in I$, where I is a totally ordered set. Let $J_1 := I$, $J_2 := \{(j_1, j_2) \in I^2 \mid j_1 \neq j_2\}$ and, in general, for $k \geq 1$,

$$J_k := \{(j_1, j_2, \ldots, j_k) \in I^k \mid j_i \neq j_{i+1},\ 1 \leq i < k\}.$$

Let now $J := \cup_{k \geq 1} J_k$ and, for $j = (j_1, j_2, \ldots, j_k) \in J$, set $\tau_j := \sigma_{j_1} \sigma_{j_2} \cdots \sigma_{j_k}$. Then for any $w \in A^*$, $j \in J$,

$$(\tau_j, w) = \sum_{x_1 x_2 \cdots x_k = w} (\sigma_{j_1}, x_1)(\sigma_{j_2}, x_2) \cdots (\sigma_{j_k}, x_k). \tag{1}$$

We show that the family $(\tau_j)_{j \in J}$ is locally finite. Indeed $\forall w \in A^*$, the set $F(w)$ of factors of w is finite. Since the family $(\sigma_i)_{i \in I}$ is locally finite, for each $x \in A^*$, the set $I_x := \{i \in I \mid (\sigma_i, x) \neq 0\}$ is finite. Put $T_w := \cup_{x \in F(w)} I_x$, from (1), it follows that if $(\tau_j, w) \neq 0$, then $\{j_1, j_2, \ldots, j_k\} \subseteq T_w$. Moreover since, by hypothesis, $(\sigma_i, 1) = 0$, $\forall i \in I$, then $j \in \cup_{1 \leq k \leq |w|} J_k$. So there are only finitely many indexes $j \in J$ such that $(\tau_j, w) \neq 0$: for each $k \geq 1$, the number of $j \in J_k$ such that $(\tau_j, w) \neq 0$ is bounded by $|T_w|(|T_w| - 1)^{k-1}$. Indeed if $(j_1, j_2, \ldots, j_k) \in J_k$ then we have at most $|T_w|$ choices for j_1 and, since $j_i \neq j_{i+1}$, $1 \leq i < k$, we have at most $|T_w| - 1$ choices for the subsequent indexes, if any. Then we can define

$$Fr_{i \in I}\, \sigma_i := \sum_{j \in J} \tau_j.$$

A first use of the free product of formal series concern a way to characterize a partition of a code.

Definition 1. *Given a family $(X_i)_{i \in I}$ of nonempty subsets of A^*, we say that the family is locally finite if the corresponding family $(\underline{X_i})_{i \in I}$ is locally finite; that is for any $w \in A^*$, the set $\{i \in I \mid w \in X_i\}$ is finite.*

Theorem 5. *Let X be a code and let $P = \{X_i \mid i \in I\}$ be a locally finite family of nonempty subsets of A^+. Then P is a partition of X if and only if $\underline{X}^+ = Fr_{i \in I}\, \underline{X_i}^+$.*

Proof. Let us first remember that if $w \in A^*$ then $(\underline{X}^+, w)$ counts the distinct X-factorizations of w.

Suppose first that P is a partition of X. If $w \notin X^+$ then $w \notin \left(\bigcup_{i \in I} X_i\right)^+$. Hence $\{x_1 x_2 \cdots x_k = w \;:\; k \geq 1, x_i \in X_{j_i}^+, j_i \in I, 1 \leq i \leq k\} = \emptyset$, and so $(\underline{X}^+, w) = (Fr_{i \in I}\, \underline{X_i}^+, w) = 0$. Let now $w \in X^+$. Since each code word belongs exactly in one class of the partition then, to each X-factorization of a message, corresponds one and only one P-factorization. Indeed if $(x_1, x_2, \ldots, x_s)$ is an X-factorization into code words of a message $w \in X^*$, we form the corresponding P-factorization $w = z_{j_1} z_{j_2} \cdots z_{j_k}$, placing in a same "block" z_{j_i}, $1 \leq i \leq k$, words belonging in the same class of the partition. So we can count all the X-factorizations counting, for each P-factorization, the distinct X-factorizations that give rise to that P-factorization: if $z_{j_1} z_{j_2} \cdots z_{j_k}$ is a P-factorization of a message w, then set

$$j = (j_1, j_2, \ldots, j_k) \text{ and } \tau_j = \underline{X_{j_1}}^+ \underline{X_{j_2}}^+ \cdots \underline{X_{j_k}}^+,$$

the distinct X-factorizations associated with this P-factorization are exactly (τ_j, w). From this follows that $\underline{X}^+ = Fr_{i \in I}\, \underline{X_i}^+$.

Conversely, suppose that $\underline{X}^+ = Fr_{i \in I}\, \underline{X_i}^+$. Let $Z := \cup_{i \in I} X_i$ and let us first show that $X^+ = Z^+$. We have $w \in X^+$ if and only if $(\underline{X}^+, w) > 0$, that is, by

hypothesis, if and only if $(Fr_{i\in I}\ \underline{X_i}^+, w) > 0$ and, by definition of free product, this is true if and only if $w \in Z^+$. Indeed if $(Fr_{i\in I}\ \underline{X_i}^+, w) > 0$, then there exist $x_1, x_2, \ldots, x_h \in A^+$, $(j_1, j_2, \ldots, j_h) \in I^h$, with $h \geq 1$, $j_{i+1} \neq j_i$, $1 \leq i < h$, such that $w = x_1 x_2 \cdots x_h$, and $(\underline{X_{j_1}}^+, x_1)(\underline{X_{j_2}}^+, x_2) \cdots (\underline{X_{j_h}}^+, x_h) > 0$. Then $x_1 \in X_{j_1}^+, x_2 \in X_{j_2}^+, \ldots, x_h \in X_{j_h}^+$, and so $w \in Z^+$; conversely, if $w \in Z^+$, it is clear that $(Fr_{i\in I}\ \underline{X_i}^+, w) > 0$.
To prove that $P = \{X_i \mid i \in I\}$ is a partition of X, we need to show that

i) $\forall w \in X$, there exists a unique $i \in I$ such that $w \in X_i$,
ii) $\forall w \in Z = \cup_{i\in I} X_i$, then $w \in X$.

Let Y be the base of $X^* = Z^*$, so $Y \subseteq Z$ and $Y \subseteq X$, and let $k(w)$ denote the maximum length of a Y-factorization of w, with $w \in X \cup Z$. We will proceed by induction on $k(w)$. Let $k(w) = 1$ so $w \in Y$; then $w \in X$ if and only if $w \in Z$, and we just need to prove only the i). Since $w \in Y \subseteq Z$, there exists $i_0 \in I$ such that $w \in X_{i_0}$ and thus $(Fr_{i\in I}\ \underline{X_i}^+, w) \geq 1$; moreover, by Remark 1, $(\underline{X}^+, w) = 1$. Then $(Fr_{i\in I}\ \underline{X_i}^+, w) = (\underline{X}^+, w) = 1$, implies that there exists a unique $i \in I$ such that $w \in X_i$.

Let now $w \in X \cup Z$ with $k(w) = n + 1$, $n \geq 1$. We observe that we can write

$$Fr_{i\in I}\ \underline{X_i}^+ = \sum_{i\in I} \underline{X_i}^+ + \sum \underline{X_{j_1}}^+ \underline{X_{j_2}}^+ \cdots \underline{X_{j_t}}^+, \tag{2}$$

where the second sum runs over all $(j_1, j_2, \ldots, j_t) \in I^t$, with $t \geq 2$, $j_{i+1} \neq j_i$, $1 \leq i < t$. Since $\forall i \in I, \underline{X_i}^+ = \underline{X_i} + \underline{X_i}\,\underline{X_i}^+$, using this relation everywhere in the right hand side of Eq. (2), we have

$$(Fr_{i\in I}\ \underline{X_i}^+, w) = \sum_{i\in I} (\underline{X_i}, w) + \sum (\underline{X_{j_1}}, x_1)(\underline{X_{j_2}}, x_2) \cdots (\underline{X_{j_h}}, x_h), \tag{3}$$

where the second sum runs over all $(j_1, j_2, \ldots, j_h) \in I^h$, such that $h \geq 2$, $x_1 x_2 \cdots x_h = w$, $x_i \in X_{j_i}$ for $1 \leq i \leq h$, and without the constraint $j_{i+1} \neq j_i$. On the other hand $\underline{X}^+ = \underline{X} + \underline{X}\,\underline{X}^+$ implies

$$(\underline{X}^+, w) = (\underline{X}, w) \quad + \sum_{x_1 x_2 \cdots x_h = w} (\underline{X}, x_1)(\underline{X}, x_2) \cdots (\underline{X}, x_h), \quad h \geq 2. \tag{4}$$

Now $h \geq 2$ and $x_1 x_2 \cdots x_h = w$ implies $k(x_i) \leq n$, for all $1 \leq i \leq h$, otherwise $k(w) > n + 1$. Since, by the i) of the induction hypothesis, for each x_i that appears in Eq. (4) there exists a unique $j_i \in I$ such that $x_i \in X_{j_i}$, then each term of the sum in Eq. (4) is a term of the second sum in Eq. (3) and, vice versa, by the ii) of the induction hypothesis, if x_i appears in Eq. (3) then x_i appears in Eq. (4). So the second sum in Eq. (3) and the sum in Eq. (4) are equal. Since, by hypothesis, $\underline{X}^+ = Fr_{i\in I}\ \underline{X_i}^+$, then

$$(\underline{X}, w) = \sum_{i\in I} (\underline{X_i}, w). \tag{5}$$

Now $w \in X$ if and only if $(\underline{X}, w) = 1$, then, by Eq. (5), this is true if and only if there exists a unique $i \in I$ such that $w \in X_i$. On the other hand, if $w \in Z$ then, by Eq. (5), $w \in X$. This concludes the induction and the proof is complete. □

The next theorem characterize a coding partitions $P = \{X_i \mid i \in I\}$ of a code X in terms of the formal series $\underline{X_i{}^+}$.

Theorem 6. *Let $P = \{X_i \mid i \in I\}$ be a partition of a code X and let $\sigma_i := \underline{X_i{}^+}$. Then P is a coding partition if and only if $\underline{X^+} = Fr_{i\in I}\ \sigma_i$.*

Proof. Suppose first $P = \{X_i \mid i \in I\}$ be a coding partition of a code X. If $w \notin X^+$ then, arguing as at the beginning of the proof of the previous Theorem 5, $0 = (\underline{X^+}, w) = (Fr_{i\in I}\ \sigma_i,\ w)$. Otherwise $\forall w \in X^+$ there exists a unique P-factorization $w = z_1 z_2 \cdots z_k$, $k \geq 1$, $z_i \in X_{j_i}{}^+$, $j_i \in I$, $j_i \neq j_{i+1}$, $\forall\, 1 \leq i < k$. Then, put $j_0 = (j_1, j_2, \ldots, j_k) \in I^k$ and $\tau_{j_0} := \sigma_{j_1}\sigma_{j_2}\cdots\sigma_{j_k}$, the uniqueness of the P-factorization shows that j_0 is the only $j \in J$ such that $(\tau_j, w) \neq 0$ and so $1 = (\underline{X^+},\ w) = (\tau_{j_0}, w) = (\sum_{j\in J}\tau_j,\ w) = (Fr_{i\in I}\ \sigma_i,\ w)$.
Let now $\underline{X^+} = Fr_{i\in I}\ \sigma_i$ and let

$$w = z_1 z_2 \cdots z_s = u_1 u_2 \cdots u_t,$$

where $z_1 z_2 \cdots z_s$, $u_1 u_2 \cdots u_t$ are P-factorizations of $w \in X^+$. To each P-factorization corresponds a $j \in J$ such that $(\tau_j, w) \neq 0$. If, by contradiction, the two P-factorizations were different, then $(Fr_{i\in I}\ \sigma_i,\ w) > 1$ and $(\underline{X^+}, w) = 1$, against the hypothesis. Thus the proof is complete. □

From the proof of the previous theorem we can easily deduce the following

Corollary 3. *Let $P = \{X_i \mid i \in I\}$ be a partition of a code X and let $\sigma_i := \underline{X_i{}^+}$. Then the number of P-factorization of a message $w \in X^+$ is given by $(Fr_{i\in I}\ \sigma_i,\ w)$.* □

The natural algebraic interpretation is given by the next characterization of the free product of monoids.

Theorem 7. *Let $M \subseteq A^*$ be a monoid and let $(N_i)_{i\in I}$ be a family of submonoids of A^*. Then M is the free product of the N_i's if and only if $\underline{M} - 1 = Fr_{i\in I}\ (\underline{N_i} - 1)$.* □

Finally we observe that, by Remark 3, $X = \{x_i \mid i \in I\}$ is a UD code if and only if $P := \{\{x_i\} \mid i \in I\}$ is a coding partition of X; thus, by Theorem 6, we can conclude with the following

Corollary 4. *Let $X = \{x_i \mid i \in I\}$ be a code. Then X is UD if and only if $\underline{X^+} = Fr_{i\in I}\ \underline{x_i^+} = Fr_{i\in I}\ \underline{x_i}{}^+$.* □

References

1. Berstel, J., Perrin, D., Reutenauer, C.: Codes and Automata. Encyclopedia of Mathematics and its Applications, vol. 129. Cambridge University Press, Cambridge (2010)
2. Burderi, F.: Full monoids and maximal codes. Int. J. Found. Comput. Sci. **23**(8), 1677–1690 (2012)
3. Burderi, F., Restivo, A.: Coding partitions. Discrete Math. Theoret. Comput. Sci. **9**(2), 227–239 (2007)
4. Burderi, F., Restivo, A.: Varieties of codes and Kraft inequality. Theor. Comput. Syst. **40**, 507–520 (2007)
5. Fliess, M.: Sur le plongement de lalgèbre des séries rationnelles non commutatives dans un corps gauche. C. R. Acad. Sci. Paris Ser. A **271**, 926–927, 1970
6. Kleene, S.C.: Representation of events in nerve nets and finite automata. In: Automata Studies. Annals of Mathematics Studies, vol. 34, pp. 3–41. Princeton University Press, Princeton (1956)
7. Schützenberger, M.P.: On the definition of a family of automata. Inf. Control **4**, 245270 (1961)
8. Schützenberger, M.P.: Certain elementary families of automata. In: Proceedings of Symposia in Mathematics, on the Theory of Automata, New York, 1962, pp. 139–153. Polytechnic Press of Polytechnic Institute of Brooklyn, New York (1963)

Linear Recurrence Sequence Automata and the Addition of Abstract Numeration Systems

Olivier Carton[1], Jean-Michel Couvreur[2], Martin Delacourt[2], and Nicolas Ollinger[2(✉)]

[1] Université Paris Cité, CNRS, IRIF, 75013 Paris, France
[2] Université d'Orléans, INSA CVL, LIFO, UR 4022, Orléans, France
Nicolas.Ollinger@univ-orleans.fr

Abstract. Abstract numeration systems encode natural numbers using radix ordered words of an infinite regular language and linear recurrence sequences play a key role in their valuation. Sequence automata, which are deterministic finite automata with an additional linear recurrence sequence on each transition, are introduced to compute various $\mathbb{Z}$-rational non commutative formal series in abstract numeration systems. Under certain Pisot conditions on the recurrence sequences, the support of these series is regular. This property can be leveraged to derive various synchronized relations including a deterministic finite automaton that computes the addition relation of various Dumont-Thomas numeration systems and deterministic finite automata converting between various numeration systems. A practical implementation for Walnut is provided.

Calculators are handy tools to accelerate tedious computations. A particular type of computation arise from the interplay between formal languages, automata and logic initiated by Büchi [6] and described in the comprehensive survey by Bruyère *et al.* [5]. Practical tools implement these ideas, such as the Walnut proof assistant developed by Mousavi, Shallit *et al.* [16,22]. Walnut particularly shines at the task to check first-order arithmetical properties on automatic sequences, provided the computational complexity is tractable on the examples considered. Indeed, the decision procedure is TOWER-complete for these arithmetics [21].

The present work is motivated by the study of first-order properties of infinite sequences constructed as fixpoints $\tau^\omega(a) = \lim_{k\to\infty} \tau^k(a)$ of some non-uniform substitutions, such as $\tau : a \mapsto aab,\ b \mapsto cab,\ c \mapsto a$ for which $\tau^\omega(a) = aabaabcabaabaabcabaaabcabaabaabcabaabaabcabaaabcabaabaabaab\ldots$. To answer a question on this sequence in a mechanized way, such as *What are the possible lengths of words appearing as cubes inside* $\tau^\omega(a)$*?*, one could express the question as a first-order predicate over $\langle \mathbb{N}, +, \leqslant, \tau^\omega(a)\rangle$:

A preliminary version of this work was presented under the title *Addition in Dumont-Thomas Numeration Systems in Theory and Practice* in the 2024 edition of *Journées Montoises d'Informatique Théorique*.

G. Gamard and J. Leroy (Eds.): WORDS 2025, LNCS 15729, pp. 70–82, 2025.
https://doi.org/10.1007/978-3-031-97548-6_7

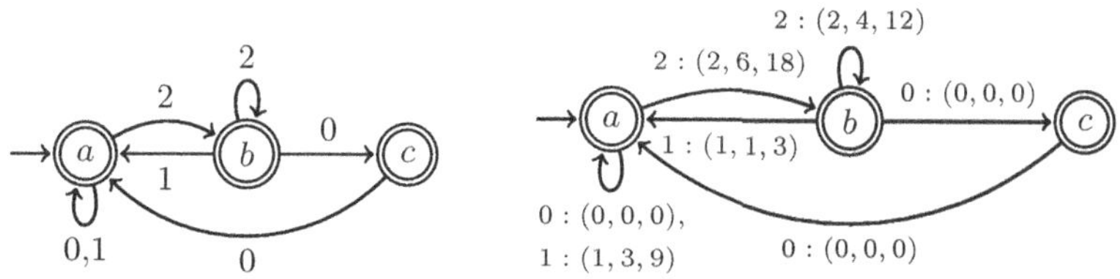

(a) Addressing automaton $\mathcal{S}_\tau$ (b) Sequence automaton $\mathcal{A}_\tau$ ($P = X^3 - 3X^2 + X - 1$)

Fig. 1. Addressing and sequence automata for $\tau : a \mapsto aab,\ b \mapsto cab,\ c \mapsto a$.

$$\text{FactorEq}(i,j,n) \equiv \forall k\ k < n \rightarrow \tau^\omega(a)[i+k] = \tau^\omega(a)[j+k]$$
$$\text{Cubes}(n) \equiv n \geqslant 1 \wedge \exists i\ \text{FactorEq}(i, i+n, n) \wedge \text{FactorEq}(i, i+2n, n)$$

A required step to mechanize these computations is to identify a numeration system in which the infinite word $\tau^\omega(a)$ is automatic as well as the addition relation $\{(x,y,z) \mid x+y=z\}$ and the order relation $\{(x,y) \mid x \leqslant y\}$. From there, the calculator computes the answer, here the possible length correspond to the linear recurrence sequence [23, A098182] given by initial values $1, 3, 9$ and recurrence relation $u_{n+3} = 3u_{n+2} - u_{n+1} + u_n$.

An immediate candidate abstract numeration system [13,19] to describe the fixpoint as an automatic word is the associated Dumont-Thomas numeration system [9,19,20]. In this numeration system, the appearance of a symbol is explained by the address of that symbol, the path taken in the substitution tree generated by τ starting from a. The language of valid labels is recognized by the addressing automaton of $\tau^\omega(a)$ depicted on Fig. 1a

To the best of the authors' knowledge, the regularity of addition has only been extensively studied in the context of positional numeration systems. Let $(\mathbf{u}_i)$ be an increasing sequence of positive integers starting from $\mathbf{u}_0 = 1$. Each number is represented greedily as $n = \sum_{i=0}^{m} \alpha_i \mathbf{u}_i$ where $\alpha_m \cdots \alpha_0$ is the representation of n in the positional numeration system $\mathbf{u}$. When the sequence $\mathbf{u}$ is a linear recurrence sequence whose characteristic polynomial is the minimal polynomial of a Pisot number, the associated numeration system is a Pisot numeration system. Bruyère and Hansel [4] and Frougny and Solomyak [10] have established the regularity of addition for Pisot numeration systems, including their canonical representative which are Bertrand numeration systems.

Dumont-Thomas numeration systems are typically neither positional nor greedy. However, given a current state in the addressing automaton the choice for the next symbol in the address is greedy among available transitions. In the literature, to study a fixpoint of a substitution, one would search for a Pisot numeration system whose minimal polynomial is the characteristic polynomial of the incidence matrix of the substitution and express the fixpoint as an automatic sequence in this positional numeration system. In the present paper, we leverage the techniques used for Pisot numeration systems to obtain a regularity of the language of the addition relation for some abstract numeration systems. In contrast to the positional numeration systems case where the normalization [11] relations are convenient relations to study in order to obtain results on the addition relation, there is no clear equivalent for abstract numeration systems.

Theorem 1. *The addition relation of the abstract numeration system associated with a Pisot substitution, a substitution with an incidence matrix whose characteristic polynomial is the minimal polynomial of a Pisot number, is synchronized.*

1 Addition Relation for Abstract Numeration Systems

An *abstract numeration system with zeros (ANSZ)* $\mathcal{S}$ is a tuple $(L, A, <, 0)$ where $(A, <)$ is an ordered alphabet of minimal element $0 \in A$ and L is a regular language over A containing ε and such that $w \in L \Leftrightarrow 0w \in L$ for all $w \in A^*$. It is a slight variation on the classical regular abstract numeration systems [13,19] where padding with 0 on the left is allowed [8,15].

The *representation*$\mathrm{rep}_{\mathcal{S}}(n)$ of an integer $n \in \mathbb{N}$ is defined by the canonical bijection between $\mathbb{N}$ and $L \setminus 0^+L$ ordered in radix order—words are compared first by length and then by lexicographic order.

The *valuation function*$\mathrm{val}_{\mathcal{S}}$ maps L to $\mathbb{N}$ so that $\mathrm{val}_{\mathcal{S}}(w)$ is the only natural number $n \in \mathbb{N}$ such that $w \in 0^* \mathrm{rep}_{\mathcal{S}}(n)$.

1.1 Automatic, Synchronized and Regular Sequences

Let $\mathcal{S}$ be an ANSZ $(L, A, <, 0)$. An infinite word $\mathbf{x} \in B^{\mathbb{N}}$ is $\mathcal{S}$*-automatic* [2,20] if there exist a deterministic finite automaton producing the output $\mathbf{x}[\mathrm{val}_{\mathcal{S}}(w)]$ for all input word $w \in L$.

Let $\langle .,. \rangle$ denote the canonical bijection between $\bigcup_{n \geqslant 0} (A^n \times A'^n)$ and $(A \times A')^*$ and extend it to tuples of arbitrary size by letting $\langle x, y, z \rangle = \langle x, \langle y, z \rangle \rangle$, *etc.*. A relation $R \subseteq \mathbb{N}^m$ is $\mathcal{S}$*-synchronized* [7] if $\{\langle y_1, \dots, y_m \rangle \mid (\mathrm{val}_{\mathcal{S}}(y_1), \dots, \mathrm{val}_{\mathcal{S}}(y_m)) \in R\}$ is regular. The *addition relation* of an ANSZ $\mathcal{S} = (L, A, <, 0)$ is the language

$$L_{\mathcal{S}}^{+} = \left\{ \langle x, y, z \rangle \;\middle|\; x, y, z \in L \wedge \mathrm{val}_{\mathcal{S}}(x) + \mathrm{val}_{\mathcal{S}}(y) = \mathrm{val}_{\mathcal{S}}(z) \right\} .$$

A sequence of natural numbers $\mathbf{a} : \mathbb{N} \to \mathbb{N}$ is $\mathcal{S}$*-regular* [1,8,20] if the function $f : A^* \to \mathbb{N}$, defined by $f(w) = \mathbf{a}(\mathrm{val}_{\mathcal{S}}(w))$ on L and $f(w) = 0$ otherwise, is a $\mathbb{Z}$-rational formal power series. That is, by Schützenberger's theorem [3], if it is recognizable by a *linear representation* (λ, μ, ρ) consisting of a row vector λ, a matrix-valued morphism $\mu : A^* \to \mathbb{Z}^{m \times m}$ and a column vector ρ, such that $\mathbf{a}(n) = \lambda\mu(\mathrm{rep}_{\mathcal{S}}(n))\rho$ for all $n \geqslant 0$ and $\lambda\mu(w)\rho = 0$ if w is not in L.

1.2 ANSZ and Dumont-Thomas Numeration Systems

Let $\mathcal{S}$ and $\mathcal{S}'$ be two ANSZ. The *conversion relation* between $\mathcal{S}$ and $\mathcal{S}'$ is the language $\{\langle x, y\rangle \mid \mathrm{val}_{\mathcal{S}}(x) = \mathrm{val}_{\mathcal{S}'}(y)\}$. Two ANSZ are said to be *equivalent* if their conversion relation is a regular language.

Proposition 1. *Equivalence between ANSZ preserve automaticity, synchronicity and regularity of sequences and relations.*

A *substitution* $\tau : A \to A^*$ generates a fixpoint $\tau^\omega(a) \in A^{\mathbb{N}}$ from a letter $a \in A$ as $\tau^\omega(a) = a \prod_{n\in\mathbb{N}} \tau^n(w)$ provided that $\tau(a) = aw$ for some $w \in A^+$ such that no $\tau^k(w)$ is the empty word. Dumont and Thomas [9] proved that every prefix p of $\tau^\omega(a)$ can be represented using a unique sequence $(p_i, a_i)_{i=0}^k \in (A^* \times A)^*$ as $p = \prod_{i=0}^k \tau^{k-i}(p_i)$ where $p_{i+1}a_{i+1}$ is a prefix of $\tau(a_i)$ for all $i < k$ and p_0a_0 is a prefix of $\tau(a)$. The sequence of the lengths of the p_i, the *address* of letter $\tau^\omega(a)_{|p|}$, completely characterizes the prefix p.

The *addressing automaton* $\mathcal{S}_\tau$ associated with the fixpoint $\tau^\omega(a)$ is the deterministic finite automaton with state set A, alphabet $B = \{0, 1, \ldots, n-1\}$ where $n = \max_{b\in A} |\tau(b)|$, initial state a, final states A and whose transitions are defined by τ as $\delta(b, i) = \tau(b)_i$ for all $b \in A$ and $i \in \{0, \ldots, |\tau(b)| - 1\}$. The *Dumont-Thomas numeration system* associated with $\tau^\omega(a)$ is the ANSZ $(L(\mathcal{S}_\tau), B, <, 0)$. The fixpoint $\tau^\omega(a)$ is automatic in this numeration system, it is recognized by $\mathcal{S}_\tau$ where the output of a state $b \in A$ is the letter b.

By taking only a subset $F \subseteq A$ as accepting states for $\mathcal{S}_\tau$, one defines the *Generalized Dumont-Thomas numeration system* associated with $\pi(\tau^\omega(a))$ where π is the erasing substitution mapping every element of F to itself and all the other elements to ε. It relates with label reduction and surminimisation in [14].

Proposition 2. *Every ANSZ is equivalent to a generalized Dumont-Thomas numeration system.*

1.3 ℕ-Rational Valuation of ANSZ

Let $\mathcal{S} = (L, A, <, 0)$ be an ANSZ with $A = \{0, \ldots, m-1\}$ and $<$ the classical order on $\mathbb{N}$. Let $(Q, A, \delta, 0, F)$ be a DFA recognizing L with $Q = \{0, \ldots, n-1\}$ and $\delta : Q \times A \to Q$ its partial transition map. The identity sequence $\mathbb{N} \to \mathbb{N}$, $n \mapsto n$ is $\mathcal{S}$-regular [18, Prop. 29]. The valuation function $\mathrm{val}_{\mathcal{S}}$ coincides on L with the $\mathbb{N}$-rational formal power series given by the linear representation (λ, μ, ρ) of dimension $2n$, that is $\lambda\mu(w)\rho = n'$ for all $n' \in \mathbb{N}$ and $w \in 0^*\,\mathrm{rep}_{\mathcal{S}}(n')$:

$$\lambda = \left(1\,0\cdots 0\middle|0\cdots 0\right) \qquad \mu : k \in A \mapsto \left(\begin{array}{c|c} \Delta_k & V_k \\ \hline 0 & M_\delta \end{array}\right) \qquad \rho = \left(0\cdots 0\middle|\chi_F\right)^T \tag{1}$$

where, for all $i, j \in Q$ and for all $k \in A$:

$$M_\delta[i,j] = \#\{a \in A | \delta(i,a) = j\} \qquad V_k[i,j] = \#\{a < k | \delta(i,a) = j\}$$

$$\chi_F[i] = \begin{cases} 1 & \text{if } i \in F \\ 0 & \text{otherwise} \end{cases} \qquad \Delta_k[i,j] = \begin{cases} 1 & \text{if } \delta(i,k) = j \\ 0 & \text{otherwise} \end{cases}$$

Thus, it is $\mathbb{N}$-rational. The linear representation simulates the DFA with Δ_k and combines the incidence matrix M_δ with the vector contributions V_k and vector ρ to count the number of accepting path of the same length that precede the input word in radix order.

1.4 Addition of ANSZ

The *synchronized addition* $f \oplus g$ between two series $f : A^* \to \mathbb{K}$ and $g : A'^* \to \mathbb{K}$ is the series satisfying $(f \oplus g)(\langle w, w' \rangle) = f(w) + g(w')$ for all pair of words of the same length $(w, w') \in A^* \times A'^*$. The synchronized addition of two $\mathbb{K}$-rational series is $\mathbb{K}$-rational by running both series at the same time in a block composition of their matrices.

The *support* $\mathrm{supp}(f)$ of a series f on A is the language $A^* \setminus f^{-1}(0)$ of inputs with non-zero image. As the characteristic function of a regular language is $\mathbb{N}$-rational and as rational series are closed by external multiplication and by Cauchy product, we obtain the following characterization.

Proposition 3. *The addition relation $L^+_{\mathcal{S}}$ of an ANSZ $\mathcal{S}$ is regular if and only if* $\mathrm{supp}(\mathrm{val}_{\mathcal{S}} \oplus \mathrm{val}_{\mathcal{S}} \oplus -\mathrm{val}_{\mathcal{S}})$ *is regular.*

Unfortunately, the above series is a $\mathbb{Z}$-series and testing for the regularity of $\mathbb{Z}$-series support is known to be undecidable in the general case [3]. We introduce, through linear recurrence sequence automata, a subfamily of $\mathbb{Z}$-series on which regularity can sometimes be effectively obtained.

2 Linear Recurrence Sequence Automata

An *integer sequence* $\mathbf{u} \in \mathbb{Z}^{\mathbb{N}}$ is *linear recurrence* if there exist a *recurrence order* m and *recurrence coefficients* $\alpha_i \in \mathbb{Z}$ such that $\mathbf{u}_{n+m} = \sum_{i=0}^{m-1} \alpha_i \mathbf{u}_{n+i}$ for all $n \geqslant 0$. The *recurrence polynomial* of this sequence is $P_{\mathbf{u}} = X^m - \sum_{i=0}^{m-1} \alpha_i X^i$.

Let the *zero sequence* be denoted as (0) and let the *shift operator* $\sigma : \mathbb{Z}^{\mathbb{N}} \to \mathbb{Z}^{\mathbb{N}}$ remove the first element of a sequence, *i.e.* $(\sigma\mathbf{u})_n = \mathbf{u}_{n+1}$ for all $\mathbf{u} \in \mathbb{Z}^{\mathbb{N}}$ and $n \in \mathbb{N}$. The *vector space* E_P of the linear recurrence sequences of recurrence polynomial $P = X^m - \sum_{i=0}^{m-1} \alpha_i X^i$ has dimension m and each linear recurrence sequence $\mathbf{u} \in E_P$ is completely defined by its *initial vector* $V_{\mathbf{u}} = (\mathbf{u}_0, \ldots, \mathbf{u}_{m-1})$. The initial vector of the shifted sequence $\sigma\mathbf{u} \in E_P$ can be computed using the *companion matrix* M_P of P as $V_{\sigma\mathbf{u}} = V_{\mathbf{u}} M_P$ where

$$M_P = \begin{pmatrix} 0 & 0 & 0 & \ldots & 0 & \alpha_0 \\ 1 & 0 & 0 & \ldots & 0 & \alpha_1 \\ 0 & 1 & 0 & \ldots & 0 & \alpha_2 \\ \vdots & \vdots & \ddots & \ddots & \vdots & \vdots \\ \vdots & \vdots & & \ddots & \ddots & \vdots \\ 0 & 0 & 0 & \ldots & 1 & \alpha_{m-1} \end{pmatrix}$$

The construction of the linear representation from Eq. 1 uses the incidence matrix M_δ to keep track of the contribution of each digit in the representation $\mathrm{rep}_{\mathcal{S}}(n)$ of a natural number n. Taking the equivalent generalized Dumont-Thomas numeration system point of view, the matrix M_δ is indeed the incidence matrix M_τ of the associated τ substitution where $M_\tau[i,j] = |\tau(i)|_j$, the number of occurrences of letter j in the word $\tau(i)$. The product $\lambda\mu(\text{rep}_\mathcal{S}(n))\rho$ is a linear combination of several $|\tau^k(i)|$. The contribution of a digit to the final value depends on the current state of the automaton and the position of the digit in the number. The sequence of its contributions is a linear recurrence sequence!

Indeed, by the Cayley-Hamilton theorem, the matrix M_τ verifies the linear recurrence relation given by its monic characteristic polynomial P. As $M_\tau^k[i,j]$ counts the number of occurrences of letter j in $\tau^k(i)$, we have $|\pi(\tau^k(i))| = \sum_{\pi(j)\neq\varepsilon} M_\tau^k[i,j] = \lambda_i M_\tau^k \rho_\pi$ for some vectors λ_i and ρ_π. All the $|\pi(\tau^k(i))|$ are linear recurrence of polynomial P. This is also the case for their linear combinations.

From there, we deduce an alternative linear representation (λ, μ', ρ') for the valuation function $\text{val}_\mathcal{S}$, with coefficients in $\mathbb{Z}$, where each digit adds its linear recurrence sequence contribution and ρ' simply extracts the first value of the initial vector of the current sequence. Let the matrix R be such that $R[i,j] = |\pi(\tau^j(i))|$ for all $i, j \in Q$. By construction, $RM_P = M_\delta R$. Now, let $W_k = V_k R$ for all $k \in A$. Use it to construct the new representation:

$$\lambda = \begin{pmatrix} 1\,0\cdots 0 | 0\cdots 0\end{pmatrix} \quad \mu' : k \in A \mapsto \left(\begin{array}{c|c} \Delta_k & W_k \\ \hline 0 & M_P \end{array}\right) \quad \rho' = \begin{pmatrix} 0\cdots 0 | 1\,0\cdots 0\end{pmatrix}^T$$

where, for all $i, j \in Q$ and for all $k \in A$:

$$W_k[i,j] = \left|\pi\left(\tau^j\left(\tau(i)_0 \cdots \tau(i)_{k-1}\right)\right)\right| \tag{2}$$

In this linear representation, a vector $\lambda\mu'(w)$ memorizes the current state of the DFA together with the initial vector of a current linear recurrence sequence after reading w. A transition applies a shift to the current vector and adds the contribution of the current digit. Linear recurrence sequence automata formalize this particular construction.

2.1 Sequence Automata

A sequence automaton is a partial deterministic finite automaton (Q, A, δ, q_0, F) equipped with *weight* sequences on its edges, given by a partial vector map π.

Definition 1. *A* sequence automaton *is a tuple* $(Q, A, \delta, q_0, F, \pi)$ *where* Q *is the* finite set of states, A *is the* finite alphabet of symbols, $\delta : Q \times A \to Q$ *is the* partial transition map, $q_0 \in Q$ *is the* initial state, $F \subseteq Q$ *is the set of* accepting states *and* $\pi : Q \times A \to \mathbb{Z}^\mathbb{N}$ *is the* partial vector map *of the automaton. The transition and the vector map share the same domain.*

The transition map and vector map are inductively extended from symbols to words as follows, for all $q \in Q$, $w \in A^*$ and $a \in A$:

$$\begin{aligned} \delta(q,\varepsilon) &= q & \pi(q,\varepsilon) &= (0) \\ \delta(q,wa) &= \delta(\delta(q,w),a) & \pi(q,wa) &= \sigma\pi(q,w) + \pi(\delta(q,w),a) \quad . \end{aligned}$$

The intuition behind a *weight* sequence associated with a transition $\delta(q,a)$ is that symbol a contributes for $\pi(q,a)_n$ to the total weight of the word when it appears in position n in the word, counting from right to left—the same way that digit 3 counts for 3×10^n in decimal.

Lemma 1. *For all $q \in Q$ and $w, w' \in A^*$, when $\delta(q, ww')$ is defined, by construction $\pi(q, ww') = \sigma^{|w'|}\pi(q,w) + \pi(\delta(q,w), w')$.*

Definition 2. *A sequence automaton is* linear recurrence (LRSA) *if every sequence appearing in the vector map is a linear recurrence sequence. The* recurrence polynomial *of a LRSA is the least common multiple of the characteristic polynomials of every linear recurrence sequence in the image of its vector map π.*

Definition 3. *A sequence automaton is* scalar *if there exists a* reference sequence $\mathbf{u}$ *of which every sequence appearing in the vector map is a multiple, i.e. for all q and a such that $\pi(q,a)$ is defined, there exists $\alpha \in \mathbb{Z}$ such that $\pi(q,a) = \alpha\mathbf{u}$.*

2.2 Combining Sequence Automata

The *scalar product* $\alpha\mathcal{A}$ of a sequence automaton $\mathcal{A}$ by $\alpha \in \mathbb{Z}$ is the same sequence automaton where the vector map π is replaced by $\alpha\pi : \mathbf{u} \mapsto \alpha\mathbf{u}$.

The *addition* $\mathcal{A}_1 + \mathcal{A}_2$ of two sequence automata $\mathcal{A}_i = (Q_i, A_i, \delta, s_i, F_i, \pi_i)$ with $i \in \{1,2\}$ is the product sequence automaton combining the vector maps. Formally, $(Q_1 \times Q_2, A_1 \times A_2, \delta, (s_1, s_2), F_1 \times F_2, \pi)$ where $\delta((q_1,q_2),(a_1,a_2)) = (\delta_1(q_1,a_1), \delta_2(q_2,a_2))$ and $\pi((q_1,q_2),(a_1,a_2)) = \pi_1(q_1,a_1) + \pi_2(q_2,a_2)$ for all $q_1 \in Q_1$, $q_2 \in Q_2$, $a_1 \in A_1$ and $a_2 \in A_2$.

Lemma 2. *The recurrence polynomial of a linear combination of LRSA divides the least common multiple of the recurrence polynomials of each automaton of the combination.*

2.3 Series of Sequence Automata and Their Support

Definition 4. *The* series$_{\mathcal{A}}$ *of a sequence automaton $\mathcal{A} = (Q, A, \delta, q_0, F, \pi)$ maps every word $w \in A^*$ to $\pi(q_0, w)[0]$.*

Definition 5.
The language$L_0(\mathcal{A})$ *of a sequence automaton $\mathcal{A} = (Q, A, \delta, q_0, F, \pi)$ is the set of words recognized by the automaton with value 0, i.e.*

$$L_0(\mathcal{A}) = \{w \in A^* \mid \delta(q_0, w) \in F \wedge s_{\mathcal{A}}(w) = 0\}$$

A simple idea, to transform a sequence automaton into a deterministic automaton recognizing the same language, is to remove the vectors from the transitions and insert them into the states. Alas, the induced flattening deterministic automaton generated this way is potentially infinite.

Definition 6. *The* flattening *of a sequence automaton* $(Q, A, \delta, q_0, F, \pi)$ *is the infinite deterministic automaton* $(Q', A, \delta', q_0', F')$ *where*

$$\begin{aligned} Q' &= Q \times \mathbb{Z}^{\mathbb{N}} \\ \delta'\left((q, \mathbf{u}), a\right) &= (\delta(q, a), \sigma\mathbf{u} + \pi(q, a)) \quad \forall q \in Q, \mathbf{u} \in \mathbb{Z}^{\mathbb{N}}, a \in A \\ q_0' &= (q_0, (0)) \\ F' &= \{(q, \mathbf{u}) \mid q \in F \wedge \mathbf{u}_0 = 0\} \end{aligned}$$

To *trim* a deterministic automaton, one only keeps states accessible from the initial state and co-accessible from the accessible accepting states.

Lemma 3. *When its trimmed flattening is finite, the language of a sequence automaton is regular.*

Remark 1. The converse is not true: the language of a LRSA can be regular without the trimmed flattening being finite.

Lemma 4. *When the support of its series is regular, the language of a sequence automaton is regular.*

3 The Pisot Condition

For LRSA with Pisot recurrence polynomial, the trimmed flattening is always finite. In the scalar case, it is obtained through an adaptation of bounds established by Bruyère and Hansel [4] to study the addition of Bertrand numeration systems. In the general LRSA case the scalar bound can be adapted by combining the polynomial representation from Frougny and Solomyak [10] with the change of basis technique from Frougny [12].

A *Pisot-Vijayaraghavan number* θ is an algebraic integer that is the dominant root of its minimal monic polynomial $P(X)$ with integer coefficients, where $P(X)$ is irreducible over $\mathbb{Z}$ and admits n complex roots $\theta_1, \ldots, \theta_n$, all distinct, satisfying $\theta = \theta_1 > 1 > |\theta_2| \geqslant \ldots \geqslant |\theta_n| > 0$. The set of power sequences of the roots $\left(\theta_i^k\right)_{k \in \mathbb{N}}$ is a base of the vector space E_P.

A linear recurrence sequence $\mathbf{u}$ is *Pisot* if its characteristic polynomial is the minimal polynomial $P(X)$ of a Pisot-Vijayaraghavan number. It is *Pisot, up to shift,* if $\sigma^k\mathbf{u}$ is Pisot for some $k \geqslant 0$, that is if its polynomial is $X^k P(X)$. By extension, such a polynomial is said to be *Pisot, up to shift.*

3.1 Scalar Pisot LRSA

Theorem 2. *The language of a scalar LRSA with a reference sequence that is Pisot, up to shift, is regular.*

Proof. Let $(Q, A, \delta, q_0, F, \pi)$ be a scalar LRSA with reference sequence $\mathbf{u}$. Assume that $\mathbf{u}$ is Pisot, up to shift, and let $X^k P(X)$ be the characteristic polynomial of $\mathbf{u}$. Let $\theta = \theta_1, \ldots, \theta_n$ be the roots of P satisfying $\theta = \theta_1 > 1 > |\theta_2| \geqslant \ldots \geqslant |\theta_n| > 0$. Let $\omega : Q \times A \to \mathbb{Z}$ retrieve the scalar coefficient of the weight so that $\pi(q,a) = \omega(q,a)\mathbf{u}$ for all $q \in Q$ and $a \in A$. Extend ω inductively to $\omega : Q \times A^* \to \mathbb{Z}^*$, where $(\mathbb{Z}^*, \cdot)$ denotes the monoid of words on $\mathbb{Z}$, by $\omega(q,\varepsilon) = \varepsilon$ and $\omega(q, wa) = \omega(q,w) \cdot \omega(\delta(q,w), a)$ for all $q \in Q$, $w \in A^*$ and $a \in A$.

As the number of transitions of the sequence automaton is finite, the scalar coefficients are bounded. Let $C = \max_{(q,a)\in Q\times A} |\omega(q,a)|$. The coefficients all lie in the interval $\mathcal{C} = \{-C, \dots, C\}$.

Let $\pi_{\mathbf{u}} : \mathbb{Z}^* \to \mathbb{Z}$ assign a weight to every finite sequence of integers according to $\mathbf{u}$ by $\pi_{\mathbf{u}}(a_m \cdots a_0) = \sum_{i=0}^{m} a_i \mathbf{u}_i$ for all $a_m \cdots a_0 \in \mathbb{Z}^*$. By construction, $(\pi(q,w))_i = \pi_{\mathbf{u}}\left(\omega(q,w)0^i\right)$, or equivalently $(\pi(q,w))_i = \pi_{\sigma^i\mathbf{u}}\left(\omega(q,w)\right)$, for all $i \in \mathbb{N}$, $q \in Q$ and $w \in A^*$.

The key argument, from [4], is to approximate $\pi_{\mathbf{u}}$ using powers of θ. As $\sigma^k\mathbf{u} \in E_P$ there exist $\gamma_1, \dots, \gamma_n \in \mathbb{C}$ such that $\mathbf{u}_{m+k} = \sum_{i=1}^{n} \gamma_i \theta_i^m$ for all $m \in \mathbb{N}$. Moreover $\gamma = \gamma_1$ is a real number. Let $\pi_\theta : \mathbb{Z}^* \to \mathbb{R}$ assign a weight to every finite sequence $a_m \cdots a_0 \in \mathbb{Z}^*$ by $\pi_\theta(a_m \cdots a_0) = \gamma \sum_{i=k}^{i=m} a_i \theta^{i-k}$, ignoring the first k values. First, let us bound the approximation error by C times a constant K. Let $a_m \cdots a_0 \in \mathcal{C}^*$,

$$\begin{aligned}
\pi_{\mathbf{u}}(a_m \cdots a_0) - \pi_\theta(a_m \cdots a_0) &= \sum_{i=0}^{k-1} a_i \mathbf{u}_i - \sum_{i=k}^{m} a_i \sum_{j=2}^{n} \gamma_j \theta_j^{i-k} \\
&= \sum_{i=0}^{k-1} a_i \mathbf{u}_i - \sum_{j=2}^{n} \gamma_j \sum_{i=k}^{m} a_i \theta_j^{i-k} \\
\text{thus} \quad |\pi_{\mathbf{u}}(a_m \cdots a_0) - \pi_\theta(a_m \cdots a_0)| &\leqslant C \left(\sum_{i=0}^{k-1} |\mathbf{u}_i| + \sum_{j=2}^{n} |\gamma_j| \sum_{i=k}^{m} |\theta_j|^{i-k} \right) \\
|\pi_{\mathbf{u}}(a_m \cdots a_0) - \pi_\theta(a_m \cdots a_0)| &\leqslant C \underbrace{\left(\sum_{i=0}^{k-1} |\mathbf{u}_i| + \sum_{j=2}^{n} \frac{|\gamma_j|}{1 - |\theta_j|} \right)}_{=K}
\end{aligned}$$

Consider now an accepting path of the automaton. Let $w, w' \in A^*$ such that $\pi(q_0, ww')$ is defined and verifies $(\pi(q_0, ww'))_0 = 0$. Let $m = |w'|$. Let $q = \delta(q_0, w)$. As $\pi_{\mathbf{u}}(\omega(q_0, ww')) = 0$, we have

$$\begin{aligned}
-CK \leqslant & \quad \pi_\theta(\omega(q_0, ww')) \leqslant CK \\
\text{thus} \quad -CK \leqslant & \pi_\theta(\omega(q_0, w)0^m) + \pi_\theta(\omega(q, w')) \leqslant CK \\
-C\left(K + \frac{\gamma\theta}{\theta - 1}\theta^m\right) \leqslant & \quad \pi_\theta(\omega(q_0, w))\theta^m \leqslant C\left(K + \frac{\gamma\theta}{\theta - 1}\theta^m\right) \\
\text{then diving by } \theta^m, & \quad |\pi_\theta(\omega(q_0, w))| \leqslant C\left(K + \frac{\gamma\theta}{\theta - 1}\right)
\end{aligned}$$

And finally we bound $\pi_{\mathbf{u}}\left(\omega(q_0, w)0^i\right)$ for all i to prepare the bound for $V_{\pi(q_0,w)}$:

$$\begin{aligned}\left|\pi_{\mathbf{u}}\left(\omega(q_0, w)0^i\right)\right| &\leqslant \left|\pi_{\mathbf{u}}\left(\omega(q_0, w)0^i\right) - \pi_\theta\left(\omega(q_0, w)0^i\right)\right| + \left|\pi_\theta\left(\omega(q_0, w)0^i\right)\right| \\ &\leqslant C\left(K + \left(K + \frac{\gamma\theta}{\theta - 1}\right)\theta^i\right)\end{aligned}$$

$V_{\pi(q_0,w)} = \left(\pi_{\mathbf{u}}\left(\omega(q_0, w)\right), \pi_{\sigma\mathbf{u}}\left(\omega(q_0, w)\right), \cdots, \pi_{\sigma^{m-1}\mathbf{u}}\left(\omega(q_0, w)\right)\right)$ is bounded thus the trimmed flattening of the sequence automaton is finite. ∎

Note that the whole process is constructive. Given a scalar LRSA, one can effectively construct the bounded flattening by encoding sequences as initial vectors and by using the companion matrix to compute transitions.

3.2 Pisot LRSA

Consider a sequence $\mathbf{u} \in E_P$ verifying the recurrence relation given by some monic polynomial P of degree m. The set of shifted sequences $\{\sigma^i\mathbf{u}\}_{i\in\mathbb{N}}$ generates a subspace of E_P. It generates the whole space E_P if and only if the following associated Hankel matrix is invertible:

$$H_{\mathbf{u}} = \begin{pmatrix} \mathbf{u}_0 & \mathbf{u}_1 & \dots & \mathbf{u}_{m-1} \\ \mathbf{u}_1 & \mathbf{u}_2 & \dots & \mathbf{u}_m \\ \vdots & \vdots & \ddots & \vdots \\ \mathbf{u}_{m-1} & \mathbf{u}_m & \dots & \mathbf{u}_{2m-2} \end{pmatrix}$$

In this case, for every sequence $\mathbf{v} \in E_P$, we have $\mathbf{v} = \sum_{i=0}^{m-1} \alpha_i\sigma^i\mathbf{u}$ where $(\alpha_0, \dots, \alpha_{m-1}) = V_{\mathbf{v}}H_{\mathbf{u}}^{-1}$. By letting $R = \sum_{i=0}^{m-1} \alpha_i X^i$, one can write $\mathbf{v} = R(\sigma)(\mathbf{u})$ and see every sequence of E_P as a polynomial in $\mathbf{u}$ with rational coefficients. The choice of the sequence $\mathbf{u}$ is free. One might for example choose the sequence of initial vector $(0, \dots, 0, 1)$ for which $H_{\mathbf{u}}$ is always invertible, as it is anti-triangular, with a determinant of value 1 or -1, ensuring integer coefficients for the polynomials.

Theorem 3. *The language of a LRSA with a recurrence polynomial that is Pisot, up to shift, is regular.*

Proof. Let $\mathcal{A}$ be a LRSA $(Q, A, \delta, q_0, F, \pi)$ with recurrence polynomial P of degree m. Assume that P is Pisot, up to shift, and let $\mathbf{u} \in E_P$ be a sequence such that the set of its shifted sequences $\{\sigma^i\mathbf{u}\}_{i\in\mathbb{N}}$ generates E_P. Let $\omega : Q\times A \to \mathbb{Q}[X]$ retrieve a polynomial representation, of degree at most $m-1$, of the weight so that $\pi(q, a) = \omega(q, a)(\sigma)(\mathbf{u})$ for all $q \in Q$ and $a \in A$. Extend ω inductively to $\omega : Q\times A^* \to \mathbb{Q}[X]$ by $\omega(q, \varepsilon) = 0$ and $\omega(q, wa) = \omega(q, w)X + \omega(\delta(q, w), a)$ for all $q \in Q$, $w \in A^*$ and $a \in A$. By construction, $(\pi(q, w))_i = \left(\omega(q, w)X^i\right)(\sigma)(\mathbf{u})[0]$ for all $i \in \mathbb{N}$, $q \in Q$ and $w \in A^*$.

As the number of transitions of the automaton is finite, the polynomial representations have bounded coefficients. Let $C = \max_{\sum_{i=0}^{m-1}\alpha_i X^i\in\pi(Q,A)} |\alpha_i|$. The coefficients all lie in the interval $\mathcal{C} = \{-C, \dots, C\}$. Moreover, the coefficients of

$(\pi(q,w))_i$ all lie in the interval $mC = \{-mC,\ldots,mC\}$ for all $i \in \mathbb{N}$, $q \in Q$ and $w \in A^*$.

When the automaton $\mathcal{A}$ admits a loop $\delta(q_0,a) = q_0$ with weight $\pi(q_0,a) = 0$ for some symbol a, one can construct a scalar LRSA $\mathcal{B}$ with reference sequence $\mathbf{u}$ so that $\mathcal{A}$ and $\mathcal{B}$ recognize the same language. To do so, $\mathcal{B}$ anticipates, with a sliding windows on tuples of letters, the future coefficients of the polynomials: instead of associating the weight $\sum_{i=0}^{m-1} \alpha_i X^i$ to a transition $\delta(q,b)$, the weights α_i for $i > 0$ are added to the weights of the $m-1$ previous symbols. By Theorem 2, the language of $\mathcal{B}$ is regular.

If the sequence automaton admits no loop of weight zero on its initial state, consider adding to the automaton $\mathcal{A}$ a new padding symbol $\# \notin A$ and a loop $\delta(q_0,\#) = q_0$ of weight $\pi(q_0,\#) = 0$ to obtain a sequence automaton $\mathcal{A}^{\#}$. By construction, $L_0(\mathcal{A}) = \{u \in A^* | \exists k < m, \#^k u \in L_0(\mathcal{A}^{\#})\}$. By previous argument, the language of $\mathcal{A}^{\#}$ is regular, and so is the language of $\mathcal{A}$. ■

Note that the whole process is still constructive and that the proof provides an effective method to transform the sequence automaton into an automaton of a single sequence.

4 Applications

Using the expressivity of LRSA, we can answer our initial question in the Pisot case and handle several other constructions.

Addition of Pisot ANSZ. Following Eq. 2, the *addressing sequence automaton* $\mathcal{A}_\tau$ associated with the fixpoint $\pi(\tau^\omega(a))$ of a generalized Dumont-Thomas numeration system is obtained by adding a vector map π to $\mathcal{S}_\tau$ as $\pi(b,k) = (|\pi(\tau^n(\tau(b)_0 \cdots \tau(b)_{k-1}))|)_{n\in\mathbb{N}}$ for all $b \in A$ and $k \in B$. The series of this sequence automaton is the valuation function of the numeration system. It is a LRSA whose recurrence polynomial divides the characteristic polynomial of the incidence matrix of τ (see for example Fig. 1b). By combining Proposition 3 with Theorems 1 and 3 is proved.

Conversion Between ANSZ. Let $\mathcal{S}$ and $\mathcal{S}'$ be two ANSZ whose valuation addressing sequence automata share a same Pisot recurrence polynomial, up to shift. Both ANSZ are equivalent: the series $\mathrm{val}_{\mathcal{S}} \oplus - \mathrm{val}_{\mathcal{S}'}$ has a regular support so the conversion between both ANSZ is regular. In particular, the equivalence to the canonical Bertrand numeration system [4] is effectively constructed.

Parikh Vectors of Fixpoint Prefixes. Using the same techniques, the Parikh vectors of prefixes of the fixpoint $\pi(\tau^\omega(a))$ of a generalized Dumont-Thomas numeration system can be obtained by constructing, for each letter $b \in B$, the erasing coding π_b that erases every letter but b and computing the series $\mathrm{val}_{\mathcal{S}} \oplus - \mathrm{val}_{\mathcal{S}_b}$ where $\mathcal{S}$ is the generalized Dumont-Thomas numeration system of $\pi(\tau^\omega(a))$ and $\mathcal{S}_b$ the one of $(\pi_b \circ \pi)(\tau^\omega(a)))$.

5 Practical Considerations

We have developed a prototype tool, `licofage` [17]. Given a substitution, `licofage` can produce its addressing sequence automaton and compute linear combinations of their series. In the Pisot case, or when given a manually selected bound, it produces the trimmed flattening deterministic finite automaton that computes a relation defined using linear combinations of sequence automata series (addition, conversion, *etc.*). The generated output can be seamlessly integrated into Walnut [16]. A companion notebook is available on arXiV to showcase the capabilities of the proposed method, accessible via the following URL: https://arxiv.org/src/2406.09868/anc.

References

1. Allouche, J.P., Shallit, J.: The ring of k-regular sequences. Theoret. Comput. Sci. **98**(2), 163–197 (1992)
2. Allouche, J.P., Shallit, J.: Automatic Sequences: Theory, Applications, Generalizations. Cambridge University Press (2003)
3. Berstel, J., Reutenauer, C.: Noncommutative Rational Series with Applications, No. 137, Cambridge University Press (2011)
4. Bruyère, V., Hansel, G.: Bertrand numeration systems and recognizability. Theoret. Comput. Sci. **181**(1), 17–43 (1997)
5. Bruyère, V., Hansel, G., Michaux, C., Villemaire, R.: Logic and p-recognizable sets of integers. Bull. Belgian Math. Soc. Simon Stevin **1**(2), 191–238 (1994)
6. Büchi, J.R.: Weak second-order arithmetic and finite automata. Z. Math. Logik Grundlag. Math. **6**(1–6), 66–92 (1960)
7. Carpi, A., D'Alonzo, V.: On factors of synchronized sequences. Theoret. Comput. Sci. **411**(44–46), 3932–3937 (2010)
8. Charlier, É., Cisternino, C., Stipulanti, M.: Regular sequences and synchronized sequences in abstract numeration systems. Eur. J. Comb. **101**, 103475 (2022)
9. Dumont, J.M., Thomas, A.: Systemes de numération et fonctions fractales relatifs aux substitutions. Theoret. Comput. Sci. **65**(2), 153–169 (1989)
10. Frougny, C., Solomyak, B.: On representation of integers in linear numeration systems. In: Pollicott, M., Schmidt, K. (eds.) Ergodic Theory of $\mathbb{Z}^d$ Actions, London Mathematical Society Lecture Note Series, vol. 228, pp. 345–368. Cambridge University Press (1996)
11. Frougny, C.: Numeration systems. In: Lothaire, M. (ed.) Algebraic Combinatorics on words, chap. 7. Cambridge University Press (2002)
12. Frougny, C.: On multiplicatively dependent linear numeration systems, and periodic points. RAIRO Theor. Inf. Appl. Informatique Théorique et Appl. **36**(3), 293–314 (2002)
13. Lecomte, P., Rigo, M.: Abstract numeration systems. In: Berthé, V., Rigo, M. (eds.) Combinatorics, Automata and Number Theory, pp. 123–178. Cambridge University Press (2010)
14. Marsault, V.: Surminimisation of automata. In: International Conference on Developments in Language Theory, pp. 352–363. Springer (2015)
15. Marsault, V., Sakarovitch, J.: The signature of rational languages. Theoret. Comput. Sci. **658**, 216–234 (2017)

16. Mousavi, H.: Automatic theorem proving in Walnut. arXiv preprint arXiv:1603.06017 (2016)
17. Ollinger, N.: Licofage software tool (2024). https://pypi.org/project/licofage/
18. Rigo, M.: Numeration systems on a regular language: arithmetic operations, recognizability and formal power series. Theoret. Comput. Sci. **269**(1–2), 469–498 (2001)
19. Rigo, M.: Formal Languages, Automata and Numeration Systems 2: Applications to Recognizability and Decidability, vol. 2. John Wiley & Sons (2014)
20. Rigo, M., Maes, A.: More on generalized automatic sequences. J. Autom. Lang. Comb. **7**(3), 351–376 (2002)
21. Schmitz, S.: Complexity hierarchies beyond elementary. ACM Trans. Comput. Theory (TOCT) **8**(1), 1–36 (2016)
22. Shallit, J.: The Logical Approach to Automatic Sequences: Exploring Combinatorics on Words with Walnut, vol. 482. Cambridge University Press (2022)
23. Sloane, N.J.A., et al.: The On-Line Encyclopedia of Integer Sequences. https://oeis.org/

Words Avoiding Half-Flips

James Currie(✉) and Narad Rampersad

The University of Winnipeg, Winnipeg, MB R3B 2E9, Canada
{j.currie,n.rampersad}@uwinnipeg.ca

Abstract. We say that a word w contains a *half-flip* if it contains non-empty factors u and vu where $|u| = |v|$. Fici reports a non-constructive proof of the existence of an infinite word over a finite alphabet avoiding half-flips and asks the size of the smallest alphabet over which half-flips may be avoided. Half-flips are unavoidable over a 4-letter alphabet. Over an 8-letter alphabet we conjecture a 3-uniform D0L avoiding half-flips; over a 5-letter alphabet we conjecture a messier HD0L construction.

Keywords: Words avoiding patterns · avoiding conjugates · morphic words · method of templates

1 Introduction

We say that a word w contains a *half-flip* if it contains non-empty factors uv and vu where $|u| = |v|$. Fici [7] reports a non-constructive proof of the existence of an infinite word over a finite alphabet avoiding half-flips and asks the size of the smallest alphabet over which half-flips may be avoided.

Several tools exist for studying avoidance problems, but half-flips do not fit neatly into existing categories:

- Cassaigne [3] gave an algorithm to decide whether an HD0L avoids formulas with constants. However, a half-flip is not simply an instance of the formula $uv \cdot vu$ because of the additional length restriction that $|u| = |v|$. (In fact $uv \cdot vu$ is unavoidable.)
- There is a family resemblance between avoiding half-flips and avoiding a given critical exponent, since a k-power is an instance of pattern uvu with the length restriction $(2-k)|u| = (k-1)|v|$. There is a large literature on words avoiding factors with large critical exponent. (See, for example [4,6,8].) Clearly words containing squares contain half-flips. It would be interesting to find a more precise connection between half-flips and critical exponent.
- Since vu is a conjugate of uv, another problem which can be linked to avoiding half-flips is that of finding an infinite word $\mathbf{w}$ over a finite alphabet such that for every non-empty factor u of $\mathbf{w}$, at least one conjugate of u is not a factor of $\mathbf{w}$. This was done over a 5-letter alphabet in [2]. Note that the word constructed in [2] does not avoid half-flips; in fact it contains both $abcd$ and $cdab$.

G. Gamard and J. Leroy (Eds.): WORDS 2025, LNCS 15729, pp. 83–90, 2025.
https://doi.org/10.1007/978-3-031-97548-6_8

In Sect. 2 we show that half-flips are unavoidable over a 4-letter alphabet. In Sect. 3 we give a highly structured 3-uniform D0L construction on an 8-letter alphabet giving an infinite word which avoids half-flips. In section 4 we give a messier HD0L construction over a 5-letter alphabet. The prefix of length 30,000 of the HD0L word avoids half-flips. We explore how the method of templates [1,5,9] introduced by the authors, and extended by Rao and Rosenfeld, could be used to analyse half-flip avoidance by this HD0L. However, a few technical issues remain to be resolved before this method could successfully verify that the HD0L avoids half-flips.

2 Unavoidability on 4 Letters

Definition 1. *For a positive integer k, let Σ_k be the alphabet $\{0, 1, 2. \dots, k-1\}$.*

Theorem 1. *Let* $\mathbf{w}$ *be an ω-word over Σ_4. Then* $\mathbf{w}$ *contains a half-flip.*

Proof. This can be verified by brute force tree search. However an analysis by hand is not difficult:

To get a contradiction, assume that $\mathbf{w}$ contains no half-flips. Since a non-empty square xx can be written as $xx = uv = vu$, where $x = u = v$, $\mathbf{w}$ contains no squares. Thus there are at least three distinct letters in $\mathbf{w}$. Thue observed that every square-free word on a three letter alphabet contains a palindrome uvu where u and v are letters. Any final segment of $\mathbf{w}$ contains no half-flips, and hence no factor uvu. We conclude that all four letters of Σ_4 appear in every final segment of $\mathbf{w}$.

For $a, b \in \Sigma_4$, say that letter a precedes b if ab is a factor of $\mathbf{w}$. Equivalently, we say that b follows a. Since $\mathbf{w}$ contains no half flips, letter a cannot both precede and follow letter b.

If each letter of Σ_4 is followed by exactly one letter, then the letters of $\mathbf{w}$ must repeat in a cycle. This is impossible by square-freeness. Thus some letter is followed by two different letters. Without loss of generality up to renaming letters, assume that 0 is followed by 1 and 2. Then 0 is not preceded by 1 or 2, and is only preceded by 3. We consider 2 cases, depending on whether one of 1 and 2 follows 3:

Case 1: One of 1 and 2 follows 3: Say without loss of generality that 1 follows 3. In this case

- Since 1 follows 0 and 3, only 2 follows 1.
- Since 2 follows 0, and 1, only 3 follows 2.
- Letter 2 cannot follow 3, since 3 follows 2.

In total, in the present case the two letter factors of $\mathbf{w}$ are 01, 02, 12, 23, 30, and 31. A final segment of $\mathbf{w}$ is concatenated from blocks starting with 0 and containing no other 0. These blocks must end in 3. Taking into account length 2 factors, these blocks must be among

$$a = 0123$$
$$b = 023$$
$$c = 023123.$$

(Here the attempt to right-extend a by 1 leads to an extension by 123, and hence the square 123123, which is impossible.)

Neither a nor c can be followed by 0230, so a final segment of $\mathbf{w}$ is just a periodic alternation of a, and c, which is impossible.

Case 2: Neither of 1 and 2 follows 3: Since we cannot have both 1 following 2 and 2 following 1, assume without loss of generality that 1 never follows 2.

The length 2 factors of $\mathbf{w}$ are thus among 01, 02, 12, 13, 23, and 30. The cyclic permutation

$$3 \to 2 \to 1 \to 0 \to 3$$

transforms this set of length 2 factors to those of the previous case, and we again get a contradiction. □

3 A D0L on Σ_8 Avoiding Half-Flips

Definition 2. *Let $g : \Sigma_8^* \to \Sigma_8^*$ be the morphism generated by*

$$\begin{aligned} g(0) &= 026 \\ g(1) &= 127 \\ g(2) &= 036 \\ g(3) &= 137 \\ g(4) &= 046 \\ g(5) &= 147 \\ g(6) &= 056 \\ g(7) &= 157. \end{aligned}$$

Let $\mathbf{w}$ be the fixed point $g^\omega(0)$ of g.

We will show that $\mathbf{w}$ contains no half-flips.

Remark 1. In principle this could be verified using `Walnut`[10], but although the relevant predicate is straightforward to write down, the computation quickly leads to an `Out of Memory` error for the implementation of `Walnut` used by the authors.

Lemma 1. *Let $x \in \Sigma_8$ and let px and $p'x$ be prefixes of $\mathbf{w}$. Then $|p| \equiv |p'|$ (mod 3).*

Proof. We see that

- If $x \in \{0, 1\}$, then $|p|, |p'| \equiv 0 \pmod 3$
- If $x \in \{6, 7\}$, then $|p|, |p'| \equiv 2 \pmod 3$
- If $x \in \{2, 3, 4, 5\}$, then $|p|, |p'| \equiv 1 \pmod 3$.

□

Corollary 1. *Let u be a non-empty factor of* $\mathbf{w}$, *and let pu and $p'u$ be prefixes of* $\mathbf{w}$. *Then $|p| \equiv |p'|$ (mod 3).*

Proof. Let x be the first letter of u. The result follows from Lemma 1. □

Lemma 2. *Let $u \in \Sigma_8^+$ be a factor of* $\mathbf{w}$ *with $|u| \equiv 0$ (mod 3). One of the following holds:*

1. *Whenever pu is a prefix of* $\mathbf{w}$ *we have $p = g(q)$, $u = g(u')$ for some prefix qu' of* $\mathbf{w}$.
2. *There exist letters $x_u, y_u \in \Sigma_8$ such whenever pu is prefix of* $\mathbf{w}$ *we have $p = g(p')x_u$, $x_u u = g(u')y_u$ for some prefix $p'u'$ of* $\mathbf{w}$.
3. *There exist letters $x_u, y_u \in \Sigma_8$ such whenever pu is prefix of* $\mathbf{w}$ *we have $px_u = g(p')$, $uy_u = x_u g(u')$ for some prefix $p'u'$ of* $\mathbf{w}$.

Proof. Since u is a non-empty factor of $\mathbf{w}$, let pu be a prefix of $\mathbf{w}$. By Lemma 1, the length of p modulo 3 is fixed by u. We make cases:

1. If $|p| \equiv 0$ (mod 3), then we have $p = g(p')$, $u = g(u')$ for some prefix $p'u'$ of $\mathbf{w}$.
2. If $|p| \equiv 1$ (mod 3), then u has a prefix of the form $z6$ or $z7$ where $z \in \{2, 3, 4, 5\}$. In the first case, $x_u = 0$ must be a suffix of p, while in the second case, $x_u = 1$ is a suffix. Letting y_u be the last letter of u, the result of the lemma is seen to hold.
3. If $|p| \equiv 2$ (mod 3), then u has a suffix of the form $0z$ or $1z$ where $z \in \{2, 3, 4, 5\}$. In the first case, $y_u = 6$ must follow pu in $\mathbf{w}$, while in the second case, $y_u = 7$ follows. Letting x_u be the first letter of u, the result of the lemma holds.

□

Lemma 3. *Suppose that* $\mathbf{w}$ *contains non-empty factors uv and vu with $|u| = |v| = r$. Then $r \equiv 0$ (mod 3).*

Proof. Let puv and qvu be prefixes of $\mathbf{w}$. By Corollary 1 applied to u, we have $|p| \equiv |qv|$ (mod 3). By Corollary 1 applied to v, we have $|q| \equiv |pu|$ (mod 3). Then

$$\begin{aligned} 2r &\equiv |uv| \\ &\equiv |puv| - |p| \\ &\equiv |pu| + |v| - |p| \\ &\equiv |q| + |v| - |p| \\ &\equiv |qv| - |p| \\ &\equiv 0 \pmod 3. \end{aligned}$$

Thus $r \equiv 0$ (mod 3). □

Lemma 4. *Suppose that* $\mathbf{w}$ *contains non-empty factors uv and vu with $|u| = |v| = r$. Then* $\mathbf{w}$ *contains factors $u'v'$ and $v'u'$, where $|u'| = |v'| = r/3$.*

Proof. Let puv and qvu be prefixes of $\mathbf{w}$. By Lemma 3, $r \equiv 0 \pmod 3$. Then

$$\begin{aligned}|p| &\equiv |pu| \\ &\equiv |q| \text{ by Corollary 1 applied to } v \\ &\equiv |qv| \pmod 3.\end{aligned}$$

We make cases based on $|p|$:

1. If $|p| \equiv 0 \pmod 3$, then by the proof of Lemma 2 applied to u and prefix pu of $\mathbf{w}$, we have $p = g(p')$, $u = g(u')$ for some prefix $p'u'$ of $\mathbf{w}$. By the proof of Lemma 2 applied to v and prefix puv of $\mathbf{w}$, we have $v = g(v')$, with $p'u'v'$ a prefix of $\mathbf{w}$. Working similarly with qv and qvu, we find that $\mathbf{w}$ has a prefix $q'v'u'$.
In total, $\mathbf{w}$ has factors $u'v'$ and $v'u'$, where $|u'| = |v'| = r/3$.
2. If $|p| \equiv 1 \pmod 3$, then by the proof of Lemma 2 applied to prefixes pu and qvu of $\mathbf{w}$, we have $p = g(p')x_u$, $x_u u = g(u')y_u$ for some prefix $p'u'$ of $\mathbf{w}$ and letters $x_u, y_u \in \Sigma_8$, and $qv = g(P')x_u$, for some prefix P' of $\mathbf{w}$. We note that

$$u = x_u^{-1} g(u') y_u.$$

By the proof of Lemma 2 applied to prefixes puv and qv of $\mathbf{w}$, we have $q = g(q')x_v$, $x_v v = g(v')y_v$, with $q'v'$ a prefix of $\mathbf{w}$ and letters $x_v, y_v \in \Sigma_8$, and $pu = g(Q')x_v$, for some prefix Q' of $\mathbf{w}$. Note that

$$v = x_v^{-1} g(v') y_v.$$

Since $qv = g(P')x_u$ and $x_v v = g(v')y_v$, we conclude that the last letter of v is $x_u = y_v$. Similarly, $x_u u = g(u')y_u$ and $pu = g(Q')x_v$ so that the last letter of u is $y_u = x_v$. (Since v follows u in $\mathbf{w}$, the last letter of u is the letter x_v which precedes every occurrence of v; since u follows v in $\mathbf{w}$, the last letter of v is the letter x_u which precedes every occurrence of u.)
It follows that

$$\begin{aligned}puv &= (g(p')x_u)(x_u^{-1} g(u') y_u)(x_v^{-1} g(v') y_v) \\ &= g(p')\cancel{x_u x_u^{-1}} g(u')\cancel{y_u x_v^{-1}} g(v') y_v \\ &= g(p'u'v') y_v,\end{aligned}$$

and $p'u'v'$ is a prefix of $\mathbf{w}$. Similarly,

$$\begin{aligned}qvu &= (g(q')x_v)(x_v^{-1} g(v') y_v)(x_u^{-1} g(u') y_v) \\ &= g(q')\cancel{x_v x_v^{-1}} g(v')\cancel{y_v x_u^{-1}} g(u') y_u \\ &= g(q'v'u') y_u,\end{aligned}$$

and $q'v'u'$ is a prefix of $\mathbf{w}$. Again, $\mathbf{w}$ has factors $u'v'$ and $v'u'$, where $|u'| = |v'| = r/3$.

3. If $|p| \equiv 2 \pmod 3$, the case is symmetrical to the previous one. We write

$$px_u = g(p'), x_u^{-1}uy_u = g(u'), y_u^{-1}vy_v = g(v')$$
$$qx_v = g(q'), x_v^{-1}vy_v = g(v'), y_v^{-1}uy_u = g(u').$$

Since v follows u in $\mathbf{w}$, the first letter of v is the unique letter y_u which follows every occurrence of u; since u follows v in $\mathbf{w}$, the first letter of u is the letter y_v which follows every occurrence of v.
We thus find that $p'u'v'$ and $q'v'u'$ are prefixes of $\mathbf{w}$ with $|u'| = |v'| = r/3$.

□

Theorem 2. *Word* $\mathbf{w}$ *contains no half-flips.*

Proof. This follows from the previous lemma by induction, since r must be a non-zero integer. □

We have seen that any infinite word that avoids half-flips must avoid squares, so the word $\mathbf{w}$ of Definition 2 must be squarefree. It is natural to ask what the critical exponent of $\mathbf{w}$ is. A word $x = x_1x_2\cdots x_n$ has *period* p if $x_i = x_{i+p}$ for $i = 1, \ldots, n-p$. The *exponent* of x is the fraction n/p, where p is the *least period* of x. The *critical exponent* of an infinite word $\mathbf{u}$ is the supremum of the set of all exponents e such that x^e is a factor of $\mathbf{u}$.

Theorem 3. *The critical exponent of* $\mathbf{w}$ *is* $16/9$.

Proof. We use the Walnut prover to verify the claim by computer. See the book by Shallit [10] for more details. The claim can be verified by the following commands:

```
morphism fi8 "0 -> 026, 1 -> 127, 2 -> 036, 3 -> 137, 4 -> 046,
  5 -> 147, 6 -> 056, 7 -> 157":
promote FICI8 fi8:
eval fici8_ce "?msd_3 Ei,n n >= 1 & Aj (j>=i & 9*j<=9*i+7*n) =>
  FICI8[j] = FICI8[j+n]":
```

The last command returns FALSE, confirming that the word $\mathbf{w}$ contains no factor with exponent greater than 16/9. Running the last eval command again with the strict inequality

```
9*j<9*i+7*n
```

returns TRUE, confirming the presence of factors with exponent 16/9 in $\mathbf{w}$. □

4 An HD0L over Σ_5

Consider the morphisms $f, h : \Sigma_5^* \to \Sigma_5^*$ generated by

$$\begin{aligned} f(0) &= 012 & h(0) &= 012432412013043 \\ f(1) &= 0342 & h(1) &= 012432412013243 \\ f(2) &= 0143 & h(2) &= 012430412013243 \\ f(3) &= 0343 & h(3) &= 0120132412 \\ f(4) &= 014342 & h(4) &= 012430412013043. \end{aligned}$$

Conjecture 1. Word $\mathcal{W} = h(f^\omega(0))$ avoids half-flips.

Remark 2. The word 012 only appears in $\mathcal{W}$ as the prefix of the image of a letter under h, and is a prefix of the image of each letter of Σ_5. The lengths of the images of letters under h are 10 and 15, so that every factor u of $\mathcal{W}$ of length at least 17 contains the factor 012. Thus u can be parsed as

$$u = sh(U)p$$

where s is a suffix of $h(i)$, some letter i, with $|s| < |h(i)|$, and p is a prefix of $h(j)$, some letter j, with $|p| < |h(j)|$, and iUj is a factor of $f^\omega(0)$. We allow s and p to be empty. The word U is uniquely determined by these conditions since h is injective on letters. The letters i and j are not uniquely determined since the images of different letters can have long common prefixes or suffixes.

Lemma 5. *Consider a word $uv \in \Sigma_5^*$ where $|u| = |v| \leq 16$. Then uv is a factor of $\mathcal{W}$ if and only if uv is a factor of $h(f^3(120))$.*

Proof. Since 0120 is a prefix of $f^\omega(0)$, the if direction is clear.

For the other direction, suppose that uv is a factor of $\mathcal{W} = h(f^\omega(0))$. We have $|uv| \leq 32$. The shortest image of a letter under h is 10. Thus uv is a factor of $h(U)$ for some factor U of $f^\omega(0)$ where $|U| \leq 5$.

The shortest image of a letter under f is 3. Thus U is a factor of $f(U')$ for some factor U' of $f^\omega(0)$ where $|U'| \leq 3$. Similarly, U' is a factor of $f(U'')$ for some factor U'' of $f^\omega(0)$ where $|U''| \leq 2$. The length 2 factors of $f^\omega(0)$ are observed to be

$$01, 03, 12, 14, 20, 30, 34, 42, \text{ and } 43,$$

which are all factors of $03420143012 = f(120)$. In total then, uv is a factor of $h(f^3(120))$. □

Lemma 6. *Word $\mathcal{W}$ contains no half-flip of length r with $r \leq 16$.*

Proof. A short computation verifies that $h(f^3(120))$ contains no half-flips. (This takes a few seconds in the SAGE programming environment.) Our result follows by Lemma 5. □

Disclosure of Interests. The authors have no competing interests to declare that are relevant to the content of this article.

Acknowledgments. The work of James Currie was supported by an NSERC grant, DDG-2024-00005. The work of Narad Rampersad is supported by an NSERC Grant, 2019-04111.

References

1. Aberkane, A. ,Currie, J. ,Rampersad, N.: The number of ternary words avoiding abelian cubes grows exponentially. J. Integer Seq. **7** (2004). Article 04.2.7, 13 pp
2. Badkobeh, G., Ochem, P.: Avoiding conjugacy classes on the 5-letter alphabet. RAIRO ITA **54** (2020). Article no. 2
3. Cassaigne, J.: Motifs évitables et régularités dans les mots. Ph.D. thesis, Paris VI (1994)
4. Currie, J., Rampersad, N.: A proof of Dejean's conjecture. Math. Comput. **80**(274), 1063–1070 (2011)
5. Currie, J., Rampersad, N.: Fixed points avoiding Abelian k-powers. J. Combin. Theor. Ser. A **119**(5), 942–948 (2012)
6. Dejean, F.: Sur un théorème de Thue. J. Combin. Theor. Ser. A **13**, 90–99 (1972)
7. Fici, G.: Personal communication
8. Rao, M.: Last cases of Dejean's conjecture. Theoret. Comput. Sci. **412**(27), 3010–3018 (2011)
9. Rao, M., Rosenfeld, M.: Avoiding Two Consecutive Blocks of Same Size and Same Sum over $\mathbb{Z}^2$. SIAM J. Disc. Math. **32**(4), 2381–2397 (2018)
10. Shallit, J.: The Logical Approach to Automatic Sequences: Exploring Combinatorics on Words with `Walnut`. Cambridge (2023)

Digital Convexity and Combinatorics on Words

Alessandro De Luca[1(✉)], Gabriele Fici[2], and Andrea Frosini[3]

[1] DIETI, Università di Napoli Federico II, Naples, Italy
alessandro.deluca@unina.it
[2] Dipartimento di Matematica e Informatica, Università di Palermo, Palermo, Italy
gabriele.fici@unipa.it
[3] Dipartimento di Matematica e Informatica, Università di Firenze, Florence, Italy
andrea.frosini@unifi.it

Abstract. An upward (resp. downward) digitally convex word is a binary word that best approximates from below (resp. from above) an upward (resp. downward) convex curve in the plane. We study these words from the combinatorial point of view, formalizing their geometric properties and highlighting connections with Christoffel words and finite Sturmian words. In particular, we study from the combinatorial perspective the operations of inflation and deflation on digitally convex words.

1 Introduction

Combinatorics on words and digital geometry have a long history of interactions. In particular, digital approximations of lines in the plane have a natural encoding as binary words. In this paper, we fix the binary alphabet $\{0, 1\}$, where 0 (resp. 1) represents a horizontal (resp. vertical) unary step in the grid $\mathbb{N} \times \mathbb{N}$.

For example, there is a clear and well-understood correspondence between finite words and approximations of digital segments, given by Christoffel words. For every pair of positive integers (a, b) there is one lower Christoffel word with a occurrences of 0 and b occurrences of 1, which is the digital approximation from below of the Euclidean segment from $(0, 0)$ to (a, b). Christoffel words are detailed in many classical references [2,12,16]. Relationships between Christoffel words and convex digital shapes have been investigated in several papers [5,8,9,18].

In this paper, we focus on digital approximations of convex curves in the plane. A binary word is called (upward) digitally convex if it is the digital approximation from below of a convex curve joining $(0, 0)$ to (a, b) and contained in the rectangle whose opposite vertices are $(0, 0)$ and (a, b). Every such word has a occurrences of 0 and b occurrences of 1. In particular, all the digitally convex words with a occurrences of 0 and b occurrences of 1 lie between the lower

This version has been shortened due to page limit constraints. The proofs and additional results are contained in the full version of the paper.

G. Gamard and J. Leroy (Eds.): WORDS 2025, LNCS 15729, pp. 91–103, 2025.
https://doi.org/10.1007/978-3-031-97548-6_9

Christoffel word with a occurrences of 0 and b occurrences of 1 and the word 1^b0^a, the upper left contour of the rectangle.

In [5], the authors gave a purely combinatorial characterization of digitally convex words: a word w is digitally convex if and only if all the Lyndon words in the Lyndon factorization of w are balanced (hence primitive lower Christoffel words). A primitive word (i.e., a word that cannot be written as a concatenation of copies of a shorter word) over $\{0, 1\}$ is a Lyndon word if it is lexicographically smaller (for the order $0 < 1$) than all its proper suffixes. Every nonempty word w has a unique factorization in non-increasing Lyndon words, which is called the Lyndon factorization of w.

A binary word w is balanced (or Sturmian) if for every length i, the difference between the number of occurrences of 0's and 1's in any two factors of length i of w is at most 1. Since a word over $\{0, 1\}$ is a primitive lower Christoffel word if and only if it is balanced and Lyndon, the previous characterization can be seen as a natural decomposition of digitally convex words in straight segments.

In this paper, we give combinatorial results on digitally convex words that shed light on their combinatorial properties and are related to the geometry of convex curves that these words approximate. Our main result is that every digitally convex word with Parikh vector (a, b) can be obtained with a sequence of inflation (resp. deflation) operations from the lower Christoffel word with the same Parikh vector (resp. from the word 1^b0^a).

The paper is organized as follows: In Sect. 2 we fix the notation and give preliminary results on words, and in particular on Christoffel words, that we will need in the sequel; in Sect. 3 we recall the notion of digitally convex words and provide some new results and characterizations. Finally, in Sect. 4 we study from the combinatorial perspective the operations of inflation and deflation on digitally convex words.

2 Preliminaries

A *word* over a finite alphabet Σ is a concatenation of letters from Σ. The *length* of a word w is denoted by $|w|$. The empty word ε has length 0. The set of all words over Σ is denoted Σ^* and is a free monoid with respect to concatenation. For a letter $x \in \Sigma$, $|w|_x$ denotes the number of occurrences of x in w. If $\Sigma = \{x_1, \ldots, x_\sigma\}$ is ordered, the vector $(|w|_{x_1}, \ldots, |w|_{x_\sigma})$ is the *Parikh vector* of w.

Let $w = uv$, with $u, v \in \Sigma^*$. We say that u is a *prefix* of w and that v is a *suffix* of w. A *factor* of w is a prefix of a suffix (or, equivalently, a suffix of a prefix) of w. If $w = w_1w_2 \cdots w_n$, with $w_k \in \Sigma$ for all k, we let $w[i...j]$ denote the nonempty factor $w_iw_{i+1} \cdots w_j$ of w, whenever $1 \leq i \leq j \leq n$. A factor u of a word $w \neq u$ is a *border* of w if u is both a prefix and a suffix of w; in this case, w has *period* $|w| - |u|$. A word w is *unbordered* if its longest border is ε; i.e., if its smallest period is $|w|$. For a word w, the *n-th power* of w is the word w^n obtained by concatenating n copies of w.

Two words w and w' are *conjugates* if $w = uv$ and $w' = vu$ for some words u and v. The conjugacy class of a word w contains $|w|$ elements if and only if w is *primitive*, i.e., $w = v^n$ for some v implies $n = 1$.

A nonempty word $w = w_1 w_2 \cdots w_n$, $w_k \in \Sigma$ for all k, is a *palindrome* if it coincides with its *reversal* $w^R = w_n w_{n-1} \cdots w_1$. The empty word is also assumed to be a palindrome.

The following proposition, whose proof is straightforward, is well known (cf. [4,6]).

Proposition 1. *A word is a conjugate of its reversal if and only if it is a concatenation of two palindromes. Moreover, these two palindromes are uniquely determined if and only if the word is primitive.*

2.1 Lyndon Words

A primitive word over an ordered alphabet is a *Lyndon word* if it is lexicographically smaller than all its conjugates (or, equivalently, lexicographically smaller than all its proper suffixes).

In the binary case, we have the class of Lyndon words for the order $0 < 1$ and the class of Lyndon words for the reversed order $1 < 0$. Whenever not otherwise specified, we assume the order $0 < 1$.

The following theorem, originally due to Chen, Fox, and Lyndon (1958), is a classical result in combinatorics on words (see [11]):

Theorem 1. *Every word w has a unique non-increasing factorization in Lyndon words.*

The factorization from the previous theorem is called the *Lyndon factorization* (or sometimes the Chen–Fox–Lyndon factorization) of w.

Example 1. Let $w = 0101001001$. The Lyndon factorization of w is

$$01 \cdot 01 \cdot 001 \cdot 001.$$

Example 2. Let $w = 1100$. The Lyndon factorization of w is

$$1 \cdot 1 \cdot 0 \cdot 0.$$

A way to obtain the Lyndon factorization is by using the following:

Property 1. If u and v are Lyndon words, then $u < v$ if and only if uv is a Lyndon word.

Starting from the trivial factorization $w = w_1 \cdots w_n$, with $|w_i| = 1$ for every i, one repeatedly applies Property 1 until the factors occur in non-increasing order.

Every Lyndon word w of length $|w| \geq 2$ has a *standard factorization* $w = uv$, where v is the lexicographically least proper suffix of w (or, equivalently, the longest proper suffix of w that is a Lyndon word), see [11].

2.2 Balanced and Christoffel Words

Definition 1. *A word over* $\{0, 1\}$ *is* balanced *if the number of* 0*'s (or, equivalently,* 1*'s) in every two factors of the same length differs by at most* 1.

We let Bal denote the set of balanced words over the alphabet $\{0, 1\}$. Balanced words are precisely the finite factors of infinite Sturmian words [12]. Balance is a *factorial property*, i.e., every factor of a balanced word is itself balanced.

Christoffel words are powers of balanced Lyndon words. But let us introduce them from the point of view of digital geometry.

In what follows, we fix the alphabet $\{0, 1\}$ and represent a word over $\{0, 1\}$ as a path in $\mathbb{N} \times \mathbb{N}$ starting at $(0, 0)$, where 0 (resp., 1) encodes a horizontal (resp., vertical) unit segment. For every pair of nonnegative integers (a, b) (not both 0), every word w that encodes a path from $(0, 0)$ to (a, b) must have exactly a zeroes and b ones, i.e., it has Parikh vector $(a, b) = (|w|_0, |w|_1)$. We define the *slope* of a word w with Parikh vector (a, b) as the rational number b/a if $a \neq 0$, or ∞ otherwise.

Definition 2. *Given a pair of nonnegative integers* (a, b) *(not both* 0*), the* lower (resp., upper) Christoffel word $w_{a,b}$ *(resp.,* $W_{a,b}$*), of slope* b/a*, is the word encoding the path from* $(0, 0)$ *to* (a, b) *that is closest from below (resp., from above) to the Euclidean segment, without crossing it.*

In other words, $w_{a,b}$ (resp. $W_{a,b}$) is the digital approximation from below (resp., from above) of the Euclidean segment joining $(0, 0)$ to (a, b). For example, the lower Christoffel word $w_{7,4}$ is the word 00100100101 (see Fig. 1).

By construction, we have that the upper Christoffel word $W_{a,b}$ is the reversal of the lower Christoffel word $w_{a,b}$ (and the two words are conjugates, see below).

Notice that by definition we allow a or b (but not both) to be 0, that is, we have $w_{n,0} = W_{n,0} = 0^n$ and $w_{0,n} = W_{0,n} = 1^n$, so that the powers of words of length 1 are both lower and upper Christoffel words.

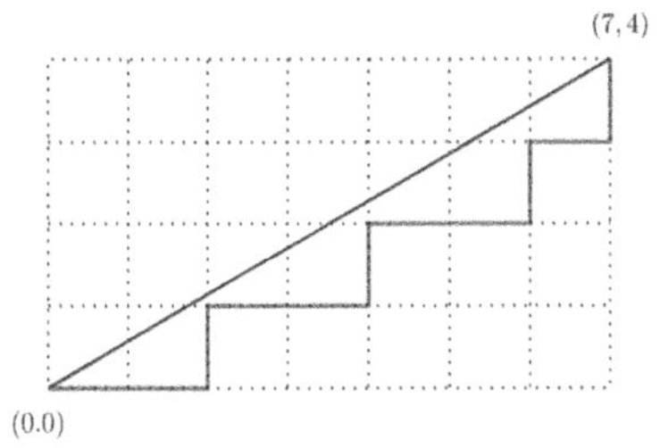

Fig. 1. The lower Christoffel word $w_{7,4} = 00100100101$. The upper Christoffel word $W_{7,4} = 10100100100$ is the reversal of $w_{7,4}$.

If a and b are coprime, the Christoffel words $w_{a,b}$ and $W_{a,b}$ do not intersect the Euclidean segment joining $(0, 0)$ to (a, b) (other than at the end points) and are primitive words. If instead $a = n\alpha$ and $b = n\beta$ for some $n > 1$, then $w_{a,b} = (w_{\alpha,\beta})^n$ (resp., $W_{a,b} = (W_{\alpha,\beta})^n$). Hence, $w_{a,b}$ (resp., $W_{a,b}$) is primitive if and only if a and b are coprime.

Therefore, letting ϕ denote the Euler totient function, for every $n > 1$ there are $2\phi(n)$ primitive Christoffel words of length n (in particular, there are $\phi(n)$ primitive lower Christoffel words and $\phi(n)$ primitive upper Christoffel words).

Theorem 2 *([1]). Let w be a word over $\{0,1\}$. Then w is a primitive (lower or upper) Christoffel word if and only if it is balanced and unbordered.*

In particular, the set of primitive lower Christoffel words is precisely the set of balanced Lyndon words over the alphabet $\{0,1\}$ for the order $0 < 1$.

Proposition 2 *([2]). For every coprime positive integers a, b, the primitive lower Christoffel word $w_{a,b}$ is the greatest (for the lexicographic order) word among all Lyndon words having Parikh vector (a,b).*

Proposition 3 *([7]). For all coprime positive integers a, b, the lower Christoffel word $w_{a,b}$ is the smallest (in the lexicographic order) word among all balanced words having Parikh vector (a,b).*

Theorem 3 *([2]). Let (a,b) and (c,d) be pairs of nonnegative integers, both distinct from $(0,0)$, such that $b/a \neq d/c$. Then*

$$w_{a,b} < w_{c,d} \iff \frac{b}{a} < \frac{d}{c}.$$

Since primitive lower Christoffel words are Lyndon words, every primitive lower Christoffel word of length $|w| \geq 2$ has a standard factorization.

On the other hand, a primitive lower Christoffel word is a conjugate of its reversal (the corresponding upper Christoffel word); hence, by Proposition 1, every primitive lower Christoffel word of length $|w| \geq 2$ has a unique *palindromic factorization*.

From the geometric point of view, the standard factorization divides $w_{a,b}$ in two shorter Christoffel words, and it determines the point S of the encoded path that is *closest* to the Euclidean segment from $(0,0)$ to (a,b); the palindromic factorization, instead, divides $w_{a,b}$ in two palindromes and determines the point S' that is *farthest* from the Euclidean segment (see [3,17]). An example is given in Fig. 2.

3 Digitally Convex Words

We now introduce the main object of study of this paper.

Definition 3. *A binary word with a occurrences of* 0 *and b occurrences of* 1 *is an* upward (resp. downward) digitally convex word *if it is the best approximation from below (resp. above) of an upward (resp. downward) convex curve that joins $(0,0)$ and (a,b) and is contained in the rectangle whose opposed vertices are $(0,0)$ and (a,b).*

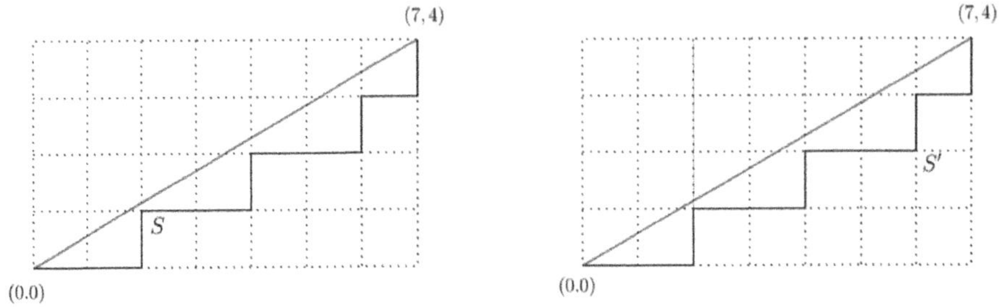

Fig. 2. The standard factorization $0Q1 \cdot 0P1 = 001 \cdot 00100101$ (left) and the palindromic factorization $0P0 \cdot 1Q1 = 00100100 \cdot 101$ (right) of the lower Christoffel word $w_{7,4}$. The point S determined by the standard factorization is the closest to the Euclidean segment, while the point S' determined by the palindromic factorization is the farthest.

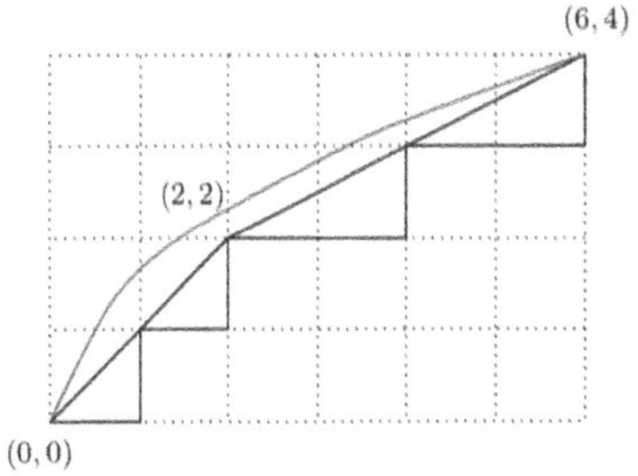

Fig. 3. The (upward) digitally convex word $w = 0101001001$ and its decomposition in two Christoffel words 0101 and 001001. In red, one of the possible approximated upward convex curves that join $(0,0)$ and $(6,4)$. (Color figure online)

For example, the word $w = 0101001001$ of Example 1 is (upward) digitally convex, as shown in Fig. 3.

Proposition 4. *We have:*

1. *A word is upward (resp. downward) digitally convex if and only if its reverse is downward (resp. upward) digitally convex;*
2. *A word is upward (resp. downward) digitally convex if and only if its binary complement is downward (resp. upward) digitally convex.*

Geometrically, given fixed a and b positive integers, the upward digitally convex words with Parikh vector (a, b) are all above the Christoffel word $w_{a,b}$ and below the word 1^b0^a. Of course, the downward digitally convex words with Parikh vector (a, b) are all below the Christoffel word $W_{a,b}$ and above the word 0^a1^b. This will be stated formally in Theorem 9.

Every balanced word is, as we will see, both upward and downward digitally convex. On the contrary, digitally convex words are not necessarily balanced, as shown by the word $w = 0101001001$ of Example 1. We have the following characterization.

Proposition 5. *Let w be a binary word, $\widetilde{w}$ its reverse, and $\overline{w}$ its binary complement. The following are equivalent:*

1. *w is balanced;*
2. *w and $\widetilde{w}$ are both upward (resp. downward) digitally convex;*
3. *w and $\overline{w}$ are both upward (resp. downward) digitally convex;*
4. *w is upward digitally convex and downward digitally convex.*

The Lyndon factorization can be used to characterize digitally convex words, as it was shown by Brlek et al. in the following

Theorem 4 *([5]). A word w is upward digitally convex if and only if all the Lyndon words in the Lyndon factorization of w are balanced (hence primitive lower Christoffel words).*

An analogous result characterizes downward digitally convex words. In fact, by a simple symmetry argument, we have that a binary word is downward digitally convex if and only if all the Lyndon words for the order $1 < 0$ in the Lyndon factorization of w are balanced (hence primitive upper Christoffel words).

Remark 1. The union of upward and downward digitally convex words is a factorial language that is also closed under reversal and binary complement, and includes the language of balanced words.

Because of the symmetries between upward and downward digitally convex words, from now on we will focus on upward digitally convex words only, which we will simply call digitally convex words. Analogous results hold, of course, for downward digitally convex words.

3.1 Minimal Forbidden Words

Minimal forbidden words are a useful tool for characterizing factorial languages.

Definition 4. *Let L be a factorial language. A word w is a minimal forbidden word for L if w does not belong to L but all proper factors of w do.*

Let $\mathcal{MF}(L)$ denote the set of minimal forbidden words of the language L. A word $w = xvy$, x, y letters, belongs to $\mathcal{MF}(L)$ if and only if

1. $xvy \notin L$;
2. $xv, vy \in L$.

Theorem 5. *([14]) There is a bijection between factorial languages and their sets of minimal forbidden words.*

As a consequence, $\mathcal{MF}(L)$ uniquely determines L.

Let Bal be the (factorial) language of balanced words over $\{0, 1\}$. The next theorem gives a characterization of the minimal forbidden words for the language of binary balanced words.

Theorem 6 *([10]).* $\mathcal{MF}(\mathrm{Bal}) = \{yvx \mid \{x,y\} = \{0,1\}, xvy$ *is a non-primitive Christoffel word*$\}$.

Example 3. The word 000101 is not balanced, but all its proper factors are. Indeed, 100100 is the square of the primitive upper Christoffel word 100.

Corollary 1 *([10]). For every* $n > 0$*, there are exactly* $n - \phi(n) - 1$ *words of length* n *in* $\mathcal{MF}(\mathrm{Bal})$ *that start with* 0*, and they are all Lyndon words. Here* ϕ *is the Euler totient function.*

In 2011, Provençal [15] studied the minimal forbidden words of the set $\mathcal{DC}$ of digitally convex words. Indeed, the set of digitally convex words is a factorial language. In his paper, Provençal attributes this result to a private communication of C. Reutenauer .

As a consequence, a word is digitally convex if and only if all its Lyndon factors are balanced.

Theorem 7 *([15]).* $\mathcal{MF}(\mathcal{DC}) = \{u(uv)^k v \mid k \geq 1, uv$ *is the standard factorization of a primitive lower Christoffel word*$\}$.

We now give an alternative description:

Theorem 8. $\mathcal{MF}(\mathcal{DC})$ *is the set of words in* $\mathcal{MF}(\mathrm{Bal})$ *that start with* 0*. Hence,* $\mathcal{MF}(\mathcal{DC}) = \{0w1 \mid 1w0$ *is a non-primitive Christoffel word*$\}$*. In particular, all words in* $\mathcal{MF}(\mathcal{DC})$ *are Lyndon words.*

By Corollary 1, we have:

Corollary 2. $\mathcal{MF}(\mathcal{DC})(n) = n - 1 - \phi(n)$.

3.2 Dominance Order

Definition 5. *Over* $\{0,1\}^n$*, we consider the dominance order defined by* $u \sqsubseteq v$ *if for every* $1 \leq i \leq n$*,* $|u[1...i]|_1 \leq |v[1...i]|_1$ *(or, equivalently,* $|u[1...i]|_0 \geq |v[1...i]|_0$*).*

Notice that the dominance order is a partial order, and that the lexicographic order is a linear extension of it.

Theorem 9. *Let* $a, b > 0$ *and* $n = a + b$*. For every digitally convex word* u *with Parikh vector* (a, b)*, and for every* $1 \leq k \leq n$*, we have that* $w_{a,b}[1...k]$ *is lexicographically smaller than or equal to* $u[1...k]$*, hence in particular* $w_{a,b}[1...k] \sqsubseteq u[1...k]$.

Corollary 3. *For every pair* (a, b)*, the lower Christoffel word* $w_{a,b}$ *is the smallest (in the lexicographic order) digitally convex word having Parikh vector* (a, b).

The previous theorem essentially says that all the paths encoded by upward digitally convex words with Parikh vector (a, b) stay above the path encoded by the lower Christoffel word $w_{a,b}$ (and below the path encoded by 1^b0^a).

By symmetry, we also have that all the paths encoded by downward digitally convex words with Parikh vector (a, b) stay above the path encoded by 0^a1^b and below the upper Christoffel word $W_{a,b}$.

Definition 6. *To each binary word w it is associated an integer sequence s_w such that $s_w[i] = |w[1...i]|_1$. The* meet *(resp.* join*) of two binary words u and v is defined as the binary word $w = u \wedge v$ (resp. $w = u \vee v$) whose associated sequence is $s_w[i] = \min\{s_u[i], s_v[i]\}$ (resp. $s_w[i] = \max\{s_u(i), s_v(i)\}$).*

In other words, $u \wedge v$ (resp. $u \vee v$) is precisely the maximum lower bound (resp. minimum upper bound) of the set $\{u, v\}$ with respect to the dominance order. From a geometrical perspective, the meet (resp. join) of u and v turns out to be the word delimiting the intersection (resp. union) of the areas below them.

Proposition 6. *Let u and v be two digitally convex words. Then $u \wedge v$ is a digitally convex word.*

Hence, the set $\mathcal{DC} \cap \{0, 1\}^n$ is a *meet-semilattice* for all n, as is the set $\mathcal{DC}_{a,b}$ of all digitally convex words with Parikh vector (a, b).

As one can expect, the join of two digitally convex words is not, in general, a digitally convex word, according to the fact that the union of convex polygons does not preserve convexity. As an example, consider $u = 010101010110001001$ and $v = 011110001001001001$. The word $u \vee v = 011110001010001001$, whose Lyndon factorization is $01111 \cdot 000101 \cdot 0001 \cdot 001$, contains a non-balanced Lyndon factor 000101, so it is not digitally convex by Theorem 4.

4 Inflation/Deflation of Digitally Convex Words

Borrowing the terminology from [19], we call *deflation* the operation that changes a digitally convex word $w = u10v$ into a digitally convex word $w' = u01v$. The name suggests the geometrical interpretation of the operation, i.e., the removal of a point from the (discrete) digitally convex set related to w to obtain a new (discrete) digitally convex set related to w'. Similarly, the addition of a point to a digitally convex set preserving the convexity is called *inflation*, and it is realized by changing a 01 occurrence into a 10 while preserving the convexity of a digitally convex word.

Lemma 1 *(see [18,19]). Let $w_1 = u1$ and $w_2 = 0v$ be lower Christoffel words such that $w_1 > w_2$. Then $u01v$ is digitally convex.*

Lemma 2. *Let $w = u10v$ be a lower Christoffel word. Then $u01v$ is* not *digitally convex.*

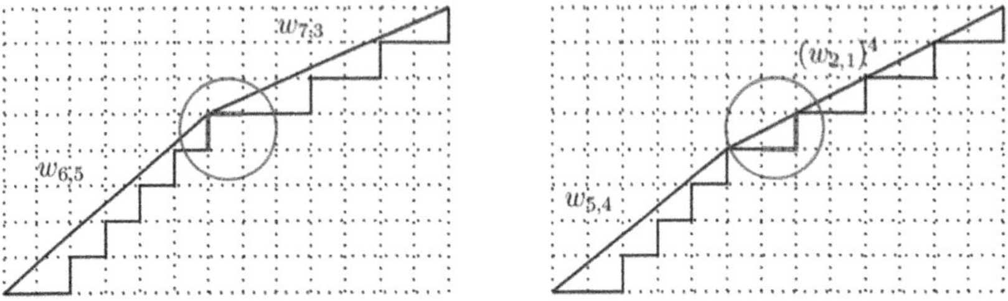

Fig. 4. On the left, the two Christoffel words with decreasing slopes $w_{6,5}$ and $w_{7,3}$. Their common point is highlighted. On the right, the deflation performed on that point produces new Christoffel words $w_{5,4}$ and $w_{8,4}$ preserving the decreasing of their slopes, and so the digital convexity.

Theorem 10 *(see [18,19]). Let $w = u10v = w_1 \cdots w_k$ be a digitally convex word, where $w_1 > \ldots > w_k$ $(k \geq 2)$ are Christoffel words. Then $u01v$ is digitally convex if and only if $0v = w_i w_{i+1} \cdots w_k$ for some $1 < i \leq k$.*

Remark 2. The deflation of a digitally convex word w is a *local* operation, in that all elements of the Lyndon factorization of $w = u10v$ are inherited in the Lyndon factorization of $w' = u01v$, except for the two elements that overlap the given occurrence of 10. By contrast, the converse inflation operation can change an arbitrarily large number of elements in the Lyndon factorization, as shown in the next example.

Example 4. Let $f = 0100101001001\cdots$ be the Fibonacci infinite word, fixed point of the substitution $0 \mapsto 01, 1 \mapsto 0$, and consider the Sturmian word $s = 001f$. As shown in [13], the Lyndon factorization of f is

$$\prod_{n\geq 1} \ell_{2n+1} = (01)(00101)(0010010100101)\cdots$$

where $\ell_1 = 1$, $\ell_2 = 0$, $\ell_{2n+1} = \ell_{2n}\ell_{2n-1}$, and $\ell_{2n+2} = \ell_{2n}\ell_{2n+1}$ for $n \geq 1$ give the sequence of all lower Christoffel factors of f (note that $|\ell_n| = F_n$, the nth Fibonacci number). It follows that for all $k > 1$, the Lyndon factorization of the prefix p_k of s of length $3+\sum_{n=1}^{k} F_{2n+1} = F_{2k+2}+2$ is $p_k = (00101)^2 \cdot \prod_{n=3}^{k} \ell_{2n+1}$.

Now, the inflated infinite word $s' = 010f$ is still Sturmian, so that its prefixes are all digitally convex; in particular, for all $k > 1$, the Lyndon factorization of its prefix p'_k such that $|p'_k| = |p_k|$ is $p'_k = 01 \cdot \ell_{2k+2}$, thus showing that an arbitrarily large number of elements in the Lyndon factorization of a digitally convex word can give rise to a constant number (2, in this case) of such elements after inflation.

In the next lemma, we analyze how the inflation operation acts on a single Christoffel word.

Lemma 3. *Let $w = v01u$ be a primitive lower Christoffel word. Then $v10u$ is digitally convex if and only if $w = 0uv1$, i.e., $(v0, 1u)$ is the palindromic factorization of w and $(0u, v1)$ is its standard factorization.*

The previous lemma essentially states that the only possible inflation point in a Christoffel word is the point S' determined by the palindromic factorization, which is the point on the path encoding the Christoffel word that is the farthest from the Euclidean segment (see Fig. 2).

A deeper investigation reveals that the inflation operation performed as in Lemma 3 cannot be applied to any Christoffel factor of a digitally convex word while preserving the digital convexity property. The following example is from [8].

Example 5. Let w be the digitally convex word $w = w_{5,3}w_{20,11}$. The application of the inflation to $w_{5,3}$, according to Lemma 3, produces $w' = w_{3,2}w_{2,1}w_{20,11}$, which is not digitally convex any more. To gain back the digital convexity, a second inflation in $w_{20,11}$ is required, thus obtaining $w'' = w_{3,2}w_{2,1}w_{9,5}w_{11,6}$. Note that $w_{2,1}w_{9,5}$ is the standard factorization of $= w_{11,6}$; so, finally, $w'' = w_{3,2}w_{11,6}w_{11,6}$.

Remark 3. Example 5 suggests that an order on the inflation operations can be established to avoid the loss of the digital convexity property. In fact, performing on w a first inflation in its Christoffel factor $w_{20,11}$ produces the digitally convex word $w_{5,3}w_{9,5}w_{11,6}$. Now, the second inflation on $w_{5,3}$ leads to w''.

The following lemma expresses what was observed in the previous remark.

Lemma 4. *Let $w = w_1 \cdots w_k$ be a digitally convex word. where $w_1 > \ldots > w_k$ $(k \geq 2)$ are Christoffel words. There exists an index $1 \leq i \leq k$ such that w_i has palindromic factorization $(u0, 1v)$ and $w' = w_1 \cdots w_{i-1}\, u10v\, w_{i+1} \cdots w_k$ is digitally convex.*

Proposition 7. *Let w be a digitally convex word with Parikh vector (a, b). There exists a sequence of applications of inflation that leads from the word $w_{a,b}$ to w.*

Proposition 8. *Let w be a digitally convex word with Parikh vector (a, b). There exists a sequence of applications of deflation that leads from the digitally convex word 1^b0^a to w.*

We therefore arrive at the main result of this section, whose proof directly follows from Propositions 7 and 8.

Theorem 11. *Iterated inflation (resp. deflation) in the word $w_{a,b}$ (resp. 1^b0^a) produces all digitally convex words with the same Parikh vector (a, b).*

Now, we are ready to define the order relation $\leq_I$ on the set $\mathcal{DC}_{a,b}$ of all the digitally convex words with Parikh vector (a, b) such that, provided $u, v \in \mathcal{DC}_{a,b}$, $u \leq_I v$ if there exists a sequence of applications of the inflation operation leading from u to v. It is worthwhile proving that the structure $\mathcal{W}_{a,b} = (\mathcal{DC}_{a,b}, \leq_I)$ is a partial order, having the words 1^b0^a and $w_{a,b}$ as maximum and minimum element, respectively.

Let us indicate the partial order provided by the dominance order on the same set $\mathcal{DC}_{a,b}$ as $\mathcal{D}_{a,b} = (\mathcal{DC}_{a,b}, \sqsubseteq)$.

Theorem 12. *The partial orders $\mathcal{W}_{a,b}$ and $\mathcal{D}_{a,b}$ define the same structure on the ground set $\mathcal{DC}_{a,b}$.*

Acknowledgments. We warmly thank Lama Tarsissi for useful discussions on Digitally Convex words.

References

1. Berstel, J., de Luca, A.: Sturmian words, Lyndon words and trees. Theor. Comput. Sci. **178**(1–2), 171–203 (1997)
2. Borel, J.-P., Laubie, F.: Quelques mots sur la droite projective réelle. J. de théorie des nombres de Bordeaux **5**(1), 23–51 (1993)
3. Borel, J.-P., Reutenauer, C.: On Christoffel classes. RAIRO Theor. Inform. Appl. **40**(1), 15–27 (2006)
4. Brlek, S., Hamel, S., Nivat, M., Reutenauer, C.: On the palindromic complexity of infinite words. Int. J. Found. Comput. Sci. **15**(2), 293–306 (2004)
5. Brlek, S., Lachaud, J.-O., Provençal, X., Reutenauer, C.: Lyndon + Christoffel = digitally convex. Pattern Recogn. **42**(10), 2239–2246 (2009)
6. de Luca, A.: On some combinatorial problems in free monoids. Discret. Math. **38**(2), 207–225 (1982)
7. De Luca, A., Fici, G.: Some results on digital segments and balanced words. Theoret. Comput. Sci. **1021**, 114935 (2024)
8. Dulio, P., Frosini, A., Rinaldi, S., Tarsissi, L., Vuillon, L.: First steps in the algorithmic reconstruction of digital convex sets. In: Brlek, S., Dolce, F., Reutenauer, C., Vandomme, É. (eds.) WORDS 2017. LNCS, vol. 10432, pp. 164–176. Springer, Cham (2017). https://doi.org/10.1007/978-3-319-66396-8_16
9. Dulio, P., Frosini, A., Rinaldi, S., Tarsissi, L., Vuillon, L.: Further steps on the reconstruction of convex polyominoes from orthogonal projections. J. Comb. Optim. **44**(4), 2423–2442 (2022)
10. Fici, G.: On the structure of bispecial Sturmian words. J. Comput. Syst. Sci. **80**(4), 711–719 (2014)
11. Lothaire, M.: Combinatorics on Words. Cambridge Mathematical Library. Cambridge University Press, Cambridge (1997)
12. Lothaire, M.: Algebraic Combinatorics on Words. Encyclopedia of Mathematics and its Applications. Cambridge University Press, Cambridge (2002)
13. Melançon, G.: Lyndon factorization of infinite words. In: Puech, C., Reischuk, R. (eds.) STACS 1996. LNCS, vol. 1046, pp. 147–154. Springer, Heidelberg (1996). https://doi.org/10.1007/3-540-60922-9_13
14. Mignosi, F., Restivo, A., Sciortino, M.: Forbidden factors in finite and infinite words. In: Karhumäki, J., Maurer, H., Paun, G., Rozenberg, G. (eds.) Jewels are Forever, Contributions on Theoretical Computer Science in Honor of Arto Salomaa, pp. 339–350. Springer (1999). https://doi.org/10.1007/978-3-642-60207-8_30
15. Provençal, X.: Minimal non-convex words. Theor. Comput. Sci. **412**(27), 3002–3009 (2011)
16. Reutenauer, C.: Studies on finite Sturmian words. Theor. Comput. Sci. **591**, 106–133 (2015)
17. Tarsissi, L.: Balance properties on Christoffel words and applications. Ph.D. thesis, Université Grenoble Alpes (2017)

18. Tarsissi, L., Coeurjolly, D., Kenmochi, Y., Romon, P.: Convexity preserving contraction of digital sets. In: Palaiahnakote, S., Sanniti di Baja, G., Wang, L., Yan, W. (eds) 5th Asian Conference on Pattern Recognition, ACPR 2019, Auckland, New Zealand, 26–29 November 2019, Revised Selected Papers, Part II, volume 12047 of LNCS, pp. 611–624. Springer (2019). https://doi.org/10.1007/978-3-030-41299-9_48
19. Tarsissi, L., Kenmochi, Y., Romon, P., Coeurjolly, D., Borel, J.-P.: Convexity preserving deformations of digital sets: characterization of removable and insertable pixels. Discrete Appl. Math. **341**, 270–289 (2023)

Clustering of Return Words in Languages of Interval Exchanges

Francesco Dolce(✉) and Christian B. Hughes

Czech Technical University in Prague, Prague, Czech Republic
{dolcefra,hughechr}@fit.cvut.cz

Abstract. A word over an ordered alphabet is said to be *clustering* if identical letters appear adjacently in its Burrows-Wheeler transform. Such words are strictly related to (discrete) interval exchange transformations. We use an extended version of the well-known Rauzy induction to show that every return word in the language generated by a regular interval exchange transformation is clustering, partially answering a question of Lapointe (2021).

Keywords: Interval exchange transformations · Burrows-Wheeler transform · Clustering words · Return words

1 Introduction

Interval exchange transformations (IETs), first introduced by Oseledec [17] in 1966, are defined by first partitioning an interval into subintervals, then translating each subinterval by a fixed permutation. They form an important class of dynamical systems that are studied from different perspectives: symbolic dynamics, combinatorics on words, ergodic theory, and others. A rich body of work has since explored various structural and combinatorial properties of these transformations. One can code IETs in a natural way to obtain sequences of linear complexity, including Sturmian sequences, which have been widely studied (see, e.g., [2,8,10]).

The Burrows-Wheeler transform, introduced in [3], is a transformation used in data compression that first rearranges the letters of a word by lexicographically sorting all of its conjugates, then reads in this order the last letters of these conjugates. Clustering words are words whose Burrows-Wheeler transform consists of adjacent occurrences of identical letters. A link between clustering words and IETs has been developed in recent years (e.g., [6,12]). In particular, each clustering word can be associated with a discrete interval exchange transformation (see [9]).

Return words to w in a language are words that when preceded by w are still in the language and end with w as well (see precise definition later). In a 2021 paper [12], Lapointe asked whether return words of a symmetric IET are themselves perfectly clustering. That is, if such return words cluster in a way that corresponds to the symmetric permutation.

G. Gamard and J. Leroy (Eds.): WORDS 2025, LNCS 15729, pp. 104–115, 2025.
https://doi.org/10.1007/978-3-031-97548-6_10

In this paper, spurred by Lapointe's 2021 question, we show that all return words of interval exchange transformations satisfying the Keane condition [11], i.e., regular IETs, are clustering. Our result leverages and extends previous combinatorial and dynamical insights, particularly from a work by the first author and Perrin on a two-sided version of Rauzy induction on regular IETs [5]. The main result of this contribution is the following.

Theorem 1. *Return words in a language generated by a regular interval exchange transformation are clustering words.*

Our approach to this theorem relies on Rauzy induction, a dynamical tool introduced in its one-sided version by Rauzy [18] and subsequently extended in various ways. We consider a family of morphisms that, under certain assumptions, preserves clustering at every step of the induction. This result, together with the fact that one can obtain the cylinders of a regular IET through Rauzy induction, allows us to prove the theorem.

We conclude this contribution by extending the link between IETs (resp. DIETs) and clustering words to clustering multisets of words instead. In order to do so, we introduce the notion of alsinicity, a generalization of the more well-known concept of dendricity.

2 Preliminaries

For all undefined terms, we refer the reader to [14].

Words. An *ordered alphabet* $\mathcal{A} = \{a_1 < a_2 < \ldots < a_d\}$ is a (finite) set of symbols called *letters* together with an order of its elements. The set of *finite words* $\mathcal{A}^*$ over $\mathcal{A}$ is the free monoid with neutral element the *empty word* ε. The product of two words $u, v \in \mathcal{A}^*$ is given by their composition uv. We denote by $\mathcal{A}^+$ the free semigroup over $\mathcal{A}$, e.g., $\mathcal{A}^+ = \mathcal{A}^* \setminus \{\varepsilon\}$. The order on $\mathcal{A}$ is naturally extended to $\mathcal{A}^*$ by the lexicographic order. For a given word $w = w_0 w_1 \cdots w_{n-1}$, where each $w_i \in \mathcal{A}$, we denote by $|w|$ its length n, and by $|w|_u$ the number of times u appears a factor of w. The Parikh vector of a word $w \in \mathcal{A}^*$ is the vector $\Psi_\mathcal{A}(w) \in \mathbb{N}^d$ defined as $(\Psi_\mathcal{A}(w))_a = |w|_a$. A word $w \in \mathcal{A}^*$ is *pangrammatic* if $|w|_a > 0$ for every $a \in \mathcal{A}$. Unless stated otherwise we will always consider words pangrammatic over their alphabet. These vectors can be generalized in a natural way to multisets of words over the same ordered alphabet.

A word w is said to be *primitive* if it is not the integer power of another word, i.e., if $w = u^k$ implies $k = 1$. Two words w, w' are *conjugate* if $w = uv$ and $w' = vu$ for some $u, v \in \mathcal{A}^*$. If a word w is primitive, then it has exactly $|w|$ distinct conjugates. A *Lyndon word* is a primitive word that is minimal for the lexicographic order among its conjugates.

A (right) *infinite word* over $\mathcal{A}$ is a sequence $\mathbf{w} = w_0 w_1 w_2 \cdots$, with $w_i \in \mathcal{A}$ for all i. An infinite word $\mathbf{w}$ is *eventually periodic* if $\mathbf{w} = uv^\omega = uvvv\cdots$. An infinite word that is not eventually periodic is called *aperiodic*.

Languages. By *language* we mean a factorial and bi-extendable set $\mathcal{L} \subset \mathcal{A}^*$, i.e., such that, for every $w \in \mathcal{A}^*$, we have $v, aw, wb \in \mathcal{L}$ for every factor v of u and for certain letters $a, b \in \mathcal{A}$. The language of an infinite word $\mathbf{w}$ is the set $\mathcal{L}(\mathbf{w})$ of all its factors, while the language of a finite word w is defined as $\mathcal{L}(w^\omega)$. A language $\mathcal{L}$ is *recurrent* if, for every $v \in \mathcal{L}$, $vuv \in \mathcal{L}$ for a certain word u. It is *uniformly recurrent* if for every $v \in \mathcal{L}$, there exists $N \in \mathbb{N}$ such that v appears as a factor of every element of length N in $\mathcal{L}$. The set $\mathcal{R}_\mathcal{L}(w)$ of (right) *return words* to w in $\mathcal{L} \subset \mathcal{A}^*$ is the set of words u such that $wu \in \mathcal{L}$ has exactly two occurrences of w as factor: as a prefix and as a suffix. Formally $\mathcal{R}_\mathcal{L}(w) = \{u \in \mathcal{A}^* \mid wu \in (\mathcal{L} \cap \mathcal{A}^* w) \setminus \mathcal{A}^+ w \mathcal{A}^+\}$. When the language $\mathcal{L}$ is clear, we will simply write $\mathcal{R}(w)$.

Morphisms. A *morphism* is a map $\varphi : \mathcal{A}^* \to \mathcal{B}^*$, with $\mathcal{A}, \mathcal{B}$ alphabets, such that $\varphi(\varepsilon) = \varepsilon$ and $\varphi(uv) = \varphi(u)\varphi(v)$ for every $u, v \in \mathcal{A}^*$. Given two distinct letters $a, b \in \mathcal{A}$, let us define the morphisms $\alpha_{a,b}, \tilde{\alpha}_{a,b} : \mathcal{A}^* \to \mathcal{A}^*$ as

$$\alpha_{a,b} = \begin{cases} a \mapsto ab \\ c \mapsto c, & c \neq a \end{cases} \qquad \text{and} \qquad \tilde{\alpha}_{a,b} = \begin{cases} a \mapsto ba \\ c \mapsto c, & c \neq a \end{cases} .$$

Permutations. When the letters of the alphabet are indexed $\{a_1 < \ldots < a_d\}$, we identify $S_\mathcal{A}$ with S_d and write $a_{\pi(d)}$ instead of $\pi(a_d)$. To describe a permutation $\pi \in S_\mathcal{A}$, we will use either the one-line notation or the cyclic one. For instance, the symmetric permutation defined by $a_{\pi(i)} = a_{d-i+1}$ for every $1 \leq i \leq d$ will be denoted as either $(a_d, a_{d-1}, \ldots, a_1)$ or as the composition of the 2-cycles $(a_1 a_d)(a_2 a_{d-2}) \cdots (a_{\frac{d}{2}} a_{\frac{d}{2}+1})$ if d is even (if d is odd, the last cycle is replaced by $(a_{\frac{d+1}{2}})$). A permutation is *circular* if it has only one cycle. It is *reducible* if $\{a_1 < \ldots < a_k\}$ is invariant under π for $1 \leq k < d$.

Burrows-Wheeler Transform. The Burrows-Wheeler transform of a word $w \in \mathcal{A}^*$ is the word $\mathrm{bwt}_\mathcal{A}(w)$ obtained by concatenating the last (not necessarily distinct) letters of the $|w|$ conjugates of w, sorted lexicographically on $\mathcal{A}$.

Example 1. Consider the word $w = \texttt{sphynx}$ on the standard ordered English alphabet $\mathcal{E} = \{\texttt{a} < \texttt{b} < \ldots < \texttt{z}\}$. Then $\mathrm{bwt}_\mathcal{E}(w) = \texttt{pysxnh}$.

The following results are well known (see, e.g., [4,15,16]).

Proposition 1. *Two words u, v over the same ordered alphabet $\mathcal{A}$ are conjugate if and only if* $\mathrm{bwt}_\mathcal{A}(u) = \mathrm{bwt}_\mathcal{A}(v)$

Proposition 2. *Let $u \in \mathcal{A}^*$. A word w is a conjugate of u^p if and only if* $\mathrm{bwt}_\mathcal{A}(u) = b_1 \cdots b_{|u|}$ *and* $\mathrm{bwt}_\mathcal{A}(w) = b_1^p \cdots b_{|u|}^p$, *with $b_i \in \mathcal{A}$.*

In [15] it is shown that, given an ordered alphabet $\mathcal{A}$, an extended version of the Burrows-Wheeler transform, denoted ebwt, gives a bijection between $\mathcal{A}^*$ and the multiset of Lyndon words over $\mathcal{A}$, where the conjugates, possibly of different length, are ordered using the ω-order instead of the lexicographic one: $u \leq_\omega v$ if $u^\omega \leq v^\omega$.

Example 2. Let W be the multiset $\{\mathtt{aac}, \mathtt{ab}, \mathtt{ab}\}$ of Lyndon words over $\mathcal{A} = \{\mathtt{a} < \mathtt{b} < \mathtt{c}\}$. We have $\Psi_{\mathcal{A}}(W) = (4, 2, 1)$ and $\mathrm{ebwt}_{\mathcal{A}}(W) = \mathtt{cbbaaaa}$.

Let π be a permutation on $\mathcal{A}$. A word $w \in \mathcal{A}^*$ is said to be *π-clustering* for $\mathcal{A}$ if $\mathrm{bwt}_{\mathcal{A}}(w) = a_{a_{\pi(1)}}^{k_1} \ldots a_{a_{\pi(d)}}^{k_r}$, where $k_i = |w|_{a_{\pi(i)}}$. It is *perfectly clustering* (for its alphabet) when $\pi \in S_{\mathcal{A}}$ is symmetric. The notions of clustering and perfectly clustering can be extended to multisets of words.

Example 3. Let us consider the three alphabets $\mathcal{A} = \{\mathtt{a} < \mathtt{b} < \mathtt{n}\}$, $\mathcal{A}' = \{\mathtt{a} < \mathtt{n} < \mathtt{b}\}$, and $\mathcal{A}'' = \{\mathtt{n} < \mathtt{a} < \mathtt{b}\}$. The word $w = \mathtt{banana}$ is defined over each of the three alphabets. One has $\mathrm{bwt}_{\mathcal{A}}(w) = \mathtt{nnbaaa}$, $\mathrm{bwt}_{\mathcal{A}'}(w) = \mathtt{bnnaaa}$ and $\mathrm{bwt}_{\mathcal{A}''}(w) = \mathtt{aabnna}$. So w is perfectly clustering for $\mathcal{A}$ and $\mathcal{A}'$, but not clustering for $\mathcal{A}''$.

Over a binary alphabet (perfectly) clustering words coincide with powers of Christoffel words and their conjugates ([15]). A characterization over larger alphabets in terms of factorization into palindromes is given in [13] (see also [19]).

The following result easily follows from Propositions 1 and 2.

Proposition 3. *Let $w = u^p \in \mathcal{A}^*$ with u primitive. Then w is π-clustering for $\mathcal{A}$ if and only if u is π-clustering for $\mathcal{A}$.*

3 Interval Exchanges

By an interval, we mean a left closed and right-open interval over the real line. Let $\mathcal{A}$ be an ordered alphabet of cardinality d and π a permutation over $\mathcal{A}$. An ordered partition $(I_a)_{a\in\mathcal{A}}$ of an interval I is such that I_a is to the left of I_b when $a < b$. The *d-interval exchange transformation* (or *d-IET* or just *IET* in short) T associated with a partition $(I_a)_{a\in\mathcal{A}}$ and a permutation π is the piecewise translation on $I = [\ell, r)$ defined by $T(x) = x + \tau_a$ if $x \in I_a$, where $\tau_a = \sum_{\pi^{-1}(b)<\pi^{-1}(a)} |I_b| - \sum_{b<a} |I_b|$. Let $D(T) = \{\sum_{b<a} |I_b| \mid a \in \mathcal{A}\} \setminus \{\ell\}$ denote the set of *formal discontinuities* of T.

The *orbit* of a point $x \in I$ under T is the set $\{T^k(x) \mid k \in \mathbb{Z}\}$. The IET is *periodic* if the orbit of any point $x \in I$ is finite. It is *minimal* if the orbit of any point is dense in I. In this case, the permutation π is irreducible (see, e.g., [5]). An IET is *regular* (or satisfies *Keane condition* or *i.d.o.c.*) if the orbits of the formal discontinuities are infinite and disjoint. A regular IET is minimal and aperiodic [11], while the inverse is not true (see, e.g., [2]).

A *connection* of an IET T is a triple (x, y, n) where $x \in D(T^{-1})$, $y \in D(T)$, $n \geq 0$ and $T^n(x) = y$. When $n = 0$, we call $x = y$ a *0-connection.* A regular IET has no connection.

Given an IET T on I, for each point $x \in I$, we can assign to it an infinite word $\Omega_T(x) = w_0 w_1 w_2 \cdots$ describing its orbit, setting $w_k = a$ if $T^k(x) \in I_a$. This word is called the *trajectory* of x under T. The *language* of an IET T is $\mathcal{L}(T) = \bigcup_{x\in I} \mathcal{L}(\Omega_T(x))$. When T is minimal or has only one periodic component (i.e., we have only one possible trajectory up to a shift), $\mathcal{L}(T)$ does not depend on the choice of x. Moreover, in this case $\mathcal{L}(T)$ is uniformly recurrent (see, e.g., [5]) and thus recurrent.

Given an IET T and a word $w = w_0 w_1 \cdots w_{n-1} \in \mathcal{L}(T)$, we define the interval $I_w = I_{w_0} \cap T^{-1}(I_{w_1}) \cap \cdots \cap T^{-(n-1)}(I_{w_{n-1}})$. By convention $I_\varepsilon = [\ell, r)$. For every point $x \in I_w$, the trajectory $\Omega_T(x)$ has w as a prefix.

4 Discrete Interval Exchanges

A *discrete interval exchange* (or *DIET* in short) associated with the composition $(n_1, n_2, \ldots, n_d)$ of $n = \sum_{i=1}^{d} n_i$ and permutation $\pi \in S_d$ is the map $T(k) = k + t_i$ if $\sum_{j<i} n_j < k \leq \sum_{j \leq i} n_j$, where $t_i = \sum_{\pi^{-1}(j) < \pi^{-1}(i)} n_j - \sum_{j<i} n_j$. A DIET corresponds to an IET associated with a partition $(I_a)_{a \in \mathcal{A}}$ and π, where $\mathcal{A} = \{a_1 < \ldots < a_d\}$ and $|I_{a_i}| = n_i$. Note that each component of this corresponding IET is periodic; thus, in particular, a DIET is never minimal (nor regular).

There is a strong link between clustering multisets of primitive words and DIETs. In fact, if a multiset $W \subset \mathcal{A}^*$ is π-clustering, then its Parikh vector gives a composition of $n = \sum_{w \in W} |w|$ that, along with π, defines a DIET. Similarly to IETs, we can encode the (periodic) trajectories by encoding each integer $k \in \left[\sum_{j<i} n_j, \sum_{j \leq i} n_j\right]$ by the ith letter of the alphabet.

In a symmetric way, it is possible to show that every DIET corresponds to a unique multiset of Lyndon words, with each orbit associated to a (clustering) Lyndon word.

Example 4. Let W be the multiset of Example 2. We can define a DIET T associated with the composition $(4, 2, 1)$ of 7 and the permutation $\pi = (\mathtt{c}, \mathtt{b}, \mathtt{a})$. The action of the DIET over $\{1, 2, \ldots, 7\}$ is given by $\mu = (1, 4, 7)(2, 5)(3, 6) \in S_7$ (see left of Fig. 1). Each orbit corresponds to one of the primitive words in W. For instance, the trajectory of 4 is given by $\Omega(4) = (\mathtt{aca})^\omega$, the infinite repetition of a conjugate of $\mathtt{aac}$. One can check that $I_{\mathtt{a}} = \{1, 2, 3, 4\}, I_{\mathtt{ab}} = \{2, 3\}, I_{\mathtt{aac}} = \{1\}$. The corresponding IET is shown on the right of Fig. 1.

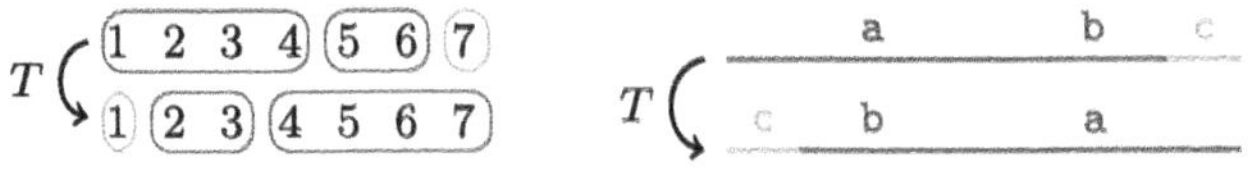

Fig. 1. A DIET (on the left) and its associated IET (on the right).

In particular, one can view every primitive π-clustering word $w \in \mathcal{A}^*$ as a DIET associated with the composition $\Psi_{\mathcal{A}}(w)$ of $|w|$ and the permutation π, the permutation μ describing the action of such a DIET being circular (see [9] for a characterization of π-clustering words in terms of trajectories in IETs or DIETs).

5 Rauzy Induction

Let $\mathcal{A} = \{a_1 < \ldots < a_d\}$ and π a permutation of $\mathcal{A}$. Let T be an IET over $[\ell, r)$ associated with $(I_a)_{a\in\mathcal{A}}$ and π. The *transformation induced* by T on a subinterval $J \subset I$ is the map $T' : J \to J$ defined by $T'(z) = T^{\nu(z)}(z)$, where $\nu(z) = \min\{n > 0 \mid T^n(z) \in J\}$ is the *first return map* of T to J. Note that $\nu(z)$ is well-defined because IETs do not have wandering intervals (see, e.g., [7]).

Rauzy induction is a procedure that associates to a regular IET T associated with an alphabet $\mathcal{A}$, a sequence of regular IETs associated with the same reordered alphabet.

The *right Rauzy step*, is the mapping ρ sending T to the induced transformation T' on $[\ell, r')$, where r' is the rightmost between the points in $D(T) \cup D(T^{-1})$. Since T is regular, it has no 0-connections and $|I_{a_d}| \neq |I_{a_{\pi(d)}}|$. Moreover, since π is irreducible we have $\pi(a_d) \neq a_d$.

If $|I_{a_d}| > |I_{a_{\pi(d)}}|$ then $\Omega(x)$ starts with $a_{\pi(d)}a_d$ for every $x \in I_{a_{\pi(d)}}$; if $|I_{a_d}| < |I_{a_{\pi(d)}}|$ then $\Omega(x)$ starts with $a_{\pi(d)}a_d$ for every $x \in T^{-1}(I_{a_d})$. We can actually give a more precise description of the obtained induced intervals: in one case the order given by the alphabet stays the same while the one given by the permutation change, while in the other case the opposite happens. Recall that we identify $S_{\mathcal{A}}$ with S_d when no confusion arises.

Lemma 1. *Let T be a regular IET associated with $\mathcal{A} = \{a_1 < \ldots < a_d\}$ and $\pi \in S_{\mathcal{A}}$. Let $h = \pi^{-1}(d)$. If $|I_{a_d}| > |I_{a_{\pi(d)}}|$, then $\rho(T)$ is the regular IET associated with $\mathcal{A}$ and $\pi' \in S_{\mathcal{A}}$ defined as*

$$\pi'(i) = \begin{cases} \pi(i) & \text{if } i \leq h \\ \pi(d) & \text{if } i = h+1 \\ \pi(i)+1 & \text{if } i > h+1 \end{cases}.$$

If $|I_{a_d}| < |I_{a_{\pi(d)}}|$, then $\rho(T)$ is the regular IET associated with

$$\mathcal{A}' = \{a_1 < \ldots < a_h < a_d < a_{h+1} < \ldots < a_{d-1}\}$$

and $\pi \in \mathcal{A}'$ defined as $\pi'(i) = \pi(i)$ for every i.

Proof. Let $S = \rho(T)$. If I_{a_d} is longer than $I_{a_{\pi(d)}}$, then the domain of S is partitioned by $(I'_a)_{a\in\mathcal{A}}$, where all $I'_a = I_a$ but I'_{a_d}, which is cut of its final part. The first return map of T into the domain of S is given by $T^2(z)$ if $z \in I_{a_{\pi(d)}}$ and $T(z)$ elsewhere. Thus, $T(I_{a_d})$ is split into $S(I_{a_d})$ and $S(I_{a_{\pi(d)}})$.

If I_{a_d} is shorter than $I_{a_{\pi(d)}}$, then S is defined as $T^2(z)$ if $z \in T^{-1}(I_{a_d})$ and $T(z)$ elsewhere. Thus, the subinterval $I_{a_{\pi(d)}}$ for T is split into $I'_{a_{\pi(d)}}$ and I'_{a_d} in the partition associated with S. The interval $S(I'_{\pi(d)})$ will remain the rightmost (even though is smaller then $T(I_{\pi(d)})$), so the permutation does not change.

The *left Rauzy step* λ is defined in a symmetric way considering the interval $[\ell', r)$, where $\ell' \neq \ell$ is the leftmost between the points in $D(T) \cup D(T^{-1})$ and the intervals considered are I_{a_1} and $I_{\pi(a_1)}$.

A symmetrical version of Lemma 1 holds.

Lemma 2. *Let T be a regular IET associated with $\mathcal{A} = \{a_1 < \ldots < a_d\}$ and $\pi \in S_\mathcal{A}$. Let $h = \pi^{-1}(1)$. If $|I_{a_1}| > |I_{a_{\pi(1)}}|$, then $\lambda(T)$ is the regular IET associated with $\mathcal{A}$ and $\pi' \in S_\mathcal{A}$ defined as*

$$\pi'(i) = \begin{cases} \pi(i) - 1 & \text{if } i < h - 1 \\ \pi(1) & \text{if } i = h - 1 \\ \pi(i) & \text{if } i \leq h \end{cases}.$$

If $|I_{a_1}| < |I_{a_{\pi(1)}}|$, then $\lambda(T)$ is the regular IET associated with

$$\mathcal{A}' = \{a_2 < \ldots < a_{h-1} < a_1 < a_h < \ldots < a_d\}$$

and $\pi \in \mathcal{A}'$ defined as $\pi'(i) = \pi(i)$ for every i.

In [5] it is shown that if T is a regular IET, then for every $w \in \mathcal{L}(T)$ the transformation induced by T on I_w is of the form $\chi(T)$, with $\chi \in \{\rho, \lambda\}^*$, where each morphism corresponds to a Rauzy step.

The following result is a consequence of Propositions 3.15, 4.12, 4.14 and Theorem 4.15 in [5].[1]

Proposition 4. *([5]) Let T be a regular IET and $w \in \mathcal{L}(T)$. The transformation induced by T on I_w is of the form $\chi(T)$, where $\chi \in \{\rho, \lambda\}^*$. Moreover, let $\chi = \chi_n \circ \cdots \circ \chi_1$. Then the morphism $\theta = \theta_1 \circ \cdots \circ \theta_n$ is an automorphism of the free group sending $\mathcal{A}$ to $\mathcal{R}(w)$, where*

$$\theta_i = \begin{cases} \alpha_{\pi(a_d^{(i)}), a_d^{(i)}} & \text{if } \chi_i = \rho \text{ and } |I_{a_d^{(i)}}| > |I_{\pi(a_d^{(i)})}| \\ \tilde{\alpha}_{a_d^{(i)}, \pi(a_d^{(i)})} & \text{if } \chi_i = \rho \text{ and } |I_{a_d^{(i)}}| < |I_{\pi(a_d^{(i)})}| \\ \alpha_{\pi(a_1^{(i)}), a_1^{(i)}} & \text{if } \chi_i = \lambda \text{ and } |I_{a_1^{(i)}}| > |I_{\pi(a_1^{(i)})}| \\ \tilde{\alpha}_{a_1^{(i)}, \pi(a_1^{(i)})} & \text{if } \chi_i = \lambda \text{ and } |I_{a_1^{(i)}}| < |I_{\pi(a_1^{(i)})}| \end{cases}$$

and $\{a_1^{(i)} < \cdots < a_d^{(i)}\}$ is the alphabet associated to $\chi_i \circ \cdots \circ \chi_1(T)$.

6 Rauzy Steps and Morphisms

In order to prove Theorem 1, we use the morphisms defined in Sect. 2 to step back from I_w to $[\ell, r)$.

Let us show that under certain additional conditions, clustering is preserved by these morphisms.

Lemma 3. *Let w be a primitive word over $\mathcal{A} = \{a_1 < \ldots < a_d\}$. Suppose w is π-clustering on $\mathcal{A}$ for some permutation π.*

[1] The result being separated in several statements in [5], the authors managed to avoid using multiple indices as it is done here.

1. *Let* $\mu \in S_{\mathcal{A}}$ *be a permutation. Then* $\mu(w) \in \mathcal{A}'^*$ *is* π'*-clustering, with* $\mathcal{A}' = \{\mu(a_1) < \ldots < \mu(a_d)\}$ *and* $\pi' \in S_{\mathcal{A}'}$.
2. *If* $b = a_1$, $\pi^{-1}(a) = a_i$ *and* $\pi^{-1}(b) = a_{i+1}$ *with* $1 \leq i < d$, *then* $\alpha_{a,b}(w) \in \mathcal{A}^*$ *is* π'*-clustering, for a certain* $\pi' \in S_{\mathcal{A}}$.
3. *If* $b = a_d$, $\pi^{-1}(b) = a_i$ *and* $\pi^{-1}(a) = a_{i+1}$ *with* $1 \leq i < d$, *then* $\alpha_{a,b}(w) \in \mathcal{A}^*$ *is* π'*-clustering, for a certain* $\pi' \in S_{\mathcal{A}}$.
4. *If* $\pi^{-1}(b) = a_1$, $a = a_i$ *and* $b = a_{i+1}$ *with* $1 \leq i < d$, *then* $\tilde{\alpha}_{a,b}(w) \in \mathcal{A}'^*$ *is* π'*-clustering, for a certain* $\pi' \in S_{\mathcal{A}'}$, *where* $\mathcal{A}' = \{a_i < a_1 < \ldots < a_{i-1} < a_{i+1} < \ldots < a_d\}$.
5. *If* $\pi^{-1}(b) = a_d$, $b = a_i$ *and* $a = a_{i+1}$ *with* $1 \leq i < d$, *then* $\tilde{\alpha}_{a,b}(w) \in \mathcal{A}'^*$ *is* π'*-clustering, for a certain* $\pi' \in S_{\mathcal{A}'}$, *where* $\mathcal{A}' = \{a_1 < \ldots < a_i < a_{i+2} < \ldots < a_d < a_{i+1}\}$.

Proof. We proceed by addressing each case.

1. Since w is π-clustering, $\mathrm{bwt}_{\mathcal{A}}(w)$ consists of contiguous blocks - possibly of length 0, if w is not pangrammatic - of each letter of $\mathcal{A}$. Since an application of μ to w simply amounts to letter renaming, we immediately obtain clustering of $\mu(w)$. Defining $\pi' = \mu \circ \pi \circ \mu^{-1} \in S_{\mathcal{A}'}$, via elementary permutation properties we see that w is π'-clustering.
2. Define $\pi' \in S_{\mathcal{A}}$ by altering π such that $\pi^{-1}(a)$ and $\pi^{-1}(b)$ are adjacent in the cycle. If $\pi^{-1}(a) = a_i$ and $\pi^{-1}(b) = a_{i+1}$, then the blocks $\pi(a)$ and $\pi(b)$ appear consecutively in $\mathrm{bwt}_{\mathcal{A}}(w)$. Replacing each a by ab in w connects the blocks $\pi(a)$ and $\pi(b)$ into a single-block adjacency in $\mathrm{bwt}_{\mathcal{A}}(\alpha_{a,b}(w))$. Thus, $\alpha_{a,b}(w)$ is π' clustering.
3. This proof is identical to that of the second argument.
4. We have $\pi(a_1) = b$. We define $\mathcal{A}' = \{a_i < a_1 < \ldots < a_{i-1} < a_{i+1} < \cdots < a_d\}$ so that a is the new smallest letter in $\mathcal{A}'$ with b appearing later. Now let $\pi' \in S_{\mathcal{A}'}$ be the permutation such that a is treated as the first letter and b follows it somewhere in the cycle. Applying $\tilde{\alpha}_{a,b}$ to w replaces every a in w with ba. In the Burrows-Wheeler transform, this forces the blocks of $\pi(a)$ and $\pi(b)$ to be contiguous. On the new order $\mathcal{A}'$ and under permutation π', we immediately obtain the clustering via elementary Burrows-Wheeler transform properties.
5. Let $\mathcal{A}' = \{a_1 < \cdots < a_i < a_{i+2} < \ldots < a_d < a_{i+1}\}$ so that a is the largest letter of $\mathcal{A}'$. We define $\pi' \in S_{\mathcal{A}'}$ so that $\pi'(a_d) = b$ and $\pi'(b)$ such that $\pi'(b)$ sits next to $\pi'(a)$. Replacing a by ba again merges the blocks of $\pi(a)$ and $\pi(b)$ contiguously in the Burrows-Wheeler transform of w over $\mathcal{A}'$. Thus, $\tilde{\alpha}_{a,b}(w)$ is π'-clustering.

We are now able to prove our main result.

Proof. (of Theorem 1) Let T be a regular IET and $w \in \mathcal{L}(T)$. By Proposition 4 there exist $\chi_1, \ldots, \chi_n \in \{\rho, \lambda\}$ such that $\chi_n \circ \cdots \circ \chi_1(T)$ is the IET induced by T on I_w. From the same proposition we obtain a morphism θ_i for each $1 \leq i \leq n$.

By Lemmata 1 and 3, θ_i sends a clustering word in $\mathcal{L}(\chi_{i-1} \circ \cdots \circ \chi_1(T))$ to a clustering word in $\mathcal{L}(\chi_i \circ \cdots \circ \chi_1(T))$. Thus, by induction on n, $\theta(u)$ is clustering for every clustering word $u \in \mathcal{L}(T)$.

Since every letter is trivially clustering and $\mathcal{R}(w) = \{\theta(a) \mid a \in \mathcal{A}\}$, we can conclude.

Example 5. Let T be the IET associated with the alphabet $\{\mathtt{a} < \mathtt{b} < \mathtt{c}\}$ and the permutation $\pi = (\mathtt{b}, \mathtt{c}, \mathtt{a})$, with $|I_\mathtt{a}| = 1 - 2\alpha$, $|I_\mathtt{b}| = |I_\mathtt{c}| = \alpha$, where $\alpha = \frac{3-\sqrt{5}}{2}$. The IET is regular, since it is just the rotation of the irrational angle α. The transformation induced to the subinterval $I_\mathtt{b}$ is $\chi(T)$, with $\chi = \lambda^2 \circ \rho^2$. We have $\theta(\{\mathtt{a}, \mathtt{b}, \mathtt{c}\}) = \mathcal{R}(\mathtt{b}) = \{\mathtt{bac}, \mathtt{b}, \mathtt{bacc}\}$, where $\theta = \alpha_{\mathtt{a},\mathtt{c}} \circ \tilde{\alpha}_{\mathtt{c},\mathtt{a}} \circ \tilde{\alpha}_{\mathtt{a},\mathtt{b}} \circ \tilde{\alpha}_{\mathtt{c},\mathtt{b}}$ (Fig. 2).

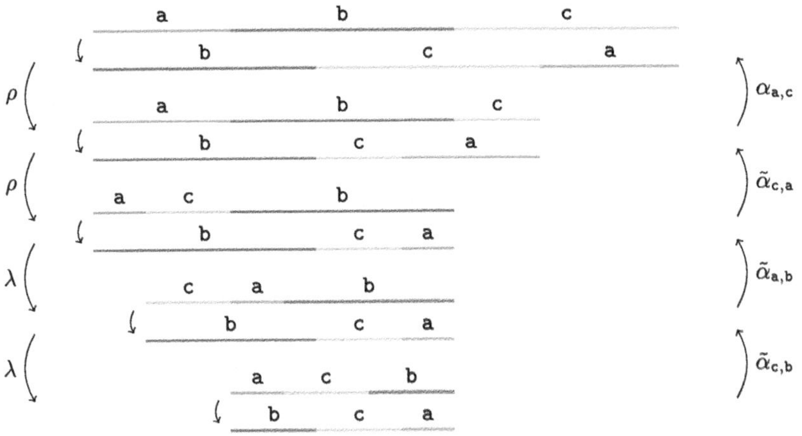

Fig. 2. Series of Rauzy steps and their associated morphisms.

7 See the Forest for the IETs

To conclude our study of clustering words in languages generated by interval exchanges, we now slightly vary our trajectory to discuss the connection between IETs (resp. DIETs) and dendricity.

Let $\mathcal{L} \subset \mathcal{A}^*$ be a language. The *extension graph* $\mathcal{G}(w)$ of a word $w \in \mathcal{L}$ is the undirected bipartite graph having as vertices the disjoint union of $L(w) = \{a \in \mathcal{A} \mid aw \in \mathcal{L}\}$ and $R(w) = \{b \in \mathcal{A} \mid wb \in \mathcal{L}\}$, and edges $B(w) = \{(a, b) \in \mathcal{A}^2 \mid awb \in \mathcal{L}\}$. The graph $\mathcal{G}(w)$ is *compatible* with two orders $<_1$ and $<_2$ on $\mathcal{A}$ if for every $(a, b), (c, d) \in B(w)$ one has the implication $a <_1 c \Longrightarrow b \leq_2 d$.

A language $\mathcal{L}$ is said to be *dendric* if the extension graph of every $w \in \mathcal{L}$ is a tree, i.e., acyclic and connected (whence the original name *tree set* in [1]). Following the same hellenophilic spirit, we call a language *alsinic* if the extension graph of every word in it is a forest, i.e., acyclic but not necessarily connected. A language $\mathcal{L}$ is *ordered dendric* (resp. *ordered alsinic*) for two orders $<_1$ and $<_2$ if every $\mathcal{G}(w)$, with $w \in \mathcal{L}$, is compatible for $<_1$ and $<_2$ (in [2] the term *planar tree* was used since the edges do not cross). Examples of dendric but not ordered dendric languages are given by Arnoux-Rauzy words on more than two letters [2].

Ordered alsinic languages are strictly linked to IETs.

Theorem 2. *([7, 8]) $\mathcal{L}$ is the language of an IET T if and only if it is a recurrent ordered alsinic language.*

$\mathcal{L}$ is the language of a minimal IET if and only if it is an aperiodic, uniformly recurrent ordered alsinic language.

$\mathcal{L}$ is the language of a regular IET if and only if it is a uniformly recurrent ordered dendric set.

As seen above, clustering words are associated with DIETs, and these can be seen as IETs where intervals have integer lengths. The following result from [6] make this link explicit. Let us denote by $<_{\mathcal{A}}$ the order on the alphabet $\mathcal{A}$ and by $<_\pi$ the order given by $a <_\pi b$ when $\pi^{-1}(a) <_{\mathcal{A}} \pi^{-1}(b)$.

Theorem 3. *([6]) A word $w \in \mathcal{A}^*$ is π-clustering if and only if for every bispecial word $v \in \mathcal{L}(w^\omega)$, the graph $\mathcal{G}(v)$ is compatible with the orders $<_\pi$ and $<_{\mathcal{A}}$.*

The following result easily follows.

Corollary 1. *A word $w \in \mathcal{A}^*$ is π-clustering if and only if $\mathcal{L}(w^\omega)$ is ordered alsinic for the orders $<_\pi$ and by $<_{\mathcal{A}}$.*

Proof. If $u \in \mathcal{L}(w^\omega)$ is not left special (resp., not right special) then $\mathcal{G}(u)$ is a tree with only one vertex to the left (resp., to the right). It is thus possible to order the vertices to the right (resp. to the left) using $<_{\mathcal{A}}$ (resp. $<_\pi$).

Following the same argument seen in Sect. 3, Theorem 3 can be generalized to multiset of words.

Example 6. Let W and T as in Example 4. Then $\mathsf{c} <_\pi \mathsf{b} <_\pi \mathsf{a}$ and $\mathsf{a} <_{\mathcal{A}} \mathsf{b} <_{\mathcal{A}} \mathsf{c}$. The extension graphs of the empty word and the letter a are shown in Fig. 3. It is easy to show that $\mathcal{G}(w)$ contains only one edge for every $w \in \mathcal{L} \setminus \{\varepsilon, \mathsf{a}\}$.

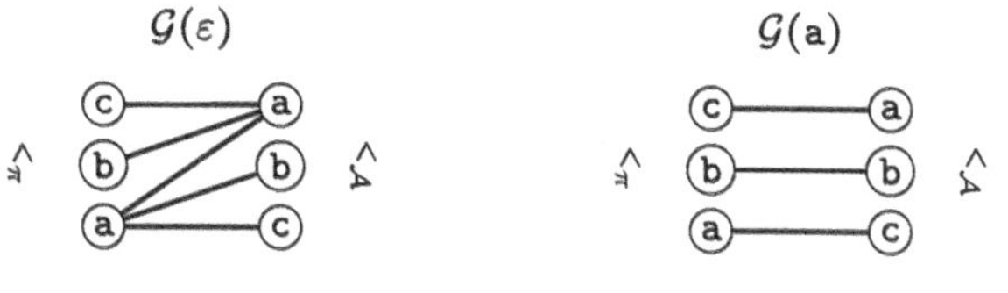

Fig. 3. Extension graphs of ε and a in $\mathcal{L}(T)$, with T as in Example 6.

Proposition 5. *If a multiset $W \subset \mathcal{A}^*$ is π-clustering, then every $w \in W$ is π_w-clustering, with π_w the restriction of π to the letters appearing in w.*

Proof. The result easily follows by considering the DIET associated with W and noticing that each word $w \in W$ corresponds to exactly one orbit of the DIET.

Note that the opposite of Proposition 5 is not true.

Example 7. Let $\mathcal{A} = \{\mathtt{a} < \mathtt{b}\}$. The words $w_1 = \mathtt{ab}, w_2 = \mathtt{aab}$ are π-clustering with $\pi = (\mathtt{ba})$. The multiset $W = \{w_1, w_2\}$ is not clustering since $\text{ebwt}_{\mathcal{A}}(W) =$ babaa. Note also that $\mathcal{L}(W^\omega) = \mathcal{L}(w_1^\omega) \cup \mathcal{L}(w_2^\omega)$ is not ordered dendric as one can easily check by considering $\mathcal{G}(\varepsilon)$, $\mathcal{G}(\mathtt{a})$ and $\mathcal{G}(\mathtt{aba})$ (see Fig. 4).

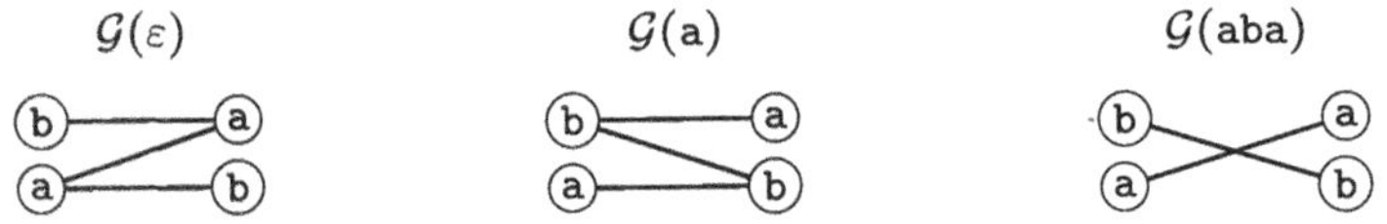

Fig. 4. Extension graphs of ε, a and aba as in Example 7.

8 Concluding Remarks

Our approach not only leverages the deep combinatorial structure inherent in Rauzy induction, but also sets the stage for potential generalizations to broader classes of interval exchange transformations in future research.

The question of whether such return words are perfectly clustering when the interval exchange transformation is symmetric remains open. More generally, are return words in languages of IETs associated with a permutation π necessarily π-clustering? It is reasonable to anticipate that the techniques developed here could aid in answering these questions in future contributions.

References

1. Berthé, V., et al.: Acyclic, connected and tree sets. Monatshefte für Mathematik **176**(4), 521–550 (2015)
2. Berthé, V., et al.: Bifix codes and interval exchanges. J. Pure Appl. Algebra **219**(7), 2781–2798 (2015)
3. Burrows, M., Wheeler, D.J.: A block-sorting lossless data compression algorithm. Syst. Res. Center - Res. Rep, 24 (1994)
4. Crochemore, M., Désarmenien, J., Perrin, D.: A note on the BurrowsWheeler transformation. Theor. Comput. Sci. **332**, 567–572 (2005)
5. Dolce, F., Perrin, D.: Interval exchanges, admissibility and branching Rauzy induction. RAIRO - Theor. Inform. Appl. **51**(3), 141–166 (2017)
6. Ferenczi, S., Hubert, P., Zamboni, L.Q.: Order conditions for languages. In: Frid, A., Mercaş, R. (eds) International Conference on Combinatorics on Words, Lecture Notes in Computer Science, pp. 155–167. Springer, Cham (2023). https://doi.org/10.1007/978-3-031-33180-0_12

7. Ferenczi, S., Hubert, P., Zamboni, L.Q.: Languages of general interval exchange transformations. Annali della Scuola Normale Superiore - Classe di Scienze, 33 (2024)
8. Ferenczi, S., Zamboni, L.Q.: Languages of k-interval exchange transformations. Bull. Lond. Math. Soc. **40**(3), 705–714 (2008)
9. Ferenczi, S., Zamboni, L.Q.: Clustering words and interval exchanges. J. Integer Sequences, 16 (2013)
10. Kanel-Belov, A.Y., Chernyat'ev, A.L.: Describing the set of words generated by interval exchange transformations. Commun. Algebra **38**, 2588–2605 (2010)
11. Keane, M.: Interval exchange transformations. Mathematische Zeitschrift **141**(1), 25–31 (1975)
12. Lapointe, M.: Perfectly clustering words are primitive positive elements of the free group. In: Combinatorics on Words, Lecture Notes in Computer Science, pp. 117–128. Springer International Publishing, Cham (2021). https://doi.org/10.1007/978-3-030-85088-3_10
13. Lapointe, M., Reutenauer, C.: Characterizations of perfectly clustering words (2024). https://arxiv.org/pdf/2407.19140.pdf
14. Lothaire, M.: Algebraic combinatorics on words. Encyclopedia of Mathematics and its Applications, vol. 90. Cambridge University Press, Cambridge (2002)
15. Mantaci, S., Restivo, A., Rosone, G., Sciortino, M.: An extension of the Burrows-Wheeler Transform. Theor. Comput. Sci. **387**, 298–312 (2007)
16. Mantaci, S., Restivo, A., Sciortino, M.: Burrows-Wheeler transform and Sturmian words. Inf. Process. Lett. **86**(5), 241–246 (2003)
17. Oseledec, V.I.: The spectrum of ergodic automorphisms. Dokl. Akad. Nauk SSSR **168**(5), 1009–1011 (1966)
18. Rauzy, G.: Changes d'intervalles et transformations induites. Acta Arithmetica, XXXIV(4), 315–328 (1979)
19. Simpson, J., Puglisi, S.J.: Words with simple Burrows-Wheeler transforms. Electron. J. Comb. **15**(R83) (2008)

Symmetries of Rich Sequences with Minimum Critical Exponent

Ľubomíra Dvořáková(✉) and Edita Pelantová

FNSPE Czech Technical University in Prague, Prague, Czech Republic
{lubomira.dvorakova,edita.pelantova}@fjfi.cvut.cz

Abstract. We state a conjecture on the repetition threshold of rich sequences over alphabet of any size. It is known to hold for binary and ternary alphabets. We provide two main contributions that may be helpful for the proof on larger alphabets. First we show that the ternary rich sequence with minimum critical exponent is a morphic image of a fixed point, i.e., an HD0L sequence. Second we draw attention to the fact that the rich sequences having the minimum critical exponent show a large degree of symmetry, i.e., they are G-rich with respect to a group G generated by more than one antimorphism. The notion of G-richness generalizes the notion of richness in palindromes which is based on one antimorphism, namely the reversal mapping.

Keywords: Repetition threshold · Palindrome · Sequences rich in palindromes · Antimorphism · Symmetry · G-rich sequences

1 Introduction

Droubay, Justin and Pirilo [6] showed that a word w of length n contains at most $n+1$ distinct palindromes. If this bound is reached, we say that w is *rich in palindromes*, or simply *rich*. They also proved that each factor of a rich word is again rich, i.e., the language of rich words is factorial. A sequence (infinite word) $\mathbf{w} = w_0w_1w_2\cdots$ is called *rich in palindromes* if each factor of $\mathbf{w}$ is rich. The best known sequence rich in palindromes is the Fibonacci sequence. A *palindrome*, which is the central notion in the definition of palindromic richness, may be formally defined as a word $w = w_0w_1\cdots w_{n-1}$ for which its reversal $R(w) = w_{n-1}w_{n-2}\cdots w_0$ is equal to w.

Another generalization of richness was introduced in [9], where instead of one antimorphism, a group G generated by involutive antimorphisms is considered. A word w is called a *G-palindrome* if $w = \Theta(w)$ for some involutive antimorphism $\Theta \in G$. The number of distinct G-palindromes occurring in a word w is bounded from above by a term depending on G and on the length of w. If this upper bound is attained, w is called *G-rich*. Analogously, if every factor of a sequence $\mathbf{w}$ is G-rich, the sequence is said to be *G-rich*. G-richness is well-understood on the binary alphabet, where we have only two involutive antimorphisms R and E and G is generated by both of them. On the one hand, the most famous G-rich

G. Gamard and J. Leroy (Eds.): WORDS 2025, LNCS 15729, pp. 116–127, 2025.
https://doi.org/10.1007/978-3-031-97548-6_11

sequence is the Thue-Morse sequence $\mathbf{t}$, which is however not rich in the classical sense [9]. On the other hand, every complementary symmetric Rote sequence[1] is G-rich and simultaneously rich in the classical sense.

Let us emphasize that the glory of the Thue-Morse sequence rests in the fact that this sequence has the smallest critical exponent among all binary sequences. Recall that $r \in \mathbb{R}$ is the *critical exponent* of a sequence $\mathbf{w}$ if no repetition of exponent $r' > r$ occurs in $\mathbf{u}$ and r is the largest number with this property.

Vesti in [15] suggested to study the critical exponent of sequences rich in palindromes. Baranwal and Shallit [1] focused on rich sequences over binary alphabet. They defined a rich sequence $\mathbf{u}$ with the critical exponent $2 + \frac{\sqrt{2}}{2}$ and conjectured that no other binary rich sequence has a smaller critical exponent. This conjecture was proven by Currie, Mol and Rampersad [5]. It is natural to call this value the *repetition threshold* of binary rich sequences. We explain in Sect. 4 that binary rich sequences reaching the repetition threshold are G-rich for the group G generated by R and E.

Recently, Currie, Mol and Peltomäki [4] constructed a sequence $\mathbf{z}$ over the alphabet $\{0, 1, 2\}$ and showed that it has the smallest critical exponent among all ternary rich sequences. More precisely, they showed that the repetition threshold of ternary rich sequences equals $1 + \frac{1}{3-\mu} \doteq 2.25876324$, where μ is the unique real root of the polynomial $x^3 - 2x^2 - 1$. The authors themselves noticed that the language of $\mathbf{z}$ is closed under the antimorphism S exchanging letters $1 \leftrightarrow 2$. We show that the sequence $\mathbf{z}$ is G-rich for the group G generated by R and S, see Theorem 14 in Sect. 6. Complexity of their article suggests that the task to find suitable candidates for rich sequences with the smallest critical exponent over larger alphabets will be certainly computationally expensive. Our result indicates that suitable candidates for rich sequences with minimum critical exponent might be found among G-rich sequences with a properly chosen group G containing more than one antimorphism.

Moreover, based on the results for binary and ternary alphabet and the computer experiments from [4] for quaternary alphabet, we formulate the following conjecture.

Conjecture 1. The repetition threshold for the class of d-ary rich sequences equals $1 + \frac{1}{3-\mu_d}$ for every $d \geq 2$, where $\mu_d \in (2, 3)$ is the root of the polynomial $x^d - 2x^{d-1} - 1$.

The structure of the paper is the following. In Sect. 3 we recall the results on G-rich sequences needed in the sequel. In Sect. 4 we show that binary rich sequences reaching the repetition threshold are G-rich for the group G generated by R and E. The two main contributions are proved in the last two sections: the sequence $\mathbf{z}$ over $\{0, 1, 2\}$ from [4], reaching the repetition threshold for ternary rich sequences, is a morphic image of a fixed point (Sect. 5) and $\mathbf{z}$ is G-rich for the group G generated by R and S, where S is the antimorphism exchanging

[1] A binary sequence is a complementary symmetric Rote sequence if its language is closed under letter exchange and it has factor complexity $C(n) = 2n$ for all $n \geq 1$ [13].

letters $1 \leftrightarrow 2$ (Sect. 6). In the last section, we present some further observations that support Conjecture 1.

2 Preliminaries

An *alphabet* $\mathcal{A}$ is a finite set, its elements are *letters*. A *word* u over $\mathcal{A}$ of *length* n is a finite sequence $u = u_0 u_1 \cdots u_{n-1}$ of letters $u_j \in \mathcal{A}$ for all $j \in \{0, 1, \ldots, n-1\}$. The length of u is denoted $|u|$. The set of all finite words over $\mathcal{A}$ is denoted $\mathcal{A}^*$. The set $\mathcal{A}^*$ equipped with concatenation as the operation forms a monoid with the *empty word* ε as the neutral element. Consider $u, p, s, v \in \mathcal{A}^*$ such that $u = pvs$, then the word p is called a *prefix*, the word s a *suffix* and the word v a *factor* of u. A *sequence* $\mathbf{u}$ over $\mathcal{A}$ is an infinite sequence $\mathbf{u} = u_0 u_1 u_2 \cdots$ of letters $u_j \in \mathcal{A}$ for all $j \in \mathbb{N}$. A *word* w over $\mathcal{A}$ is called a *factor* of the sequence $\mathbf{u} = u_0 u_1 u_2 \cdots$ if there exists $j \in \mathbb{N}$ such that $w = u_j u_{j+1} u_{j+2} \cdots u_{j+|w|-1}$. If $j = 0$, then w is a *prefix* of $\mathbf{u}$.

The *language* $\mathcal{L}(\mathbf{u})$ of a sequence $\mathbf{u}$ is the set of factors occurring in $\mathbf{u}$, the set of factors of length n is denoted $\mathcal{L}_n(\mathbf{u})$. The *factor complexity* of a sequence $\mathbf{u}$ is a mapping $\mathcal{C} : \mathbb{N} \to \mathbb{N}$, where

$$\mathcal{C}(n) = \#\mathcal{L}_n(\mathbf{u}) .$$

A factor w of a sequence $\mathbf{u}$ is *left special* if $iw, jw \in \mathcal{L}(\mathbf{u})$ for at least two distinct letters $i, j \in \mathcal{A}$. A *right special* factor is defined analogously.

A sequence $\mathbf{u}$ is *recurrent* if each factor of $\mathbf{u}$ has infinitely many occurrences in $\mathbf{u}$. Moreover, a recurrent sequence $\mathbf{u}$ is *uniformly recurrent* if for every $n \in \mathbb{N}$ there exists $N \in \mathbb{N}$ such that each factor of $\mathbf{u}$ of length N contains all factors of $\mathbf{u}$ of length n.

A *morphism* is a map $\psi : \mathcal{A}^* \to \mathcal{B}^*$ such that $\psi(uv) = \psi(u)\psi(v)$ for all words $u, v \in \mathcal{A}^*$. The morphism ψ can be naturally extended to a sequence $\mathbf{u} = u_0 u_1 u_2 \cdots$ over $\mathcal{A}$ by setting $\psi(\mathbf{u}) = \psi(u_0)\psi(u_1)\psi(u_2)\cdots$. If a morphism $\psi : \mathcal{A}^* \to \mathcal{A}^*$ satisfies $\psi(i) \neq \varepsilon$ for every letter $i \in \mathcal{A}$ and there exists $j \in \mathcal{A}$ and $w \in \mathcal{A}^*, w \neq \varepsilon$, such that $\psi(j) = jw$, then there exists a sequence $\mathbf{u}$ having the prefix $\psi^n(j)$ for every $n \in \mathbb{N}$, thus $\psi(\mathbf{u}) = \mathbf{u}$ and $\mathbf{u}$ is a *fixed point* of ψ. In the sequel, we use the notation $\psi^\omega(j)$. A morphism $\psi : \mathcal{A}^* \to \mathcal{A}^*$ is *primitive* if there exists $k \in \mathbb{N}$ such that $\psi^k(i)$ contains all letters of $\mathcal{A}$ for each letter $i \in \mathcal{A}$. It is known that a fixed point of a primitive morphism is uniformly recurrent [12].

An *antimorphism* is a map $\Theta : \mathcal{A}^* \to \mathcal{A}^*$ such that $\Theta(uv) = \Theta(v)\Theta(u)$ for all words $u, v \in \mathcal{A}^*$. The *reversal* mapping $R : \mathcal{A}^* \to \mathcal{A}^*$ is an antimorphism satisfying $R(i) = i$ for each letter $i \in \mathcal{A}$. The language $\mathcal{L}(\mathbf{u})$ is called *closed under* Θ if for each factor w, the word $\Theta(w)$ is also a factor of $\mathbf{u}$. A word $w \in \mathcal{A}^*$ is a Θ-*palindrome* if $w = \Theta(w)$. If a uniformly recurrent sequence $\mathbf{u}$ contains infinitely many Θ-palindromic factors, then its language $\mathcal{L}(\mathbf{u})$ is closed under Θ. The Θ-*palindromic complexity* of a sequence $\mathbf{u}$ is a mapping $\mathcal{P}_\Theta : \mathbb{N} \to \mathbb{N}$, where

$$\mathcal{P}_\Theta(n) = \#\{w \in \mathcal{L}_n(\mathbf{u}) \ : \ w = \Theta(w)\} .$$

3 Definition of Generalized Richness of Sequences

In this section, we mainly cite definitions and results from [11]. Throughout the paper, let G be a finite group of morphisms and antimorphisms on $\mathcal{A}^*$ such that G contains at least one antimorphism. We say that a sequence $\mathbf{u}$ over $\mathcal{A}$ is closed under G if, for every $\eta \in G$ and every factor w of $\mathbf{u}$, the word $\eta(w)$ is a factor of $\mathbf{u}$, too. On the set of factors of $\mathbf{u}$ we define an equivalence as usual

$$w \sim v \quad \Longleftrightarrow \quad w = \eta(v) \text{ for some } \eta \in G.$$

Then $[w]$ denotes the class of equivalence containing w, i.e., $[w]$ is the orbit of w under the action of the group G.

To recall the definition of G-richness, we need to introduce the graph of symmetries of $\mathbf{u}$.

Definition 2. Let $\mathbf{u}$ be a sequence with language closed under G and let $n \in N$. The undirected graph of symmetries of the sequence $\mathbf{u}$ of order n is $\Gamma_n(\mathbf{u}) = (V, E)$ with the set of vertices

$$V = \{[w] : w \in \mathcal{L}_n(\mathbf{u}), \quad w \ \text{ is left or right special}\}.$$

Two vertices $[w] \in V$ and $[v] \in V$ are connected by an edge $[e] \in E$ if there exists a factor $u \in [e] \subset \mathcal{L}(\mathbf{u})$ of length $> n$ such that

1. the prefix of u of length n belongs to $[w]$,
2. the suffix of u of length n belongs to $[v]$,
3. u contains no other special factor of length n (except the prefix and the suffix of length n).

Note that the role of $[w]$ and $[v]$ in the previous definition of edges is symmetric, as the language of $\mathbf{u}$ is closed under G.

Example 3. Let us construct $\Gamma_2(\mathbf{u}) = (V, E)$ for the group $G = \{I, R\}$, where $\mathbf{u}$ is the fixed point of φ defined in (4). The left and right special factors of $\mathbf{u}$ of length 2 are: $01, 10, 04, 40$, thus $V = \{[01], [04]\}$.
Factors of length > 2 satisfying items 1., 2. and 3. in Definition 2 are:

$$010, \ 10201, \ 10301, \ 040, \ 0440, \ 104, \ 401.$$

Hence the vertex $[01]$ has three loops, namely $[010], [10201], [10301]$, the vertex $[04]$ has two loops, namely $[0440]$ and $[040]$. The vertices $[04]$ and $[01]$ are connected by one edge $[104]$, see Fig. 1.

Definition 4. Let $\mathbf{u}$ be a sequence with the language closed under G. We say that $\mathbf{u}$ has Property G-*tls*(N), if for every $n \geq N$ the following two conditions are satisfied:

- the graph $\Gamma_n(\mathbf{u})$ after removing the loops is a tree;

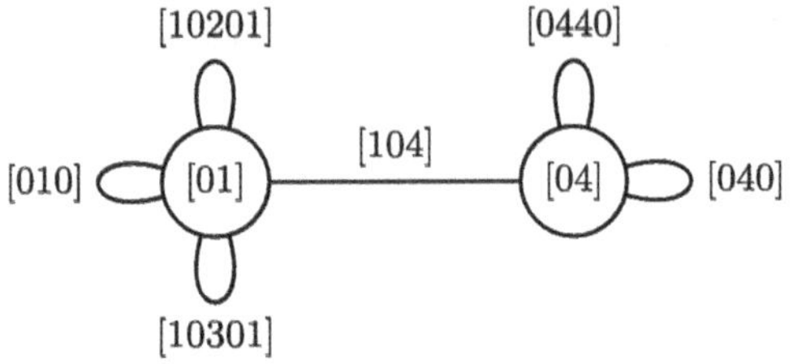

Fig. 1. The graph of symmetries $\Gamma_2(\mathbf{u})$ for $\mathbf{u} = \varphi^\omega(0)$ and the group $G = \{I, R\}$.

- if an edge $[e]$ is a loop in $\Gamma_n(\mathbf{u})$, then e is a Θ-palindrome for some antimorphism $\Theta \in G$.

Note that *tls* is an abbreviation for "tree like structure".

Definition 5. Let $\mathbf{u}$ be a sequence with the language closed under G. We say that $\mathbf{u}$ is G-rich, if $\mathbf{u}$ has Property G-$tls(1)$.

Recall that a recurrent sequence is rich if and only if it is G-rich for $G = \{I, R\}$.

Sequences rich in the classical sense may be characterized by various ways: using complete return words [8], using occurrences of the longest palindromic suffixes of words, the extensions of bispecial factors [2], etc. All these characterizations also have analogies for G-richness, see [11]. One of the tools for proving G-richness is based on the relation between palindromic and factor complexity. For the classical richness this relation was observed and proved in [3].

The set of involutive antimorphisms of G is denoted $G^{(2)}$, i.e.,

$$G^{(2)} = \{\psi \in G : \psi \text{ is an antimorphism and } \psi^2 = I\}\,.$$

Remark 6. In general, G may contain also antimorphisms that are not involutive. However, in this paper, we will deal only with groups whose antimorphisms are all involutive.

- For "classical" richness, where G contains only identity I and reversal mapping R, we have $G^{(2)} = \{R\}$.
- In Sect. 4, the considered group $G = \{I, R, E, ER\}$, where R is the reversal mapping and E is the antimorphism exchanging letters. In this case $G^{(2)} = \{R, E\}$.
- In Sect. 6, we focus on the group $G = \{I, R, S, RS\}$, where R and S are antimorphisms on the ternary alphabet $\{0, 1, 2\}$. R denotes the reversal mapping and S is defined by $S(0) = 0, S(1) = 2$ and $S(2) = 1$. In this case $G^{(2)} = \{R, S\}$.

We say that $N \in \mathbb{N}$ is G-distinguishing on $\mathbf{u}$ if, for every factor $w \in \mathcal{L}_N(\mathbf{u})$ and every pair of antimorphisms $\Theta_1, \Theta_2 \in G$, the following implication holds: $\Theta_1(w) = \Theta_2(w) \Rightarrow \Theta_1 = \Theta_2$.

Example 7. Consider the group $G = \{I, R, S, RS\}$. Any factor of length 3 occurring in $\mathbf{z}$ contains either the letter 1 or 2, see Sect. 5. Thus $R(w) \neq S(w)$ if $|w| \geq 3$ and therefore $N = 3$ is G-distinguishing on $\mathbf{z}$.

Theorem 8. *[[9]] Let* $\mathbf{u}$ *be a sequence with language closed under* G *and let* $N \in \mathbb{N}$ *be* G*-distinguishing on* $\mathbf{u}$.

1. *For every* $n \geq N$,

$$\mathcal{C}(n+1) - \mathcal{C}(n) + \#G \geq \sum_{\Theta \in G^{(2)}} \mathcal{P}_\Theta(n) + \mathcal{P}_\Theta(n+1). \qquad (1)$$

2. $\mathbf{u}$ *has Property* G*-tls*(N) *if and only if the equality is attained for all* $n \geq N$ *in* (1).

4 G-Richness of the Rich Binary Sequences with Minimum Critical Exponent

The authors of [5] showed that the language of each rich sequence with minimum critical exponent over the binary alphabet $\{0,1\}$ contains, up to the letter permutation, the language of one of the following sequences, namely $f(h^\omega(0))$ and $f(g(h^\omega(0)))$, where the morphisms $g, h : \{0,1,2\}^* \to \{0,1,2\}^*$ and $f : \{0,1,2\}^* \to \{0,1\}^*$ are defined by letter images as follows

$$f : \begin{cases} 0 \to 0 \\ 1 \to 01 \\ 2 \to 011 \end{cases} \qquad g : \begin{cases} 0 \to 011 \\ 1 \to 0121 \\ 2 \to 012121 \end{cases} \qquad h : \begin{cases} 0 \to 01 \\ 1 \to 02 \\ 2 \to 022 \end{cases}.$$

They also proved that both $f(h^\omega(0))$ and $f(g(h^\omega(0)))$ are complementary symmetric Rote sequences. Pelantová and Starosta [10] proved that every complementary symmetric Rote sequence is G-rich for the group $G = \{I, R, E, ER\}$, where E is the antimorphism defined by $E(0) = 1$ and $E(1) = 0$. Hence, the rich sequences with minimum critical exponent are G-rich.

5 The Rich Ternary Sequence z with Minimum Critical Exponent as a Morphic Image of a Fixed Point

Currie, Mol and Peltomäki [4] showed that the repetition threshold for ternary rich sequences is reached by the sequence $\mathbf{z}$, defined below. Moreover, the language of every sequence that reaches the threshold contains, up to the letter permutation, the language of $\mathbf{z}$. In the sequel, we keep notation from [4]. Let the morphisms $f, g : \{0,1,2\}^* \mapsto \{0,1,2\}^*$ be defined by letter images

$$f : \begin{cases} 0 \to 01 \\ 1 \to 022 \\ 2 \to 02 \end{cases} \quad \text{and } g : \begin{cases} 0 \to 20 \\ 1 \to 21 \\ 2 \to 2 \end{cases} .$$

Denote $\mathbf{x} = f^\omega(0)$ and $\mathbf{y} = g(\mathbf{x}) = y_0 y_1 y_2 \cdots$.

As the last step, a transducer τ is applied to $\mathbf{y}$. To describe the action of the transducer τ we denote

$$A = 00101101, \quad B = 001, \quad C = 00202202 \text{ and } D = 002. \tag{2}$$

τ maps the letters standing at even positions and odd positions in a different way, namely

$$\tau(y_{2i}) = \begin{cases} B & \text{if } y_{2i} = 0 \\ A & \text{if } y_{2i} = 1 \\ AA & \text{if } y_{2i} = 2 \end{cases} \quad \text{and} \quad \tau(y_{2i+1}) = \begin{cases} D & \text{if } y_{2i+1} = 0 \\ C & \text{if } y_{2i+1} = 1 \\ CC & \text{if } y_{2i+1} = 2 \end{cases} .$$

The sequence

$$\mathbf{z} = \tau(\mathbf{y}) = \tau\big(g(f^\omega(0))\big) \tag{3}$$

is a rich sequence with minimum critical exponent.

We will show that $\mathbf{z}$ may be obtained as a morphic image of a fixed point (also called HD0L sequence [14]). Consider two morphisms $\varphi : \{0,1,2,3,4\}^* \mapsto \{0,1,2,3,4\}^*$ and $\psi : \{0,1,2,3,4\}^* \mapsto \{0,1,2\}^*$

$$\varphi : \begin{cases} 0 \to 01 \\ 1 \to 02 \\ 2 \to 03 \\ 3 \to 04 \\ 4 \to 044 \end{cases} \quad \text{and } \psi : \begin{cases} 0 \to 0 \\ 1 \to 1 \\ 2 \to 22 \\ 3 \to 202 \\ 4 \to 20102 \end{cases} . \tag{4}$$

Denote $\mathbf{u} = \varphi^\omega(0)$, a prefix of $\mathbf{u}$ reads

$\mathbf{u} = 010201030102010401020103010201044010201030102010401020103010 2010440\cdots$

To reach our goal, i.e., to show that $\mathbf{z}$ is a morphic image of a fixed point, the following lemma is handy.

Lemma 9. *$f^n\psi = \psi\varphi^n$ for every $n \in \mathbb{N}$.*

Proof. It is easy to see that $f\psi = \psi\varphi$ as both morphisms map

$0 \mapsto 01, \quad 1 \mapsto 022, \quad 2 \mapsto 0202, \quad 3 \mapsto 020102, \quad 4 \mapsto 02010220102\,.$

Let us complete the proof by induction on $n \in \mathbb{N}$. We have just shown the statement for $n = 1$. Let $n \geq 2$. Using the induction hypothesis and the validity of the statement for $n = 1$, we get

$$f^n\psi = f(f^{n-1}\psi) = f(\psi\varphi^{n-1}) = (f\psi)\varphi^{n-1} = (\psi\varphi)\varphi^{n-1} = \psi\varphi^n\,.$$

Proposition 10. *Using words A, B, C, D from (2), we can write $\mathbf{z} = \xi(\varphi^\omega(0))$, where $\xi : \{0,1,2,3,4\}^* \to \{0,1,2\}^*$ is a morphism*

$$\xi : \begin{cases} 0 \mapsto A \\ 1 \mapsto AD \\ 2 \mapsto AC \\ 3 \mapsto ACC \\ 4 \mapsto ACCBCC \end{cases} .$$

In particular, $\mathbf{z}$ is uniformly recurrent.

Proof. By Lemma 9, $g\big(f^\omega(0)\big) = g\big(f^\omega(\psi(0))\big) = g\big(\psi\varphi^\omega(0)\big) = (g\psi)\varphi^\omega(0)$. The composition $g\psi$ maps letters of $\{0,1,2,3,4\}$ as follows

$$0 \mapsto 20, \quad 1 \mapsto 21, \quad 2 \mapsto 22, \quad 3 \mapsto 2202, \quad 4 \mapsto 22021202.$$

As $\mathbf{y} = g\big(f^\omega(0)\big) = (g\psi)\varphi^\omega(0)$ and the length of $g\psi(a)$ is even for every letter $a \in \{0,1,2,3,4\}$, we may give an explicit formula for $\tau g\psi$

$$\tau g\psi : \begin{cases} 0 \mapsto AAD \\ 1 \mapsto AAC \\ 2 \mapsto AACC \\ 3 \mapsto AACCBCC \\ 4 \mapsto AACCBCCACCBCC \end{cases} .$$

It is straightforward to check that $\tau g\psi = \xi\varphi$. Hence $\mathbf{z} = \tau(\mathbf{y}) = (\tau g\psi)\varphi^\omega(0) = (\xi\varphi)(\varphi^\omega(0)) = \xi\big(\varphi(\varphi^\omega(0)\big) = \xi\big(\varphi^\omega(0)\big)$. Uniform recurrence follows from the fact that fixed points of primitive morphisms are uniformly recurrent.

6 G-Richness of the Rich Ternary Sequence z

In the paper [4] it is proven that $\mathbf{z}$ is rich in the classical sense. Our goal is to show that $\mathbf{z}$ is also G-rich, where $G = \{I, R, S, RS\}$. Let us summarize the properties of $\mathbf{z}$ deduced in [4].

Proposition 11. *Let $\mathbf{z}$ be the sequence defined in (3). Denote by $\mathcal{C}$ and $\mathcal{P}_R$ its factor and R-palindromic complexity function, respectively. Then*

1. *$\mathcal{C}(0) = 1$, $\mathcal{C}(1) = 3$, $\mathcal{C}(2) = 7$, $\mathcal{C}(3) = 12$ and*

$$\mathcal{C}(n) = 4n + 2 \quad \text{for all } n \geq 4.$$

2. *$\mathcal{P}_R(0) = 1$, $\mathcal{P}_R(1) = \mathcal{P}_R(2) = 3$, $\mathcal{P}_R(3) = 4$ and*

$$\mathcal{P}_R(n) = \begin{cases} 2 & \text{if } n \geq 5 \text{ and } n \text{ is odd;} \\ 4 & \text{if } n \geq 4 \text{ and } n \text{ is even.} \end{cases}$$

3. *The language of $\mathbf{z}$ is closed under the antimorphism S.*

To show G-richness of $\mathbf{z}$, we have to examine S-palindromic complexity of $\mathbf{z}$. The following lemma will be helpful for this purpose.

Lemma 12. *Let w be a factor of $\varphi^\omega(0)$. If $\xi(w)00$ is an S-palindrome, then $\xi(\varphi(w))00$ is an S-palindrome, too.*

Proof. Let us prove the statement by induction on the length of $|w|$. If $w = \varepsilon$, the statement is trivial. If $|w| = 1$, then the only S-palindrome of the discussed form is $\xi(2)00 = AC00$, where A, B, C, D are given in (2), and the reader may easily check that $\xi(\varphi(2))00 = \xi(03)00 = AACC00$ is an S-palindrome, too. If $|w| = 2$, then the only S-palindrome of the discussed form is $\xi(03)00$ and the reader may again check that $\xi(\varphi(03))00 = \xi(0104)00 = AADAACCBCC00$ is an S-palindrome, too. Now, consider w of length $|w| \geq 3$. By the form of ξ and φ and since the only factors of length two of $\varphi^\omega(0)$ are $01, 02, 03, 04, 44$ and their mirror images, we observe that if $\xi(w)00$ is an S-palindrome, then there exists a non-empty factor w' of $\varphi^\omega(0)$ such that w is of one of the listed forms:

$$0104, 0102w'4, 0103w'04, 0104w'0104, 02w'3, 03w'03, 04w'0103, 2w'2, 3w'02, 4w'0102\,.$$

Let us comment on the most laborious case. If w starts in 01, then $\xi(w)$ starts in AAD. Since $\xi(w)00$ is an S-palindrome, $\xi(w)$ ends in BCC, but then w necessarily ends in 4. This implies that $\xi(w)$ starts in $AADAAC$ and ends in $ACCBCC$. We have several possibilities:

- $\xi(w) = AADAACCBCC$, which leads to $w = 0104$;
- $\xi(w) = AADAACACCBCC$, which means that $w = 01024$, but 01024 is not a factor of $\varphi^\omega(0)$;
- $\xi(w) = AADAACA\cdots CACCBCC$, which leads to the form $w = 0102w'4$ for some non-empty factor w' of $\varphi^\omega(0)$;
- $\xi(w) = AADAACCAACCBCC$, which means that $w = 010304$, but 010304 is not a factor of $\varphi^\omega(0)$;
- $\xi(w) = AADAACCA\cdots CAACCBCC$, which means that $w = 0103w'04$ for some non-empty factor w' of $\varphi^\omega(0)$;
- $\xi(w) = AADAACCBCCAADAACCBCC$, which gives $w = 01040104$, but 01040104 is not a factor of $\varphi^\omega(0)$;
- $\xi(w) = AADAACCBCC\cdots AADAACCBCC$, which means that $w = 0104w'0104$ for some non-empty factor w' of $\varphi^\omega(0)$.

In any case, it is readily seen that $\xi(w')00$ is an S-palindrome, too. Since $|w'| < |w|$, by induction assumption, $\xi(\varphi(w'))00$ is an S-palindrome. It follows then by mechanical verification that $\xi(\varphi(w))00$ is an S-palindrome, too.

Proposition 13. *Let $\mathbf{z}$ be a sequence defined in* (3). *Denote by $\mathcal{P}_S$ its S-palindromic complexity function. Then $\mathcal{P}_S(2n) \geq 2$ for every $n \geq 2$.*

Proof. According to Lemma 12, the word $x_k = \xi(\varphi^k(2))00$ for $k \in \mathbb{N}$ is an S-palindrome. It suffices to observe that $\xi(2)00 = AC00$ is an S-palindrome, where A, C are defined in (2). Moreover, x_k is obviously a factor of $\mathbf{z}$ and x_k is of even length. An S-palindrome of odd length could contain only 0 as its central factor. However none of the words 000, 102, 201 is a factor of $\mathbf{z}$. Consequently, for any $n \geq 2$, the central factor of length $2n$ of the word x_k and its reversal are distinct S-palindromes (they contain both letters 1 and 2, which guarantees their difference).

Theorem 14. *The sequence* $\mathbf{z}$ *is* G*-rich.*

Proof. By Definition 5 we have to show that for every $n \in \mathbb{N}, n \geq 1$, the graph $\Gamma_n(\mathbf{z})$ after removing loops is a tree and that every loop in $\Gamma_n(\mathbf{z})$ is an R-palindrome or an S-palindrome.

We start by showing that it is true for $n = 1$ and $n = 2$ and then we complete the proof by showing that $\mathbf{z}$ has Property G-$tls(3)$. Consider $n = 1$. All letters are left or right special factors, hence the graph has two vertices [0] and $[1] = \{1, 2\}$, there is one edge $[01] = \{01, 10, 02, 20\}$ between those two vertices, the only loop corresponding to 0 is [00] and the only loop corresponding to [1] is $[11] = \{11, 22\}$. Both loops are formed by R-palindromes. See Fig. 2 (left).

Consider $n = 2$. There are two vertices of $\Gamma_2(\mathbf{z})$: [00] and [01] and the only edge connecting them is [001]. The loops corresponding to [01] are $[010] = \{010, 020\}$, $[101] = \{101, 202\}$ and $[0110] = \{0110, 0220\}$ and they all are R-palindromes. See Fig. 2 (right).

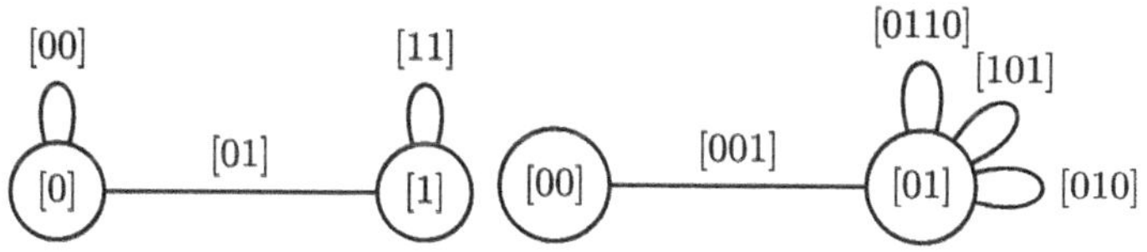

Fig. 2. The graphs of symmetries $\Gamma_1(\mathbf{z})$ (left) and $\Gamma_2(\mathbf{z})$ (right) for $\mathbf{z}$ defined in (3) and the group $G = \{I, R, S, RS\}$.

According to Example 7, the number $N = 3$ is G-distinguishing on $\mathbf{z}$. Using the second part of Theorem 8, to deduce Property G-$tls(3)$ it is necessary and sufficient to prove for every $n \geq 3$ the equality

$$\mathcal{C}(n+1) - \mathcal{C}(n) + 4 = \mathcal{P}_R(n) + \mathcal{P}_R(n+1) + \mathcal{P}_S(n) + \mathcal{P}_S(n+1). \qquad (5)$$

We will make use of Propositions 11 and 13.

Inserting $n = 3$, we have on the left hand side $\mathcal{C}(4) - \mathcal{C}(3) + 4 = 6 + 4 = 10$ and on the right hand side $\mathcal{P}_R(3) + \mathcal{P}_R(4) + \mathcal{P}_S(3) + \mathcal{P}_S(4) \geq 4 + 4 + 2 = 10$. By the first part of Theorem 8, the equality (5) follows.

If $n \geq 4$, then we have on the left hand side $\mathcal{C}(n+1) - \mathcal{C}(n) + 4 = 4 + 4 = 8$ and on the right hand side $\mathcal{P}_R(n) + \mathcal{P}_R(n+1) + \mathcal{P}_S(n) + \mathcal{P}_S(n+1) \geq 4 + 2 + 2 = 8$. Again, the first part of Theorem 8 forces the equality (5).

7 Conjecture

For every $d \in \mathbb{N}$, $d \geq 2$, let us denote by $\mathbf{v}_d$ the fixed point of the morphism φ_d defined over the alphabet $\mathcal{A}_d = \{0, 1, \ldots, 2d-2\}$ as follows

$$\varphi_d(i) = \begin{cases} 0(i{+}1) & \text{for } i = 0, 1, \ldots, 2d-3; \\ 0\,i\,i & \text{for } i = 2d-2. \end{cases}$$

In the paper [7], we studied the asymptotic critical exponent of $\mathbf{v}_d$, i.e., the value

$$E^*(\mathbf{v}_d) = \lim_{n \to \infty} \sup\{r : u^r \in \mathcal{L}(\mathbf{v}_d) \text{ and } |u| \geq n\}.$$

We have shown that $E^*(\mathbf{v}_d) = 1 + \frac{1}{3-\mu_d}$, where $\mu_d \in (2,3)$ and μ_d is the unique real root of the polynomial $x^d - 2x^{d-1} - 1$. It is interesting that the repetition threshold for binary rich sequences is attained on the sequence $\hat{\pi}(\mathbf{v}_2)$, where $\hat{\pi}$ is a morphism from $\{0,1,2\}$ to $\{0,1\}$, and its value is $E^*(\mathbf{v}_2)$. As we have shown in this paper, the repetition threshold for ternary rich sequences is attained on $\pi(\mathbf{v}_3)$, where π is a morphism from $\{0,1,2,3,4\}$ to $\{0,1,2\}$, and its value is $E^*(\mathbf{v}_3)$.

In the last section of the paper [4], the authors mention that their computer experiments suggest that the repetition threshold for 4-ary rich sequences lies between 2.117 and 2.2. Let us point out that $E^*(\mathbf{v}_4) \doteq 2.11972$. These observations somehow support Conjecture 1.

Let us repeat that a further natural question is the following one. Is the repetition threshold of rich sequences on alphabets of size larger than three reached by G-rich sequences for some groups G containing more than one antimorphism as it is the case for binary and ternary alphabet?

Disclosure of Interests. The authors have no competing interests to declare that are relevant to the content of this article.

References

1. Baranwal, A.R., Shallit, J.: Repetitions in infinite palindrome-rich words. In: Mercaş, R., Reidenbach, D. (eds.) WORDS 2019. LNCS, vol. 11682, pp. 93–105. Springer, Cham (2019). https://doi.org/10.1007/978-3-030-28796-2_7
2. Balková, L', Pelantová, E., Starosta, Š: Sturmian jungle (or garden?) on multiliteral alphabets. RAIRO - Theor. Inform. Appl. **44**, 443–470 (2010). https://doi.org/10.1051/ita/2011002
3. Bucci, M., De Luca, A., Glen, A., Zamboni, L.Q.: A connection between palindromic and factor complexity using return words. Adv. Appl. Math. **42**, 60–74 (2009). https://doi.org/10.1016/j.aam.2008.03.005
4. Currie, J. D., Mol, L., Peltomäki, J.: The repetition threshold for ternary rich words (2024). https://arxiv.org/abs/2409.12068
5. Currie, J. D., Mol, L., Rampersad, N.: The repetition threshold for binary rich words. Disc. Math. Theoret. Comput. Sci. **22**(1) (2020). https://doi.org/10.23638/DMTCS-22-1-6

6. Droubay, X., Justin, J., Pirillo, G.: Episturmian words and some constructions of de Luca and Rauzy. Theoret. Comput. Sci. **255**, 539–553 (2001). https://doi.org/10.1016/S0304-3975(99)00320-5
7. Dvořáková, L', Klouda, K., Pelantová, E.: The asymptotic repetition threshold of sequences rich in palindromes. Eur. J. Comb. **126**, 104124 (2025). https://doi.org/10.1016/j.ejc.2025.104124
8. Glen, A., Justin, J., Widmer, S., Zamboni, L.Q.: Palindromic richness. Eur. J. Combin. **30**, 510–531 (2009). https://doi.org/10.1016/j.ejc.2008.04.006
9. Pelantová, E., Starosta, Š: Languages invariant under more symmetries: overlapping factors versus palindromic richness. Discrete Math. **313**, 2432–2445 (2013). https://doi.org/10.1016/j.disc.2013.07.002
10. Pelantová, E., Starosta, Š.: Constructions of words rich in palindromes and pseudopalindromes. DMTCS **18**(3) (2016). https://doi.org/10.46298/dmtcs.655
11. Pelantová, E., Starosta, Š: Palindromic richness for languages invariant under more symmetries. Theoret. Comput. Sci. **518**, 42–63 (2014). https://doi.org/10.1016/j.tcs.2013.07.021
12. Queffélec, M.: Substitution dynamical systems - Spectral analysis. Lecture Notes in Math, vol. 1294. Springer, Berlin, Heidelberg (1987). https://doi.org/10.1007/978-3-642-11212-6
13. Rote, G.: Sequences with subword complexity $2n$. Journal of Number Theory **46**, 196–213 (1994). https://doi.org/10.1006/jnth.1994.1012
14. Rozenberg, G., Salomaa, A.: The mathematical theory of L systems. In: Tou, J.T. (eds) Advances in Information Systems Science. Springer, Boston, MA (1976). https://doi.org/10.1007/978-1-4615-8249-6_4
15. Vesti, J.: Extensions of rich words. Theoret. Comput. Sci. **548**, 14–24 (2014). https://doi.org/10.1016/j.tcs.2014.06.033

The Heinis Spectrum Has Non-empty Interior

Harold Erazo[1(✉)] and Carlos Gustavo Moreira[1,2]

[1] IMPA, Estrada Dona Castorina 110, Rio de Janeiro 22460-320, Brazil
{harold.erazo,gugu}@impa.br
[2] SUSTech International Center for Mathematics, Shenzhen, Guangdong, People's Republic of China

Abstract. The Heinis spectrum Ω is the set of all pairs (α_u, β_u) such that $\alpha_u = \liminf_{n\to\infty} \frac{p_u(n)}{n}$ and $\beta_u = \limsup_{n\to\infty} \frac{p_u(n)}{n}$ for some infinite word u. In this paper, we demonstrate that there exists a closed connected set with non-empty interior contained in Ω. Furthermore, every point in this set can be represented as the pair (α_u, β_u) for some recurrent word u. The construction is explicit, algorithmic in nature and is based on constructing certain "Cantor sets of integers", whose "gaps" correspond to blocks of zeros.

Keywords: Factor complexity · Heinis spectrum · Cantor set

1 Introduction

Let $u = u_1u_2\ldots$ be a infinite word over some finite alphabet $\mathcal{A}$. A *factor* or *subword* of length m of u is any word of the form $u_{i+1}u_{i+2}\ldots u_{i+m}$. Let $p_u : \mathbb{N}_{>0} \to \mathbb{N}_{>0}$ be the complexity function of u, where $p_u(m)$ is defined as the number of length m subwords of u. A broad class of words extensively studied is the class of words with *linear growth*, i.e., $p_u(m) = O(m)$. Examples of such words are Sturmian words, Arnoux-Rauzy words, the Thue-Morse word, paperfolding words, Rote words, automatic words, primitive substitutive words, etc. In this paper, we will work with words with linear complexity growth.

Define

$$\alpha_u := \liminf_{n\to\infty} \frac{p_u(n)}{n}, \quad \beta_u := \limsup_{n\to\infty} \frac{p_u(n)}{n}.$$

The Heinis spectrum is

$$\Omega := \{(\alpha_u, \beta_u) : u \in \mathcal{A}^{\mathbb{N}}, \mathcal{A} \text{ any finite alphabet}\} \subset (0, \infty]^2.$$

Recall that a word u is called *recurrent* if all factors of u occur infinitely often. A natural subset of Ω is

$$\Omega_R := \{(\alpha_u, \beta_u) : u \in \mathcal{A}^{\mathbb{N}}, u \text{ recurrent}, \mathcal{A} \text{ any finite alphabet}\}.$$

G. Gamard and J. Leroy (Eds.): WORDS 2025, LNCS 15729, pp. 128–138, 2025.
https://doi.org/10.1007/978-3-031-97548-6_12

The Heinis spectrum was probably introduced in [2, Page 244]. By the classical Morse–Hedlund theorem [9], the complexity grows at least linearly if u is not periodic. In particular, we have $1 \leq \alpha$ for any $(\alpha, \beta) \in \Omega$ with $(\alpha, \beta) \neq (0, 0)$. Clearly, $\alpha \leq \beta$ must hold. Interestingly, Heinis[1] showed that if $1 \leq \alpha \leq 2$ and $(\alpha, \beta) \in \Omega_R$, then the inequality

$$\beta \geq (3\alpha - 2)/\alpha$$

holds. This lower bound can be improved for $1 \leq \alpha \leq 2$, $(\alpha, \beta) \neq (3/2, 5/3)$ again under the additional assumption that u is recurrent [1].

It is also known [2, Theorem 4.9.4] that if $\alpha = \beta$ (i,e., when the limit exists), then $\alpha = \beta \in \mathbb{N}_{>0}$. These points lie in Ω, as they can be realized via codings of k-interval exchange transformations.

Furthermore, it has been shown [10] that the only points of Ω_R lying in the curve $(\alpha, (3\alpha - 2)/\alpha)$ where $1 \leq \alpha \leq 2$ are $(1, 1)$, $(3/2, 5/3)$, $(2, 2)$. Moreover, the point $(3/2, 5/3)$ is an isolated point in Ω_R. On the other hand, Aberkane [1] constructed a sequence of points in Ω converging to $(1, 1)$.

Some open problems about this set, which we attribute to J. Cassaigne, are:

1. Is Ω topologically closed?
2. Is the interior of Ω non-empty?
3. Is $\{1\} \times [1, \infty) \subset \Omega$?
4. Is Ω the closure of

$$\Omega_{PS} := \{(\alpha_u, \beta_u) : u \in \mathcal{A}^{\mathbb{N}}, u \text{ purely substitutive}, \mathcal{A} \text{ any finite alphabet}\}.$$

In fact, it seems that is not even known if Ω is uncountable. Our main objective is to show that Ω contains a connected closed set that has non-empty interior.

Theorem 1. *The Heinis spectrum Ω contains*

$$S = \left\{(x, y) : 1 \leq x \leq 2, \frac{3x}{x+1} \leq y \leq 2\right\}. \tag{1}$$

In particular Ω has non-empty interior. Moreover, for any $(\alpha, \beta) \in S$, there is a infinite recurrent word such that $(\alpha, \beta) = (\alpha_u, \beta_u)$.

More generally, one could consider the complexity functions of subshifts. Recently, the problem of determining which functions $f \colon \mathbb{N} \to \mathbb{N}$ can be asymptotically equivalent to the complexity function of a subshift Σ was solved in [4]. Recall that a function f and the complexity function ρ_Σ are *asymptotically equivalent* if $f(n) \leq C\rho_\Sigma(Dn)$ for some constants $C, D > 0$ and also $\rho_\Sigma(n) \leq Ef(Fn)$

[1] The lower bound from [6, Theorem 2] was originally stated for bi-infinite words; however, as noted in [5, Page 66], it also applies to right-infinite words, provided they are recurrent.

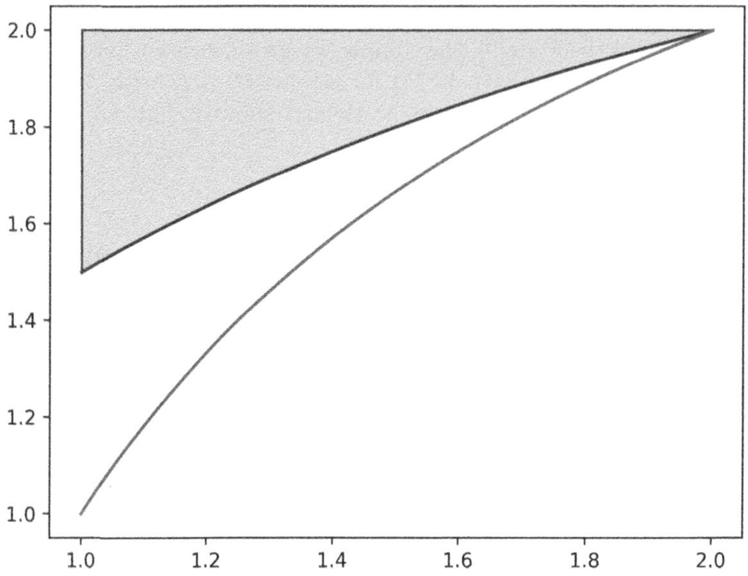

Fig. 1. The Heinis spectrum of recurrent words Ω_R contains the closed set S in blue. The blue curve is given by the function $x \mapsto 3x/(x+1)$ on the interval $[1,2]$, the top line is the line $y = 2$ and the vertical line is $x = 1$. The red curve is the graph of $x \mapsto (3x-2)/x$ which is the lower bound for Ω_R found by Heinis. (Color figure online)

for some constants $E, F > 0$. Beside characterizing such functions, they also showed that any complexity function of a subshift is asymptotically equivalent to the complexity function of a infinite recurrent word (Fig. 1).

The construction of this paper is based on the same construction of [4]. These ideas of using "positive measure Cantor sets of integers" to construct infinite words first appeared in the works [7,8] of C. Mauduit and the second author. In fact, since we are only interested in words with linear complexity, the construction of [4] is significantly simplified. However, we need to control carefully the complexity function of the words we construct, since asymptotic equivalence is not enough to control the liminf and limsup.

2 Construction

We will work exclusively with words over the binary alphabet $\{0,1\}$. The construction we present is recursive and is philosophically based on the construction of a Cantor set. Let $(n_k)_{k\geq 1}$ be a sequence of positive integers such that $n_1 = 1$ and $3n_k \leq n_{k+1}$. Define another sequence $(g_k)_{k\geq 1}$ by $g_k = n_{k+1} - 2n_k$. The sequence $(g_k)_{k\geq 1}$ is strictly increasing because $g_{k+1} = n_{k+2} - 2n_{k+1} \geq n_{k+1} > n_{k+1} - 2n_k = g_k$ and corresponds to the "sizes of gaps" in the construction. We will see that different choices of $(n_k)_{k\geq 1}$ yield various recursive words u with distinct linear complexity growth behaviors.

Let $\alpha_1 = 1$ and define recursively $\alpha_{k+1} = \alpha_k 0^{g_k} \alpha_k$ for $k \geq 1$. Since each α_k is a prefix of α_{k+1}, we can define the infinite recursive word

$$u = \lim_{k \to \infty} \alpha_k. \tag{2}$$

Note that the size of α_k is equal to $n_k = |\alpha_k|$ for all $k \geq 1$, and by construction, each word α_k begins and ends with 1.

This explicit construction of the word u defined on (2) has appeared previously in the literature [2, Page 179]. In particular, the special case where $n_{k+1} = 3n_k$ for all $k \geq 1$, yields the so-called "triadic Cantor word", studied by Anna Frid [3]. This example corresponds to a purely substitutive word, defined as the non-periodic fixed point of the morphism $0 \mapsto 000, 1 \mapsto 101$.

The asymptotic behavior of the factor complexity of u can be determined by the techniques introduced in [2]. The main novelty of this work is that we determine the **exact** complexity function of u for any sequence $(n_k)_{k \geq 1}$ with $n_k \leq 3n_{k+1}$ for all $k \geq 1$.

Although the following fact is not used in the proofs, notice that the bispecial factors of u are precisely the factors 0^k and $0^{g_k}\alpha_k 0^{g_k}$ for all $k \geq 1$ (Fig. 2).

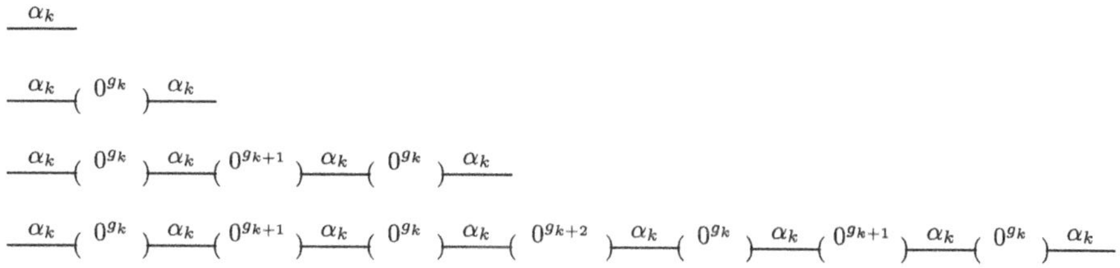

Fig. 2. "Cantor like" construction of the recursive word u.

Lemma 1. *Let $(n_k)_{k \geq 1}$ be a sequence of positive integers such that $3n_k \leq n_{k+1}$ for all $k \geq 1$ and define $g_k = n_{k+1} - 2n_k$ for all $k \geq 1$. With the convention $g_0 = n_0 = 0$, the complexity function of the word u defined in (2) is*

$$p_u(m) = \begin{cases} 2m - g_k + n_k - 1, & \text{if } \max\{n_{k-1} + 2g_{k-1}, g_k + 1\} < m \\ & \quad \leq \min\{n_k + 2g_k, g_{k+1} + 1\}; \\ m + n_k, & \text{if } n_{k-1} + 2g_{k-1} < m \leq g_k + 1; \\ 3m - g_k - n_k + 3n_{k-1} - 2, & \text{if } g_k + 1 < m \leq n_{k-1} + 2g_{k-1}. \end{cases}$$

Proof. If $n_{k-1} + 2g_{k-1} < m \leq n_k + 2g_k$, then the length-$m$ factors of $0^m \alpha_{k+1} 0^{g_k}$ starting in $0^m \alpha_k$ are all distinct: they are distinguished by the position of the first 1, the position of the first factor 1×1 of length n_{k-1} (i.e., α_{k-1}) and the position of the first 1 after this factor. Note also that the largest gap inside α_{k-1} is $0^{g_{k-2}}$ and that the sequence $(g_k)_{k \geq 1}$ is strictly increasing. On the other hand, the factors starting in $0^{g_k}\alpha_k 0^{g_k}$ have an occurrence before. We therefore have $m+n_k$ different length-m factors. The same argument applies for $1 \le m \le n_1 + 2g_1$.

The length-m factors of $0^{g_{k+1}}\alpha_{k+1}0^{g_{k+1}}$ that are not factors of $0^m\alpha_{k+1}0^{g_k}$ are precisely those of the form $v0^{1+g_k}0^a$ with v a non empty suffix of $0^{g_{k+1}}\alpha_{k+1}$ and $a = m - g_k - 1 - |v|$. In particular, they only exist if $m > g_k + 1$. They are entirely determined by $a \in [0, m - g_k - 2]$ so we have $m - g_k - 1$ such distinct factors.

Since all length-m factors appear in $0^{g_{k+1}}\alpha_{k+1}0^{g_{k+1}}$ if $m \le g_{k+1} + 1$, we conclude that:

- if $\max\{n_{k-1} + 2g_{k-1}, g_k + 1\} < m \le \min\{n_k + 2g_k, g_{k+1} + 1\}$, then $p_u(m) = 2m + n_k - g_k - 1$ as we have the two types of factors;
- if $n_{k-1} + 2g_{k-1} < m \le g_k + 1$, then $p_u(m) = m + n_k$ as we only have the first type of factors;
- if $g_k+1 < m \le n_{k-1}+2g_{k-1} (\le g_{k+1})$, then $p_u(m) = 3m - g_k + 3n_{k-1} - n_k - 2$. Indeed, we have $2m + n_{k-1} - g_{k-1} - 1$ factors in $0^m\alpha_k 0^m$, as above. Note that, since $m \le n_{k-1} + 2g_{k-1} \le n_k + 2g_k$, these factors are either in $0^{g_{k+1}}\alpha_k 0^{g_k}$ or in $0^{g_k}\alpha_k 0^{g_{k+1}}$ (and we have all such factors) so they are in $0^{g_{k+1}}\alpha_{k+1}0^{g_{k+1}}$. We are missing the factors $v_1 0^{g_k} v_2$ where v_1 is a non empty suffix of $0^{g_{k+1}}\alpha_k$ and v_2 is a non empty prefix of $\alpha_k 0^{g_{k+1}}$. They are distinct and entirely determined by $|v_1| \in [1, m - g_k - 1]$ so, in total, we have

$$\begin{aligned} 2m + n_{k-1} - g_{k-1} - 1 + m - g_k - 1 &= 3m - g_k + n_{k-1} - (n_k - 2n_{k-1}) - 2 \\ &= 3m - g_k - n_k + 3n_{k-1} - 2 \end{aligned}$$

factors.

Remark 1. Note that if $g_k + 1 \ge n_{k-1} + 2g_{k-1}$ for all $k \ge 2$, then we can write

$$p_u(m) = \begin{cases} 2m - g_{k-1} + n_{k-1} - 1, & \text{if } g_k + 1 < m \le n_k + 2g_k; \\ m + n_k, & \text{if } n_{k-1} + 2g_{k-1} < m \le g_k + 1; \end{cases}$$

This condition is satisfied for example if $4n_k \le n_{k+1}$ for all $k \ge 1$.

Denote $\lambda_k := n_k/n_{k+1}$ for all $k \ge 1$. We will now show that the range of $p_u(m)/m$ is controlled by these quotients.

Lemma 2. *Let $n_k \le m < n_{k+1}$ with $k \ge 2$.*

1. *If $g_k + 1 \ge n_{k-1} + 2g_{k-1}$, then $p_u(m)/m$ lies and attains the extremes of the interval*

$$\left[1 + \min\left\{\frac{\lambda_k}{1 - 2\lambda_k + 2n_{k+1}^{-1}}, 3\lambda_{k-1} - n_k^{-1}\right\},\right. \\ \left. 1 + \max\left\{\frac{1 - n_k^{-1}}{2 - 3\lambda_{k-1}}, \frac{1}{2 - 3\lambda_{k-1} + n_k^{-1}}, \frac{3\lambda_k - n_{k+1}^{-1}}{1 - n_{k+1}^{-1}}\right\}\right] \quad (3)$$

2. *If $g_k + 1 < n_{k-1} + 2g_{k-1}$, then $p_u(m)/m$ is increasing in $n_k \le m < n_{k+1}$. Moreover it lies and attains the extremes of the interval*

$$\left[1 + 3\lambda_{k-1} - n_k^{-1}, 1 + \frac{3\lambda_k - n_{k+1}^{-1}}{1 - n_{k+1}^{-1}},\right]. \tag{4}$$

Proof. Let us assume $g_k + 1 \ge n_{k-1} + 2g_{k-1}$ first. According to Lemma 1, the quotient $p_u(m)/m$ belongs to the following intervals:

a) If $n_k \le m \le n_{k-1} + 2g_{k-1}$, since $\max\{n_{k-2} + 2g_{k-2}, g_{k-1} + 1\} < n_k$, we have that $p_u(m) = 2m + n_{k-1} - g_{k-1} - 1$. As consequence

$$\frac{p_u(m)}{m} = \frac{2m + n_{k-1} - g_{k-1} - 1}{m} = 2 + \frac{n_{k-1} - g_{k-1} - 1}{m}.$$

Since $n_{k-1} \le g_{k-1}$, one has that $p_u(m)/m$ is increasing, so it lies in the interval

$$\left[2 + \frac{n_{k-1} - g_{k-1} - 1}{n_k}, 2 + \frac{n_{k-1} - g_{k-1} - 1}{n_{k-1} + 2g_{k-1}}\right] = \left[2 + \frac{n_{k-1} - g_{k-1} - 1}{n_k}, 1 + \frac{2n_{k-1} + g_{k-1} - 1}{n_{k-1} + 2g_{k-1}}\right]$$

Since $n_k = 2n_{k-1} + g_{k-1}$, we can write $n_{k-1} + 2g_{k-1} = 2n_k - 3n_{k-1}$ and $n_{k-1} - g_{k-1} = 3n_{k-1} - n_k$ so the interval becomes

$$\left[1 + 3\frac{n_{k-1}}{n_k} - \frac{1}{n_k}, 1 + \frac{n_k - 1}{2n_k - 3n_{k-1}}\right] = \left[1 + 3\lambda_{k-1} - n_k^{-1}, 1 + \frac{1 - n_k^{-1}}{2 - 3\lambda_{k-1}}\right]. \tag{5}$$

b) If $n_{k-1} + 2g_{k-1} < m \le g_k + 1$, then $p_u(m) = m + n_k$, so $p_u(m)/m$ is decreasing and lies in the interval

$$\left[1 + \frac{n_k}{g_k + 1}, 1 + \frac{n_k}{n_{k-1} + 2g_{k-1} + 1}\right].$$

Since $n_{k-1} + 2g_{k-1} = 2n_k - 3n_{k-1}$ and $g_k = n_{k+1} - 2n_k$, this interval can be written as

$$\left[1 + \frac{n_k}{n_{k+1} - 2n_k + 1}, 1 + \frac{n_k}{2n_k - 3n_{k-1} + 1}\right] = \left[1 + \frac{\lambda_k}{1 - 2\lambda_k + n_{k+1}^{-1}}, 1 + \frac{1}{2 - 3\lambda_{k-1} + n_k^{-1}}\right]. \tag{6}$$

c) If $g_k + 1 < m < n_{k+1}$, then $p_u(m) = 2m + n_k - g_k - 1$, hence $p_u(m)/m$ is increasing and belongs to the interval

$$\left[2 + \frac{n_k - g_k - 1}{g_k + 2}, 2 + \frac{n_k - g_k - 1}{n_{k+1} - 1}\right] = \left[1 + \frac{n_k + 1}{g_k + 2}, 1 + \frac{3n_k - 1}{n_{k+1} - 1}\right].$$

Since $g_k = n_{k+1} - 2n_k$ and $g_k = n_{k+1} - 2n_k$, this interval can be written as

$$\left[1 + \frac{\lambda_k}{1 - 2\lambda_k + 2n_{k+1}^{-1}}, 1 + \frac{3\lambda_k - n_{k+1}^{-1}}{1 - n_{k+1}^{-1}}\right]. \tag{7}$$

Taking the minimums (resp. maximums) of the left borders (resp. right borders) of the intervals (5), (6), (7) one obtains the interval (3).

Now assume $g_k + 1 < n_{k-1} + 2g_{k-1}$. According to Lemma 1, the quotient $p_u(m)/m$ belongs to the following intervals:

a) If $n_k \leq m \leq g_k + 1$, we have again that

$$\frac{p_u(m)}{m} = \frac{2m + n_{k-1} - g_{k-1} - 1}{m} = 2 + \frac{n_{k-1} - g_{k-1} - 1}{m}.$$

Since $n_{k-1} \leq g_{k-1}$, then $p_u(m)/m$ is increasing, so lies in the interval

$$\left[\frac{p_u(n_k)}{n_k}, \frac{p_u(g_k+1)}{g_k+1}\right] = \left[1 + 3\lambda_{k-1} - n_k^{-1}, \frac{p_u(g_k+1)}{g_k+1}\right]. \tag{8}$$

b) If $g_k + 1 < m \leq n_{k-1} + 2g_{k-1}$, then $p_u(m) = 3m - g_k - n_k + 3n_{k-1} - 2$ and since $n_k \geq 3n_{k-1}$, we have that $p_u(m)/m$ is also increasing. Hence $p_u(m)/m$ lies in the interval

$$\left[\frac{p_u(g_k+2)}{g_k+2}, \frac{p_u(n_{k-1}+2g_{k-1})}{n_{k-1}+2g_{k-1}}\right]. \tag{9}$$

c) If $n_{k-1} + 2g_{k-1} < m < n_{k+1}$, then $p_u(m) = 2m - g_k + n_k - 1$, so $p_u(m)/m$ is again increasing, so it lies in the interval

$$\left[\frac{p_u(n_{k-1}+2g_{k-1}+1)}{n_{k-1}+2g_{k-1}+1}, \frac{p_u(n_{k+1}-1)}{n_{k+1}-1}\right] = \left[\frac{p_u(n_{k-1}+2g_{k-1}+1)}{n_{k-1}+2g_{k-1}+1}, 1 + \frac{3\lambda_k - n_{k+1}^{-1}}{1 - n_{k+1}^{-1}}\right] \tag{10}$$

Since for $m = g_k + 1$ one has $p_u(m) = 2m + n_{k-1} - g_{k-1} - 1 = 2g_k + n_{k-1} - g_{k-1} + 1 = 3m - g_k - n_k + 3n_{k-1} - 2$, we have that $p_u(g_k+1)/(g_k+1) < p_u(g_k+2)/(g_k+2)$. Moreover, for $m = n_{k-1} + 2g_{k-1} + 1$ one has $3m - g_k - n_k + 3n_{k-1} - 2 = 2n_k - g_k + 2g_{k-1} + 1 = 2m - g_k + n_k - 1$, so $p_u(n_{k-1}+2g_{k-1})/(n_{k-1}+2g_{k-1}) < p_u(n_{k-1}+2g_{k-1}+1)/(n_{k-1}+2g_{k-1}+1)$. Note that we used again that the corresponding expressions for $p_u(m)/m$ are increasing. In particular the intervals (8), (9), (10) are *ordered*. As consequence, if $g_k < n_{k-1} + 2g_{k-1}$, the quotient $p_u(m)/m$ lies and attains the extremes of the interval

$$\left[1 + 3\lambda_{k-1} - n_k^{-1}, 1 + \frac{3\lambda_k - n_{k+1}^{-1}}{1 - n_{k+1}^{-1}}\right].$$

3 Proof of Theorem 1

Let (α, β) be such that $0 \leq \alpha \leq \beta \leq 1/3$. We will construct a sequence of positive integers $(n_k)_{k\geq 1}$ with $n_1 = 1$ and such that the quotients $\lambda_k = n_k/n_{k+1}$ satisfy the following conditions:

i. $0 \leq \lambda_k \leq 1/3$ for all $k \geq 1$.

ii. Let $V := \{k \geq 2 : \lambda_k^{-1} \geq 4 - 3\lambda_{k-1}\}$. Then $2\ell \in V$ for all $\ell \geq 1$.
iii. The following limits hold:

$$\alpha = \liminf_{k\to\infty} \lambda_k = \lim_{\ell\to\infty} \lambda_{4\ell-3} = \lim_{\ell\to\infty} \lambda_{4\ell-2}, \tag{11}$$

and

$$\beta = \limsup_{k\to\infty} \lambda_k = \lim_{\ell\to\infty} \lambda_{4\ell-1} = \lim_{\ell\to\infty} \lambda_{4\ell}. \tag{12}$$

Let $n_1 := 1$. Suppose $n_1, \ldots, n_k$, $k = 4\ell - 3$, $\ell \geq 1$ have been chosen so that the conditions i. and ii. above are satisfied for all $1 \leq i \leq k-1$. We define

$$n_{k+1} := \lceil n_k/\alpha \rceil, \qquad n_{k+2} := \lceil n_{k+1}^2/n_k \rceil,$$
$$n_{k+3} := \lceil n_{k+2}/\beta \rceil, \qquad n_{k+4} := \lceil n_{k+3}^2/n_{k+2} \rceil.$$

Let us check that the above properties hold:

i. We have $n_{k+1} \geq n_k/\alpha \geq 3n_k$, so $\lambda_k \leq 1/3$ and similarly for λ_{k+2}. Since $n_{k+2} \geq n_{k+1}^2/n_k$, one has $\lambda_{k+1} = n_{k+1}/n_{k+2} \leq n_k/n_{k+1} = \lambda_k \leq 1/3$ and similarly for λ_{k+3}.
ii. Note that if $\lambda_{i+1} \leq \lambda_i$, then $\lambda_{i+1}^{-1} \geq 4 - 3\lambda_{i+1} \geq 4 - 3\lambda_i$ because $0 \leq \lambda_{i+1} \leq 1/3$. Since $\lambda_{k+1} \leq \lambda_k$ and $\lambda_{k+3} \leq \lambda_{k+2}$, we have that $k+1, k+3 \in V$.

Once the sequence $(n_k)_{k\geq 1}$ has been constructed, we can show that the condition iii. holds. By construction, for any $k = 4\ell - 3$ where ℓ is a positive integer, one has

$$\alpha - \frac{\alpha}{n_{k+1}} < \frac{n_k}{n_{k+1}} = \lambda_k \leq \alpha, \qquad \lambda_k - \frac{\lambda_k}{n_{k+1}} < \lambda_{k+1} \leq \lambda_k,$$

and since $n_i \to \infty$ as $i \to \infty$ (because $n_{i+1} \geq 3n_i$), we conclude that

$$\alpha = \lim_{\ell\to\infty} \lambda_{4\ell-3} = \lim_{\ell\to\infty} \lambda_{4\ell-2}.$$

Analogously,

$$\beta - \frac{\beta}{n_{k+3}} < \frac{n_{k+2}}{n_{k+3}} = \lambda_{k+2} \leq \alpha, \qquad \lambda_{k+3} - \frac{\lambda_{k+3}}{n_{k+4}} < \lambda_{k+4} \leq \lambda_{k+3},$$

and since $n_i \to \infty$, we get

$$\beta = \lim_{\ell\to\infty} \lambda_{4\ell-1} = \lim_{\ell\to\infty} \lambda_{4\ell}.$$

Now that we constructed positive integers $(n_k)_{k\geq 1}$ such that $0 \leq \lambda_k := n_k/n_{k+1} \leq 1/3$ for all $k \geq 1$, we can let $g_k = n_{k+1} - 2n_k$ for all $k \geq 1$ and finally we can construct the recurrent word u given by (2). We will show that

$$\liminf_{m\to\infty} \frac{p_u(m)}{m} = 1 + \frac{\alpha}{1-2\alpha}, \quad \text{and} \quad \limsup_{m\to\infty} \frac{p_u(m)}{m} = 1 + \frac{1}{2-3\beta}.$$

Observe that the condition $g_k \geq n_{k-1} + 2g_{k-1}$ (that fits in the first case of Lemma 2) is equivalent to $n_{k+1} - 2n_k \geq 2n_k - 3n_{k-1}$ which in turn is equivalent to $\lambda_k^{-1} \geq 4 - 3\lambda_{k-1}$. In resume $g_k \geq n_{k-1} + 2g_{k-1}$ is equivalent to $k \in V$. Although the following observation is not used in the proof, note that this condition is always satisfied if $0 \leq \lambda_k \leq 1/4$, because it implies $g_k = n_{k+1} - 2n_k \geq 2n_k = 4n_{k-1} + 2g_{k-1}$ (so the choice of the sequence $(n_k)_{k\geq 1}$ and the following analysis can be simplified when $0 \leq \alpha \leq \beta \leq 1/4$).

In what follows we will use the elementary inequalities $3x(1-2x) \geq x$ and $3x(2-3x) \leq 1$ that hold for all $0 \leq x \leq 1/3$. For the infinitely many indices $k \in V$, since we are in the first case of Lemma 2, we have

$$\liminf_{\substack{m\to\infty \\ n_k \leq m < n_{k+1}, k\in V}} \frac{p_u(m)}{m} = \liminf_{k\to\infty, k\in V} \min\left\{1 + \frac{\lambda_k}{1-2\lambda_k}, 1 + 3\lambda_{k-1}\right\}.$$

Since, using (11) and the fact that V contains all even numbers, one has

$$\liminf_{k\to\infty,k\in V}\left(1 + \frac{\lambda_k}{1-2\lambda_k}\right) = 1 + \frac{1}{(\liminf_{k\to\infty,k\in V}\lambda_k)^{-1} - 2} = 1 + \frac{\alpha}{1-2\alpha} \leq 1 + 3\alpha,$$

it follows that

$$\liminf_{\substack{m\to\infty \\ n_k \leq m < n_{k+1}, k\in V}} \frac{p_u(m)}{m} = 1 + \frac{\alpha}{1-2\alpha}.$$

Analogously,

$$\limsup_{\substack{m\to\infty \\ n_k \leq m < n_{k+1}, k\in V}} \frac{p_u(m)}{m} = \limsup_{k\to\infty, k\in V} \max\left\{\frac{1}{2-3\lambda_{k-1}}, 1 + 3\lambda_k\right\}.$$

Since $1/(2-3\alpha) \leq 1/(2-3\beta)$ are the only limit points of the sequence $1/(1-2\lambda_k)$, again using that V contains all evens and (12)

$$\limsup_{k\to\infty,k\in V}\left(1 + \frac{1}{2-3\lambda_{k-1}}\right) = 1 + \frac{1}{2 - 3\limsup_{k\to\infty,k\in V}\lambda_{k-1}} = 1 + \frac{1}{2-3\beta} \geq 1 + 3\beta.$$

If the complement of V is finite then we are done. Otherwise, using the second case of Lemma 2 and (12)

$$\liminf_{\substack{m\to\infty \\ n_k \leq m < n_{k+1}, k\notin V}} \frac{p_u(m)}{m} = \liminf_{k\to\infty,k\notin V}(1 + 3\lambda_{k-1}) \geq 1 + 3\alpha \geq 1 + \frac{\alpha}{1-2\alpha},$$

and similarly

$$\limsup_{\substack{m\to\infty \\ n_k \leq m < n_{k+1}, k\notin V}} \frac{p_u(m)}{m} = \limsup_{k\to\infty,k\notin V}(1 + 3\lambda_k) \leq 1 + 3\beta \leq 1 + \frac{1}{2-3\beta}.$$

Therefore, it follows from the previous analysis that

$$\liminf_{m\to\infty}\frac{p_u(m)}{m} = 1 + \frac{\alpha}{1-2\alpha}, \quad \text{and} \quad \limsup_{m\to\infty}\frac{p_u(m)}{m} = 1 + \frac{1}{2-3\beta}.$$

Since for any choice of (α, β) such that $0 \leq \alpha \leq \beta \leq 1/3$, we can find a sequence a recurrent word $u \in \{0,1\}^{\mathbb{N}}$ that satisfies the previous equation, we conclude that the Heinis spectrum of recurrent words Ω_R contains the closed set

$$S = \left\{ \left(1 + \frac{\alpha}{1-2\alpha}, 1 + \frac{1}{2-3\beta}\right) : 0 \leq \alpha \leq \beta \leq 1/3 \right\}.$$

If we call $x = 1 + \alpha/(1-2\alpha)$ and $y = 1 + 1/(2-3\beta)$, then $\alpha = (x-1)/(2x-1)$, whence $1 \leq x \leq 2$ and $3x/(x+1) \leq y \leq 2$, so indeed S coincides with the set (1).

Remark 2. Using Lemma 2, it is easy to show that for any choice of positive integers $(n_k)_{k\geq 1}$ with $n_1 = 1$ and $3n_k \leq n_{k+1}$ for all $k \geq 1$, the corresponding pair (α_u, β_u) belongs to S. The choice of the sequence $(n_k)_{k\geq 1}$ we made above was for showing that actually one fills the whole region S.

4 Concluding Remarks

The result of this paper gives some progress on whether $\{1\} \times [1, \infty) \subset \Omega$, since we proved that $\{1\} \times [3/2, 2] \subset \Omega$. It seems that with our technique is difficult to reach the "Sturmian case" $(1, 1)$, because the words we constructed are far from being uniformly recurrent. However, we believe that using more symbols and the same construction, more regions of Ω could be studied, as for example $\{1\} \times (2, \infty)$.

As we stated in the introduction, we use a simplification of the construction done in [4] (that was in turn based on [7,8]). In fact, we just used the first case of the general construction done there. It is likely that using the full construction of [4] and a detailed case-by-case analysis, one could construct many infinite recursive words with quite different behaviors, which not need to be linear, so one could for example change $p_u(m)/n$ by $p_u(m)/n^2$ and also study the corresponding pairs of liminf and limsup.

Acknowledgments. We would like to thank the anonymous referees for their helpful comments and specially the one who simplified greatly the original proof of Lemma 1. The first author was partially supported by CAPES and FAPERJ. The second author was partially supported by CNPq and FAPERJ.

Disclosure of Interests. The authors have no competing interests to declare that are relevant to the content of this article.

References

1. Aberkane, A.: Words whose complexity satisfies lim $\frac{p(n)}{n} = 1$. Theor. Comput. Sci. **307**(1), 31–46 (2003). https://doi.org/10.1016/S0304-3975(03)00091-4
2. Cassaigne, J., Nicolas, F.: Factor complexity. In: Combinatorics, Automata and Number Theory, vol. 135, pp. 163–247. Cambridge University Press, Cambridge (2010)

3. Frid, A.E.: On the subword complexity of iteratively generated infinite words. Discrete Anal. Oper. Res. **114**(1–3), 115–120 (2001). https://doi.org/10.1016/S0166-218X(00)00364-4
4. Greenfeld, B., Moreira, C.G., Zelmanov, E.: On the complexity of subshifts and infinite words. arXiv: 2408.03403 (2024)
5. Heinis, A.: Arithmetics and combinatorics of words of low complexity. Ph. D. thesis, University of Leiden (2001)
6. Heinis, A.: The P(n)/n-function for bi-infinite words. Theor. Comput. Sci. **273**(1–2), 35–46 (2002). https://doi.org/10.1016/S0304-3975(00)00432-1
7. Mauduit, C., Moreira, C.G.: Complexity of infinite sequences with zero entropy. Acta Arith. **142**(4), 331–346 (2010). https://doi.org/10.4064/aa142-4-3
8. Mauduit, C., Moreira, C.G.: Generalized Hausdorff dimensions of sets of real numbers with zero entropy expansion. Ergodic Theory Dynam. Syst. **32**(3), 1073–1089 (2012). https://doi.org/10.1017/S0143385711000137
9. Morse, M., Hedlund, G.A.: Symbolic Dynamics. Amer. J. Math. **60**(4), 815–866 (1938). https://doi.org/10.2307/2371264
10. Turki, R.: An isolated point in the Heinis spectrum. RAIRO Theor. Inform. Appl. **50**(1), 21–38 (2016). https://doi.org/10.1051/ita/2016004

Binomial Coefficients of Multidimensional Arrays

Mehdi Golafshan and Michel Rigo(✉)

Department of Mathematics, University of Liège, Allée de la découverte 12 (B37), 4000 Liège, Belgium
{mgolafshan,M.Rigo}@uliege.be

Abstract. Motivated by Parikh matrices of picture arrays introduced in combinatorial image analysis, we propose a generalization of binomial coefficients of words to multidimensional arrays. These coefficients recursively count prescribed patterns occurring in an array. The base case is the one of binomial coefficients of words.

With our definition we extend Pascal's rule, the Chu–Vandermonde identity and therefore, the concept of Parikh matrices, in a natural way. We further present some more binomial-related identities and introduce (q,t)-deformations, i.e., multivariate polynomials whose evaluation at $(q,t) = (1,1)$ recovers the value of the classical coefficients. We explain the additional combinatorial information encoded in the coefficients of these (q,t)-polynomials compared to their integer-valued counterparts.

Keywords: Binomial coefficient of words · Parikh matrix · Combinatorial image analysis · Multidimensional array · Gaussian coefficients

1 Introduction

Combinatorial properties of multidimensional words have been thoroughly investigated by many authors: periodicity, primitiveness, Lyndon words, automaticity, etc. [6,10,14]. Much attention has been given to two-dimensional arrays in image processing. For instance, one of the main motivations for research in two-dimensional pattern matching is related to searching aerial photographs where array elements encode the color of pixels. See, for instance, [1,2].

Let A be a finite alphabet, and let $A^{m\times p}$ denote the set of $m \times p$ *arrays* (or *pictures*), i.e., functions from $\{1,\dots,m\}\times\{1,\dots,p\}$ to A. Although the concepts we develop apply to both the rows and columns, we focus on columns for concise notation; analogous definitions for rows are straightforward.

Recall that the *binomial coefficient* of two words $u = u_1 \cdots u_k$ and v (with each u_i a letter) is defined by

$$\binom{u}{v} = \#\{1 \le i_1 < \cdots < i_{|v|} \le k \mid u_{i_1} \cdots u_{i_{|v|}} = v\}. \tag{1}$$

M. Rigo—Supported by the FNRS Research grant T.196.23 (PDR).

G. Gamard and J. Leroy (Eds.): WORDS 2025, LNCS 15729, pp. 139–152, 2025.
https://doi.org/10.1007/978-3-031-97548-6_13

For further details, see, for instance, [8, Chap. 6].

Let $A = \{\mathtt{a}, \mathtt{b}\}$ and $M = (M_1 \cdots M_p)$ be an array whose columns are $M_i \in A^{m\times 1}$ for each $i \in \{1, \dots, p\}$. Recent work (see, e.g., [3,7,16]) has focused on Parikh matrices of binary picture arrays, with particular emphasis on combinatorial image analysis. These matrices capture information about the occurrence of letters and of the scattered subword ab in the array M. Precisely, the *column-Parikh matrix* associated with M is an upper-triangular matrix of the form

$$\begin{pmatrix} 1 & x & z \\ 0 & 1 & y \\ 0 & 0 & 1 \end{pmatrix} \in \mathbb{N}^{3\times 3} \quad \text{where} \quad z = \sum_{i=1}^{p} \binom{M_i^{\mathrm{T}}}{\mathtt{ab}} \quad \text{and} \quad x + y = m \cdot p.$$

Here, x (resp. y) denotes the total number of occurrences of the letter a (resp. b) in M; hence, z counts the total number of subwords ab occurring locally in the columns.

In this article, arrays of the form $m \times 1$ or $1 \times p$ are treated as conventional unidimensional words. In fact, the transpose of an $m \times 1$ array yields a word of length m. Consequently, horizontal and vertical words can be compared directly, making it legitimate to state that ab may appear as a subword of the column M_i without further notice.

Example 1. The column-Parikh matrix associated with the three arrays

$$M = \begin{pmatrix} \mathtt{a\ b\ a\ a} \\ \mathtt{b\ a\ a\ a} \\ \mathtt{a\ a\ b\ a} \\ \mathtt{a\ b\ b\ a} \end{pmatrix}, \ M' = \begin{pmatrix} \mathtt{a\ a\ a\ a} \\ \mathtt{a\ b\ b\ a} \\ \mathtt{a\ a\ b\ b} \\ \mathtt{a\ a\ a\ b} \end{pmatrix} \ \text{and} \ M'' = \begin{pmatrix} \mathtt{a\ a\ a\ a} \\ \mathtt{b\ b\ a\ a} \\ \mathtt{a\ b\ b\ a} \\ \mathtt{a\ a\ b\ a} \end{pmatrix} \ \text{is} \ \begin{pmatrix} 1 & 11 & 7 \\ 0 & 1 & 5 \\ 0 & 0 & 1 \end{pmatrix}$$

because, considering columns of M from left to right, gives

$$\binom{\mathtt{abaa}}{\mathtt{ab}} + \binom{\mathtt{baab}}{\mathtt{ab}} + \binom{\mathtt{aabb}}{\mathtt{ab}} + \binom{\mathtt{aaaa}}{\mathtt{ab}} = 1 + 2 + 4 + 0 = 7.$$

Therefore, these matrices (which could easily be extended to larger alphabets) contain information about the occurrence of specific patterns within the columns of an array. This example shows that several arrays have the same column-Parikh matrix: any permutation of columns does not affect the Parikh matrix. Replacing a column by an equivalent one also leads to the same matrix. For instance, if the fourth column of M is placed in the first position and if the column containing baab is replaced by abba which, also contains the subword ab twice and the same number of a's and b's, then the array M' has the same column-Parikh matrix as above. This is a reason why many authors are interested in M-ambiguity, i.e., an equivalence relation defined by arrays having the same Parikh matrix. This is a difficult problem of genuine interest around Parikh matrices (of unidimensional words). Thus, if one considers the problem of reconstructing an image made up of pixels, the Parikh matrix alone is insufficient to guarantee a unique reconstruction.

Classical Parikh matrices associated with unidimensional words have been generalized in [15]. With a word v over an alphabet A, for each $a \in A$, one associates a square unitriangular matrix M_a of dimension $|v|+1$ encoding the positions of a within v. For any word $u = a_1 \cdots a_{|u|} \in A^*$, compute the corresponding product of $|u|$ matrices $M_{a_1} \cdots M_{a_{|u|}}$. The upper-right element of the resulting matrix is exactly $\binom{u}{v}$. Therefore, Parikh matrices and binomial coefficients of words are intimately linked [4].

By convention, the binomial of integers $\binom{m}{p} = 0$ if $m < p$ and similarly, the binomial of words $\binom{u}{v} = 0$ whenever $|u| < |v|$. These conventions are extended to a multidimensional setting. Henceforth, it will not be necessary to specify whether one array is smaller than another.

1.1 Main Object of Study

Our aim is to introduce generalized binomial coefficients that allow counting the occurrence of patterns beyond just subwords in columns. In particular, we want to count the number of occurrences of certain (scattered) subarrays. Several definitions could be considered. To stay in line with [3,7,16], we have adopted an approach that focuses on subwords appearing in the different columns. Our choice also relies on a definition that can be recursively extended to higher dimensions. For the sake of readability, we will focus primarily on two-dimensional arrays.

Reconsider (1), which defines binomial coefficients of words. Let $u = a_1 \cdots a_k$ and $v = b_1 \cdots b_\ell$ be words where a_i and b_j are letters. We can write

$$\binom{u}{v} = \sum_{1 \leq i_1 < \cdots < i_\ell \leq k} \delta_{a_{i_1}, b_1} \cdots \delta_{a_{i_\ell}, b_\ell}.$$

The Kronecker symbol can be thought of as a zero-dimensional binomial coefficient. Roughly speaking, letters are 0-dimensional objects and words are 1-dimensional. So for any two letters a, b, $\binom{\mathsf{a}}{\mathsf{b}} = \delta_{\mathsf{a},\mathsf{b}} = 1$ if and only if $\mathsf{a} = \mathsf{b}$. Such a trivial observation permits us to introduce our object of study in a natural way. As binomial coefficients of words are built from 0-dimensional coefficients, our 2-dimensional coefficients are built from 1-dimensional ones.

Definition 1. *Let $M = (M_1 \cdots M_m)$ and $P = (P_1 \cdots P_p)$ be two-dimensional arrays. We define the* column-binomial coefficient *of the arrays M and P by*

$$\binom{M}{P} = \sum_{1 \leq i_1 < \cdots < i_p \leq m} \binom{M_{i_1}}{P_1} \cdots \binom{M_{i_p}}{P_p}$$

where on the r.h.s. we have classical binomial coefficients of (unidimensional) words, with columns M_i and P_j interpreted as words. In particular, if $m < p$ or if the number of rows of M is less than the number of rows of P, then the column-binomial coefficient is zero.

Note that in the above definition, if $m = p = 1$, then we get back the usual binomial of words.

Example 2. Consider the array M from Example 1 and the sixteen 2×2 arrays

$$\begin{pmatrix} \mathtt{a}\,\mathtt{a} \\ \mathtt{a}\,\mathtt{a} \end{pmatrix} \begin{pmatrix} \mathtt{a}\,\mathtt{a} \\ \mathtt{a}\,\mathtt{b} \end{pmatrix} \begin{pmatrix} \mathtt{a}\,\mathtt{a} \\ \mathtt{b}\,\mathtt{a} \end{pmatrix} \begin{pmatrix} \mathtt{a}\,\mathtt{a} \\ \mathtt{b}\,\mathtt{b} \end{pmatrix} \begin{pmatrix} \mathtt{a}\,\mathtt{b} \\ \mathtt{a}\,\mathtt{a} \end{pmatrix} \begin{pmatrix} \mathtt{a}\,\mathtt{b} \\ \mathtt{a}\,\mathtt{b} \end{pmatrix} \cdots \begin{pmatrix} \mathtt{b}\,\mathtt{b} \\ \mathtt{b}\,\mathtt{b} \end{pmatrix}$$

As P ranges over these sixteen arrays, the column-binomial coefficient $\binom{M}{P}$ takes the values

$$37, 22, 46, 14, 6, 7, 2, 4, 30, 20, 13, 4, 4, 6, 0, 1. \tag{2}$$

Computations show that there is exactly one other 4×4 array with the same column-binomial coefficient, the array M'' from Example 1. Unlike column-Parikh matrices, when P has two columns, permuting the columns of M alters the corresponding binomial coefficient.

Let us briefly discuss a notion of ambiguity. Out of the 2^{16} binary matrices of size 4×4, there are 49 497 pairwise distinct vectors of binomial values such as (2). There are at most 16 matrices having the same vector (and this is the single class with 16 elements). One representative is composed of four identical columns, namely $(\mathtt{a}\ \mathtt{b}\ \mathtt{b}\ \mathtt{a})^{\mathsf{T}}$. There are four equivalence classes, and each consists of 12 matrices.

Remark 1. A natural candidate for extracting subarrays is to consider submatrices A *submatrix* of a matrix is obtained by deleting any set of rows and/or columns. In this case, counting the number of occurrences of a submatrix P within a matrix M is bounded from above by our column-binomial coefficient $\binom{M}{P}$. However, choosing to count submatrices does not lead to a Pascal's rule as in Proposition 1. Indeed, the constraint of considering the same set of rows in every column is too strong. Whenever a product such as $\binom{C}{D}\binom{M}{P}$ occurs, as in (3), there is no easy way to ensure that the rows selected in C and M coincide.

Let $k \geq 1$ and M be a k-dimensional array with dimension $(d_1, \ldots, d_k)$. For each $j \in \{1, \ldots, d_k\}$, we let M_j denote the $(k-1)$-dimensional *section* (or slice) of M where the last component is set to j. Hence, M_j is a $(k-1)$-dimensional word of dimension $(d_1, \ldots, d_{k-1})$ where

$$M_j[i_1, \ldots, i_{k-1}] = M[i_1, \ldots, i_{k-1}, j].$$

In particular, if M is a 1-dimensional word, M_j denotes its jth letter and if M is a 2-dimensional array, M_j is its jth column.

Definition 2. *Let* $M \in A^{d_1 \times \cdots \times d_k}$ *and* $P \in A^{e_1 \times \cdots \times e_k}$ *be* k*-dimensional arrays. The* binomial coefficient *of* M *and* P *is given by*

$$\binom{M}{P} = \sum_{1 \leq i_1 < \cdots < i_{e_k} \leq d_k} \binom{M_{i_1}}{P_1} \cdots \binom{M_{i_{e_k}}}{P_{e_k}}$$

where, on the r.h.s., we have binomial coefficients of $(k-1)$*-dimensional section arrays.*

1.2 Our Results

One could debate our choice of Definition 1, as other counting functions could be considered. However, besides Remark 1, another argument in favor of our choice is that the classical formulas for binomials naturally extend. We list those below.

We define the *concatenation* of two arrays. Let $A \oplus B$ denote the matrix obtained by juxtaposing, side by side, the two arrays A and B that have the same number of rows. As a fundamental property, we have Pascal's rule:

Proposition 1. *Let $M \in A^{r\times m}$, $P \in A^{s\times p}$, $C \in A^{r\times 1}$ and $D \in A^{s\times 1}$.*

$$\binom{M \oplus C}{P \oplus D} = \binom{M}{P \oplus D} + \binom{C}{D}\binom{M}{P}. \tag{3}$$

This generalizes the relation for words u, v and letters $\mathtt{c}, \mathtt{d}$ [8, Prop. 6.3.2]

$$\binom{u\mathtt{c}}{v\mathtt{d}} = \binom{u}{v\mathtt{d}} + \delta_{\mathtt{c},\mathtt{d}}\binom{u}{v}.$$

It is worth noting that in (3), $\binom{C}{D}$ serves as a 1-dimensional coefficient replacing the 0-dimensional coefficient $\delta_{\mathtt{c},\mathtt{d}}$ (whereas the other coefficients are column-binomial coefficients). We also have a Chu–Vandermonde identity, [8, Cor. 6.3.7].

Proposition 2. *Let $N \in A^{r\times n}$ so that $M \oplus N$ belongs to $A^{r\times(m+n)}$*

$$\binom{M \oplus N}{P} = \sum_{P_1 \oplus P_2 = P} \binom{M}{P_1}\binom{N}{P_2}.$$

Since we have established Pascal's rule, it is not surprising that we can obtain an analogue of (generalized) Parikh matrices. The strategy is the same as in Definition 1: we replace the 0-dimensional Kronecker symbols in the classical definition [4] with 1-dimensional coefficients.

Definition 3. *Let $M = (M_1 \cdots M_m)$ and $P = (P_1 \cdots P_p)$. Let $\Psi : A^{r\times 1} \to (\mathbb{N}^{(p+1)\times(p+1)}, \cdot)$ be a morphism of monoids such that, for all $k \in \{1, \ldots, m\}$ and $i \in \{1, \ldots, p+1\}$,*

$$[\Psi(M_k)]_{i,i} = 1, \quad [\Psi(M_k)]_{i,i+1} = \binom{M_k}{P_i} \text{ for } i \le p$$

and $[\Psi(M_k)]_{i,j} = 0$ otherwise.

Hence $\Psi(M) := \Psi(M_1) \cdots \Psi(M_m)$ is a *unitriangular* matrix, i.e., an upper-triangular matrix with ones on the diagonal.

Theorem 1. *Let $M = (M_1 \cdots M_m)$ and $P = (P_1 \cdots P_p)$ be two-dimensional arrays. With the above notation, the product $\Psi(M_1) \cdots \Psi(M_m)$ is unitriangular* *matrix. For each entry above the diagonal, i.e., at position (i,j) with $i<j$, the value is given by the column-binomial coefficient*

$$\binom{M}{(P_i \cdots P_{j-1})}.$$

In particular, the upper-right element of the matrix $\Psi(M)$ is $\binom{M}{P}$.

Example 3. With the array M from Example 1 and $P = \begin{pmatrix} \mathsf{a} & \mathsf{a} \\ \mathsf{b} & \mathsf{a} \end{pmatrix}$, we obtain

$$\Psi\begin{pmatrix} \mathsf{a} \\ \mathsf{b} \\ \mathsf{a} \\ \mathsf{a} \end{pmatrix} \Psi\begin{pmatrix} \mathsf{b} \\ \mathsf{a} \\ \mathsf{a} \\ \mathsf{b} \end{pmatrix} \Psi\begin{pmatrix} \mathsf{a} \\ \mathsf{a} \\ \mathsf{b} \\ \mathsf{b} \end{pmatrix} \Psi\begin{pmatrix} \mathsf{a} \\ \mathsf{a} \\ \mathsf{a} \\ \mathsf{a} \end{pmatrix} = \begin{pmatrix} 1 & 1 & 0 \\ 0 & 1 & 3 \\ 0 & 0 & 1 \end{pmatrix}\begin{pmatrix} 1 & 2 & 0 \\ 0 & 1 & 1 \\ 0 & 0 & 1 \end{pmatrix}\begin{pmatrix} 1 & 4 & 0 \\ 0 & 1 & 1 \\ 0 & 0 & 1 \end{pmatrix}\begin{pmatrix} 1 & 0 & 0 \\ 0 & 1 & 6 \\ 0 & 0 & 1 \end{pmatrix} = \begin{pmatrix} 1 & 7 & 46 \\ 0 & 1 & 11 \\ 0 & 0 & 1 \end{pmatrix}.$$

Indeed, the entries of the matrix correspond to

$$\binom{M}{(\mathsf{a}\ \mathsf{b})^{\intercal}} = 7, \quad \binom{M}{(\mathsf{a}\ \mathsf{a})^{\intercal}} = 11 \quad \text{and} \quad \binom{M}{P} = 46.$$

In Sect. 2, we prove these results and discuss more properties. We sum the total values of the coefficients $\binom{M}{P}$ when P ranges over all arrays of a given size. We obtain a generalization of Manvel–Meyerowitz–Schwenk–Smith–Stockmeyer formula [9]. We also discuss the product $\binom{M}{P}\binom{M}{Q}$. Then in Sect. 3, we introduce multivariate deformations of these coefficients. The number of variables is equal to the dimension of the arrays. The so-called *q-analogue* of a counting function reduces to the original function when q tends to 1. For a deformation to be useful, it should preserve at least some if not all of the algebraic properties of the original function.

2 Proofs and Extra Results

The proofs of Pascal's rule, Chu–Vandermonde identity, and the structure theorem of Parikh matrices are rather straightforward. Readers familiar with this type of combinatorial identities will not be surprised. After presenting these proofs, we consider some more relations.

Proof (of Proposition 1). Let $M \in A^{r\times m}$ and $P \in A^{s\times p}$. Let $C \in A^{r\times 1}$ and $D \in A^{s\times 1}$. By definition of the column-binomial coefficient, we have

$$\binom{M \oplus C}{P \oplus D} = \sum_{1\le i_1<\cdots<i_{p+1}\le m+1} \prod_{j=1}^{p} \binom{(M\oplus C)_{i_j}}{P_j} \binom{(M\oplus C)_{i_{p+1}}}{D},$$

where $(M \oplus C)_k = M_k$ for $1 \le k \le m$ and $(M\oplus C)_{m+1} = C$. Similarly, $(P\oplus D)_j = P_j$ for $1\le j\le p$ and $(P\oplus D)_{p+1} = D$. Partitioning the sum based on whether the last column index i_{p+1} is $m+1$ yields the desired result. □

Due to space constraints, we provide only the proof of its generalization in Proposition 6, since the arguments are similar.

Proof (of Theorem 1). Define $\Phi_j = \Psi(M_1)\cdots\Psi(M_j)$. We use the same notation as in the previous proof. We show, by induction on $j \geq 1$, that for $1 \leq i < r \leq p+1$,

$$[\Phi_j]_{i,r} = \binom{M_{[1..j]}}{P_{[i..(r-1)]}}.$$

In particular, $[\Phi_j]_{i,r} = 0$ whenever $j < r-i$ because $M_{[1..j]}$ has fewer columns than $P_{[i..r-1]}$. For $j = 1$, $\Phi_1 = \Psi(M_1)$ and the conclusion follows from the definition of Ψ. For the inductive step, observe that $[\Psi(M_{j+1})]_{\ell,r}$ is nonzero only when $\ell = r$ or $\ell = r-1$. Hence the off-diagonal entry $[\Phi_{j+1}]_{i,r}$ with $i < r$, is

$$[\Phi_j]_{i,r} + [\Phi_j]_{i,r-1}\binom{M_{j+1}}{P_{r-1}} = \binom{M_{[1..j]}}{P_{[i..(r-1)]}} + \binom{M_{[1..j]}}{P_{[i..(r-2)]}}\cdot\binom{M_{j+1}}{P_{r-1}}$$

by induction hypothesis. The conclusion then follows from Proposition 1. □

Summing over all possible patterns P is equivalent to consider arrays over a unary alphabet $\{\mathtt{a}\}$. This generalizes the fact that for words, $\sum_{v\in A^k}\binom{u}{v} = \binom{|u|}{k}$.

Proposition 3. *Let $M \in A^{r\times m}$*

$$\sum_{P\in A^{s\times p}}\binom{M}{P} = \binom{\mathtt{a}^{r\times m}}{\mathtt{a}^{s\times p}} = \binom{m}{p}\cdot\binom{r}{s}^p.$$

Proof. The sum in the l.h.s. counts all possible ways to choose p columns from M and, within each chosen column, select s rows. The number of ways to select p columns is $\binom{m}{p}$. For each selected column, there are $\binom{r}{s}$ ways to choose s rows. Since the choices between columns are independent, the total number of configurations is $\binom{m}{p}\cdot\binom{r}{s}^p$. This count is independent of the entries in M because all possible subarrays P are considered. □

Applying the above proposition to Example 2, we get $\binom{4}{2}\binom{4}{2}^2 = 6^3 = 216$. This is indeed the total sum of (2). The relation can easily be extended to an arbitrary dimension. Let $k \geq 2$ and $M \in A^{d_1\times\cdots\times d_k}$. Proceeding by induction on k, we easily get

$$\sum_{P\in A^{e_1\times\cdots\times e_k}}\binom{M}{P} = \binom{d_k}{e_k}\binom{d_{k-1}}{e_{k-1}{}^{e_k}}\binom{d_{k-2}}{e_{k-2}{}^{e_k e_{k-1}}}\cdots\binom{d_1}{e_1}^{e_k e_{k-1}\cdots e_2}.$$

Similarly to the case of words, we consider the (column)-shuffle ⧢ of two arrays $P \in A^{s\times p}$, $Q \in A^{s\times q}$ having the same number s of rows. It is a formal polynomial in $\mathbb{N}\langle\langle A^{s\times 1}\rangle\rangle$. As an example, we have

$$\begin{pmatrix}\mathtt{a\ b}\\ \mathtt{a\ a}\end{pmatrix} ⧢ \begin{pmatrix}\mathtt{a\ b}\\ \mathtt{a\ b}\end{pmatrix} = \begin{pmatrix}\mathtt{a\ b\ a\ b}\\ \mathtt{a\ a\ a\ b}\end{pmatrix} + 2\begin{pmatrix}\mathtt{a\ a\ b\ b}\\ \mathtt{a\ a\ a\ b}\end{pmatrix} + 2\begin{pmatrix}\mathtt{a\ a\ b\ b}\\ \mathtt{a\ a\ b\ a}\end{pmatrix} + \begin{pmatrix}\mathtt{a\ b\ a\ b}\\ \mathtt{a\ b\ a\ a}\end{pmatrix}.$$

Note that the concatenation of the two arrays appears in the shuffle with a coefficient of at least one. If f is a polynomial or a formal series, the coefficient of an element R is denoted by $\langle f, R\rangle$. For the definition of the shuffle product on the module $\mathbb{Z}\langle\langle A\rangle\rangle$, see [8, p. 126].

Proposition 4. *Let $M \in A^{r\times m}$, $P \in A^{s\times p}$, and $Q \in A^{s\times q}$. We have:*

$$\binom{M}{P}\binom{M}{Q} = \sum_{R\in(A^{s\times 1})^*} \langle P \sqcup\!\sqcup Q, R\rangle \binom{M}{R}$$

$$+ \sum_{\substack{1\le i_1<\cdots<i_p\le m\\ 1\le j_1<\cdots<j_q\le m\\ \{i_1,\ldots,i_p\}\cap\{j_1,\ldots,j_q\}\neq\emptyset}} \binom{(M_{i_1}\cdots M_{i_p})}{P}\cdot\binom{(M_{j_1}\cdots M_{j_q})}{Q}.$$

In particular, the above result permits us to express $\binom{M}{P\oplus Q}$ occurring in the first sum of the r.h.s. in the above formula as an expression involving $\binom{M}{P}\binom{M}{Q}$ divided by $\langle P\sqcup\!\sqcup Q, P\oplus Q\rangle$. This is similar to [13, Thm. 6.4], which is used to express binomial coefficients using only Lyndon words.

Proof. By definition of the column-binomial coefficient, we have

$$\binom{M}{P}\binom{M}{Q} = \sum_{1\le i_1<\cdots<i_p\le m} \binom{M_{i_1}}{P_1}\cdots\binom{M_{i_p}}{P_p} \sum_{1\le j_1<\cdots<j_q\le m} \binom{M_{j_1}}{Q_1}\cdots\binom{M_{j_q}}{Q_q}.$$

We expand this product and focus on the $\binom{m}{p}\binom{m-p}{q}$ terms where $\{i_1,\ldots,i_p\}\cap\{j_1,\ldots,j_q\}=\emptyset$, i.e., $p+q$ pairwise distinct columns from M are selected. Since all these indices are distinct, they are linearly ordered. Each such ordering is in one-to-one correspondence with a shuffle of columns. For example, $i_1<i_2<j_1<i_3<j_2$ ($p=3$ and $q=2$) corresponds to $(P_1P_2Q_1P_3Q_2)$. Since the indices belong to $\{1,\ldots,m\}$, several choices of the $p+q$ indices lead to the same ordering. Continuing the example, if we fix one ordering, e.g., $i_1<i_2<j_1<i_3<j_2$, and consider all choices leading to this ordering, we get

$$\sum_{1\le i_1<i_2<j_1<i_3<j_2\le m} \binom{M_{i_1}}{P_1}\binom{M_{i_2}}{P_2}\binom{M_{j_1}}{Q_1}\binom{M_{i_3}}{P_3}\binom{M_{j_2}}{Q_2} = \binom{M}{(P_1P_2Q_1P_3Q_2)}.$$

Now we consider all the orderings of the $p+q$ indices and we thus see all the elements of the shuffle $P\sqcup\!\sqcup Q$. Different orderings may lead to the same resulting shuffle of columns. As an example, if $P_2=Q_1$, then $i_1<i_2<j_1<i_3<j_2$ and $i_1<j_1<i_2<i_3<j_2$ lead to the same array $(P_1P_2Q_1P_3Q_2)=(P_1Q_1P_2P_3Q_2)$. If we express the sum as a sum over these elements occurring in the shuffle, this explains the coefficient $\langle P\sqcup\!\sqcup Q, R\rangle$ in the first part of the formula. □

Note that we could refine the above formula by summing over the size of the intersection of the two sets of indices. We now extend a result from Manvel et al. [9]. See also [5].

Theorem 2. *Let $M \in A^{r\times m}$, $P \in A^{s\times p}$ and $r \geq q \geq s$, $m \geq t \geq p$. We have*

$$\sum_{T\in A^{q\times t}} \binom{M}{T}\binom{T}{P} = \binom{m-p}{t-p}\binom{r}{q}^{t-p}\binom{r-s}{q-s}^{p}\binom{M}{P}.$$

Proof. In the r.h.s., as in Proposition 3, we review all possible choices of $q \cdot t$ elements that give rise to a valid pattern T and within each such T count the occurrences of P. However, a fixed occurrence of P appears many times in this counting, as we now explain.

Fix a particular *occurrence* of $P \sqsubseteq M$. Concretely, $P \sqsubseteq M$ means selecting p columns of M (out of m) and within each column, independently select s rows (out of r), matching P. Next, to build a $q \times t$ subarray $T \sqsubseteq M$ containing that chosen copy of P:

- In each of the p columns used by P, we independently pick $q - s$ more elements (out of $r - s$ available). Hence we have $\binom{r-s}{q-s}^p$ ways to do so.
- Since P already occupies p columns, we must choose $t - p$ columns from the remaining $m - p$. There are $\binom{m-p}{t-p}$ such choices. In each of these $t - p$ columns, we independently pick q elements. Hence we have $\binom{m-p}{t-p}\binom{r}{q}^{t-p}$ ways in total. □

Corollary 1. *Let $M, M' \in A^{r\times m}$. Let q, t, s, p be such that $r \geq q \geq s$ and $m \geq t \geq p$. If $\binom{M}{T} = \binom{M'}{T}$ for all $T \in A^{q\times t}$, then $\binom{M}{P} = \binom{M'}{P}$ for all $P \in A^{s\times p}$.*

3 Deformations of these Coefficients

The q-deformations of binomial coefficients of words have been introduced in [11] as a generalization of Gaussian coefficients. As discussed in that paper, two definitions may coexist: letters can be processed either from the left or from the right. The two resulting univariate polynomials are different, but their properties are similar. Equivalently, one may consider reversals of the words. Since we index rows and columns from the upper left corner starting with index 1, we choose the recursive definition of q-binomials processed from the left, i.e., for $u, v \in A^*$ and $\mathtt{c}, \mathtt{d} \in A$

$$\binom{\mathtt{c}u}{\mathtt{d}v}_q = \binom{u}{\mathtt{d}v}_q \cdot q^{|\mathtt{d}v|} + \delta_{\mathtt{c},\mathtt{d}}\binom{u}{v}_q. \tag{4}$$

We recall a useful result in that setting, see again [11] for details.

Theorem 3. *Let u be a word over A, $k \geq 0$, and $a_1, \ldots, a_k \in A$. Then*

$$\binom{u}{a_1 \cdots a_k}_q = \sum_{\substack{u_1,u_2,\ldots,u_{k+1}\in A^* \\ u=u_1a_1\cdots u_ka_ku_{k+1}}} q^{\sum_{i=1}^{k}(k+1-i)|u_i|}.$$

As an example of classical q-binomial coefficients of words, we have

$$\binom{\mathtt{aab}}{\mathtt{ab}}_q = q^2 + q \quad \text{and} \quad \binom{\mathtt{aba}}{\mathtt{ab}}_q = 1. \tag{5}$$

For instance, the word aab can be factorized as $u_1 \mathtt{a} u_2 \mathtt{b} u_3$ with $u_1 = u_3 = \varepsilon$ and $u_2 = \mathtt{a}$, which yields the term q in the first q-binomial above (resp. $u_2 = u_3 = \varepsilon$ and $u_1 = \mathtt{a}$ giving q^2). This combinatorial interpretation is important for the next definition, because a similar idea applies to the exponent of the second variable t.

For bidimensional arrays, we consider the following definition.

Definition 4. *Let* $M = (M_1 \cdots M_m)$ *and* $P = (P_1 \cdots P_p)$ *be two-dimensional arrays. We define the* (q,t)-column-binomial coefficient *of* M *and* P *by*

$$\binom{M}{P}_{q,t} = \sum_{1 \le i_1 < \cdots < i_p \le m} t^{\sum_{j=1}^{p}(i_j - j)} \binom{M_{i_1}}{P_1}_q \cdots \binom{M_{i_p}}{P_p}_q.$$

Here, the l.h.s. we have q-binomial coefficients of words (obtained by considering the transpose of the columns).

As expected with such deformations, when q and t are equal to 1, we recover the column-binomial coefficient discussed in the first part of this paper.

Example 4. We make use of the q-binomials of the columns computed in (5). Let

$$M = \begin{pmatrix} \mathtt{a} & \mathtt{a} & \mathtt{a} \\ \mathtt{a} & \mathtt{b} & \mathtt{a} \\ \mathtt{b} & \mathtt{a} & \mathtt{b} \end{pmatrix}, \quad P = \begin{pmatrix} \mathtt{a} & \mathtt{a} \\ \mathtt{b} & \mathtt{b} \end{pmatrix}.$$

Then

$$\binom{M}{P}_{q,t} = (q^2 + q) \cdot 1 + t \cdot (q^2 + q) \cdot (q^2 + q) + t^2 \cdot 1 \cdot (q^2 + q)$$

which is equal to $q^4t + 2q^3t + q^2t^2 + q^2t + q^2 + qt^2 + q$. The three terms of the sum correspond, respectively, to the choices of columns $1,2$; $1,3$; and $2,3$.

Remark 2. If M and P both have a single row, then the polynomial $\binom{M}{P}_{q,t}$ only contains the variable t and is equal to $\binom{M}{P}_q(t)$ (i.e., replace the variable q by t). This shows the coherence of the definition. Indeed, let $M = \mathtt{m}_1 \cdots \mathtt{m}_k$ and $P = \mathtt{p}_1 \cdots \mathtt{p}_\ell$ be words. The definition reduces to

$$\binom{M}{P}_{q,t} = \sum_{1 \le i_1 < \cdots < i_\ell \le k} t^{\sum_{j=1}^{\ell}(i_j - j)} \delta_{\mathtt{m}_{i_1}, \mathtt{p}_1} \cdots \delta_{\mathtt{m}_{i_\ell}, \mathtt{p}_\ell}.$$

Observe that if P appears as a subword of M with the letters occurring in position $i_1, \ldots, i_\ell$, i.e., $M = u_1 \mathtt{p}_1 \cdots u_\ell \mathtt{p}_\ell u_{\ell+1}$ with $i_j = \sum_{i=1}^{j} |u_i| + j$ for all j, then

$$\sum_{j=1}^{\ell} (i_j - j) = \sum_{j=1}^{\ell} \sum_{i=1}^{j} |u_i| = \sum_{i=1}^{\ell} (\ell + 1 - i) |u_i|.$$

We conclude with Theorem 3.

If M and P both have a single column, then the polynomial $\binom{M}{P}_{q,t}$ only contains the variable q and it is trivially equal to $\binom{M}{P}_q$.

For classical binomial coefficients of words, the *reconstruction problem* is fundamental. Given some binomial coefficients associated with a word u, one asks whether u can be uniquely reconstructed i.e., whether the knowledge of certain coefficients uniquely determines the word [5]. As observed in [11], q-binomials contain more precise information. The same observation holds for (q,t)-binomials. For example,

$$\text{with } M = \begin{pmatrix} \mathtt{a\,a\,b} \\ \mathtt{a\,b\,b} \\ \mathtt{b\,a\,b} \end{pmatrix}, \quad \binom{M}{(\mathtt{b})}_{q,t} = q^2 + tq + t^2\left(q^2 + q + 1\right).$$

We clearly see that $q^{i-1}t^{j-1}$ occurs in the (q,t)-binomial if and only if there is a letter $\mathtt{b}$ in position (i,j) (counted from 1 in the upper-left corner). This observation is general. To be consistent with our convention of processing from the left, we concatenate a column to the left of an array. The exponent $p+1$ occurring in the statement is similar to the exponent $|\mathtt{b}v|$ in (4).

Proposition 5. *Let $M \in A^{r\times m}$, $P \in A^{s\times p}$, $C \in A^{r\times 1}$, and $D \in A^{s\times 1}$. Then*

$$\binom{C \mathbin{\text{⦶}} M}{D \mathbin{\text{⦶}} P}_{q,t} = \binom{M}{D \mathbin{\text{⦶}} P}_{q,t} \cdot t^{p+1} + \binom{C}{D}_q \binom{M}{P}_{q,t}.$$

Proof. By definition of the (q,t)-column-binomial coefficient, we have

$$\binom{C \mathbin{\text{⦶}} M}{D \mathbin{\text{⦶}} P}_{q,t} = \sum_{1\le i_1<\cdots<i_{p+1}\le m+1} t^{\sum_{j=1}^{p+1}(i_j-j)} \binom{(C \mathbin{\text{⦶}} M)_{i_1}}{(D \mathbin{\text{⦶}} P)_1}_q \cdots \binom{(C \mathbin{\text{⦶}} M)_{i_{p+1}}}{(D \mathbin{\text{⦶}} P)_{p+1}}_q.$$

We partition the selection of columns in the concatenated array $C \mathbin{\text{⦶}} M$ into two cases, depending on whether the first column C is included. If $i_1 = 1$, the contribution is $\binom{C}{D}_q \binom{M}{P}_{q,t}$. Otherwise, the first column C is not selected. All $p+1$ columns are selected from M. By reindexing with $\ell_j = i_j - 1$, the contribution becomes:

$$t^{p+1} \cdot \sum_{1\le \ell_1<\cdots<\ell_{p+1}\le m} t^{\sum_{j=1}^{p+1}(\ell_j-j)} \prod_{k=1}^{p+1} \binom{M_{\ell_k}}{(D \mathbin{\text{⦶}} P)_k}_q = t^{p+1} \cdot \binom{M}{D \mathbin{\text{⦶}} P}_{q,t}.$$

□

We also have an analogue of Chu–Vandermonde identity:

Proposition 6. *Let $M \in A^{r\times m}$, $N \in A^{r\times n}$, and $P \in A^{s\times p}$. We have*

$$\binom{M \mathbin{\text{⦶}} N}{P}_{q,t} = \sum_{P_1 \mathbin{\text{⦶}} P_2 = P} t^{(m-|P_1|)|P_2|} \binom{M}{P_1}_{q,t} \binom{N}{P_2}_{q,t},$$

where $|P|$ is the number of columns of P.

Proof. By definition of the (q,t)-binomial coefficient, we have

$$\binom{M \oplus N}{P}_{q,t} = \sum_{1 \le i_1 < \cdots < i_p \le m+n} t^{\sum_{j=1}^{m+n}(i_j - j)} \binom{(M \oplus N)_{i_1}}{P_1}_q \cdots \binom{(M \oplus N)_{i_p}}{P_p}_q .$$

Partitioning the sum by how many elements of the sequence $i_1 < \cdots < i_p$ are less than or equal to m, the above coefficient can be written

$$\sum_{k=0}^{p} \left(\sum_{1 \le i_1 < \cdots < i_k \le m} t^{\sum_{j=1}^{k}(i_j - j)} \prod_{j=1}^{k} \binom{(M \oplus N)_{i_j}}{P_j}_q \right) \cdot$$
$$\left(\sum_{m < i_{k+1} < \cdots < i_p \le m+n} t^{\sum_{j=k+1}^{p}(i_j - j)} \prod_{j=k+1}^{p} \binom{(M \oplus N)_{i_j}}{P_j}_q \right) .$$

Observe that $\left(M \oplus N\right)_j = M_j$ if $j \le m$ and $\left(M \oplus N\right)_j = N_{j-m}$ if $j > m$. Hence, the second factor in the above formula can be written

$$\sum_{m < i_{k+1} < \cdots < i_p \le m+n} t^{\sum_{j=k+1}^{p}(i_j - m - (j-k) + m - k)} \prod_{j=k+1}^{p} \binom{N_{i_j - m}}{P_j}_q$$
$$= t^{\sum_{j=k+1}^{p}(m-k)} \binom{N}{P_{[(k+1)..p]}}_{q,t} = t^{(m-|P_1|)|P_2|} \binom{N}{P_{[(k+1)..p]}}_{q,t} .$$

where $P_{[(k+1)..p]}$ is made of the last $p-k$ columns. □

Flipping the arrays around a vertical axis reverses the order of coefficients of the (q,t)-binomial, when it is viewed as a polynomial in t with coefficients in $\mathbb{N}[q]$:

$$[t^j] \binom{M_1 \cdots M_m}{P_1 \cdots P_p}_{q,t} = [t^{p(m-p)-j}] \binom{M_m \cdots M_1}{P_p \cdots P_1}_{q,t} .$$

Similarly, if $\widetilde{M}$ denotes a flip of M around a horizontal axis and $\widetilde{P}$ is defined correspondingly, then when we view the (q,t)-binomial as a polynomial in q with coefficients in $\mathbb{N}[t]$, we have

$$[q^j] \binom{\widetilde{M}}{\widetilde{P}}_{q,t} = [q^{ps(r-s)-j}] \binom{M}{P}_{q,t}$$

where $M \in A^{r \times m}$, $P \in A^{s \times p}$. For the two arrays in Example 4, with $r = m = 3$ and $s = p = 2$, we have

$$\binom{M}{P}_{q,t} = q^4 t + 2tq^3 + \left(t^2 + t + 1\right) q^2 + \left(t^2 + 1\right) q,$$

and

$$\binom{\widetilde{M}}{\widetilde{P}}_{q,t} = \left(t^2 + 1\right) q^3 + \left(t^2 + t + 1\right) q^2 + 2tq + t.$$

4 Conclusions

In this short paper, we have introduced a natural generalization of binomial coefficients of words to multidimensional arrays. Expected properties extend without much effort, thus justifying our definition. Moreover, this opens the way to further developments, as coefficients of words already appear in numerous contexts (formal language theory, algebra, theoretical computer science, combinatorics, etc.). Based on a robust definition, one can hope that other non-trivial combinatorial properties may also extend to multidimensional arrays. Furthermore, as we have shown in the two-dimensional case, one can work either rows or columns. In higher dimension d, it is necessary to specify the order in which sections are selected, potentially leading to $d!$ distinct (but related) coefficients. The multivariate polynomials we obtain are also worth to be studied further. For instance, it is straightforward to adapt [12] to this context.

References

1. Amir, A., Benson, G., Farach, M.: An alphabet independent approach to two-dimensional pattern matching. SIAM J. Comput. **23**(2), 313–323 (1994). https://doi.org/10.1137/S0097539792226321
2. Amir, A., Kapah, O., Tsur, D.: Faster two-dimensional pattern matching with rotations. Theor. Comput. Sci. **368**(3), 196–204 (2006). https://doi.org/10.1016/j.tcs.2006.09.012
3. Bera, S., Sriram, S., Nagar, A.K., Pan, L., Subramanian, K.G.: Algebraic properties of Parikh matrices of binary picture arrays. J. Math. **2020**, 3236405 (2020). https://doi.org/10.1155/2020/3236405
4. Şerbănuţă, T.-F.: Extending Parikh matrices. Theoret. Comput. Sci. **310**(1–3), 233–246 (2004). https://doi.org/10.1016/S0304-3975(03)00396-7
5. Dudík, M., Schulman, L.J.: Reconstruction from subsequences. J. Combin. Theory Ser. A **103**(2), 337–348 (2003). https://doi.org/10.1016/S0097-3165(03)00103-1
6. Gamard, G., Richomme, G., Shallit, J., Smith, T.J.: Periodicity in rectangular arrays. Inf. Process. Lett. **118**, 58–63 (2017). https://doi.org/10.1016/j.ipl.2016.09.011
7. Janaki, K., Arulprakasam, R., Paramasivan, M., Rajkumar Dare, V.: Algebraic properties of Parikh q-matrices on two-dimensional words. In: Barneva, R.P., Brimkov, V.E., Nordo, G. (eds.) Combinatorial Image Analysis. IWCIA 2022. LNCS, vol. 13348, pp. 171–188. Springer, Cham (2023). https://doi.org/10.1007/978-3-031-23612-9_11
8. Lothaire, M.: Combinatorics on Words. Cambridge Mathematical Library. Cambridge University Press (1997). https://doi.org/10.1017/CBO9780511566097
9. Manvel, B., Meyerowitz, A.D., Schwenk, A.J., Smith, K.W., Stockmeyer, P.K.: Reconstruction of sequences. Discrete Math. **94**(3), 209–219 (1991). https://doi.org/10.1016/0012-365X(91)90026-X
10. Marcus, S., Sokol, D.: On two-dimensional Lyndon words. In: Kurland, O., Lewenstein, M., Porat, E. (eds) String Processing and Information Retrieval. SPIRE 2013. LNCS, vol. 8214, pp. 206–217. Springer, Cham (2013). https://doi.org/10.1007/978-3-319-02432-5_24

11. Renard, A., Rigo, M., Whiteland, M.A.: Introducing q-deformed binomial coefficients of words. J. Algebr. Comb. **61**(2), 33 (2025). https://doi.org/10.1007/s10801-025-01384-9
12. Renard, A., Rigo, M., Whiteland, M.A.: q-Parikh matrices and q-deformed binomial coefficients of words. Discrete Math. **348**(5), 114381 (2025). https://doi.org/10.1016/j.disc.2024.114381
13. Reutenauer, C.: Free lie algebras. In: Handbook of Algebra, vol. 3, vol. 3, pp. 887–903. Elsevier/North-Holland, Amsterdam (2003). https://doi.org/10.1016/S1570-7954(03)80075-X
14. Salon, O.: Automatic sequences with multi-indices. Appendix by Shallit, J.O., Sémin. Théor. Nombres, Univ. Bordeaux I 1986-1987, Exp. No. 4, 27 p. (1987). (Appendix 29A-36A (1987))
15. Şerbănuţă, T.-F.: Extending Parikh matrices. Theor. Comput. Sci. **310**(1–3), 233–246 (2004). https://doi.org/10.1016/S0304-3975(03)00396-7
16. Subramanian, K.G., Mahalingam, K., Abdullah, R., Nagar, A.K.: Two-dimensional digitized picture arrays and Parikh matrices. Int. J. Found. Comput. Sci. **24**(3), 393–408 (2013). https://doi.org/10.1142/S012905411350010X

Circularity and Repetitiveness in Non-Injective DF0L Systems

Herman Goulet-Ouellet(✉), Karel Klouda, and Štěpán Starosta

Faculty of Information Technology, Czech Technical University in Prague, Prague, Czech Republic
{herman.goulet.ouellet,karel.klouda,stepan.starosta}@fit.cvut.cz

Abstract. We study circularity in DF0L systems, a generalization of D0L systems. We focus on two different types of circularity, called weak and strong circularity. When the morphism is injective on the language of the system, the two notions are equivalent, but they may differ otherwise. Our main result shows that failure of weak circularity implies unbounded repetitiveness, and that unbounded repetitiveness implies failure of strong circularity. This extends previous work by the second and third authors for injective systems. To help motivate this work, we also give examples of non-injective but strongly circular systems.

Keywords: D0L systems · Circularity

1 Introduction

The notion of circularity in D0L systems appeared in several different forms, starting with the work of Cassaigne on pattern avoidance [1] and Mignosi and Séébold on repetitiveness [9]. It is closely related with recognizability, a key notion from symbolic dynamics [8]. The study of circularity in D0L systems is also motivated by its connection with bispecial factors [2,4,7], in turn linked with factor complexity, return words, Rauzy graphs, etc. One of the main motivations behind this paper is to better understand under which conditions weak and strong circularity can be safely decided, with effective calculation of the circularity thresholds being the ultimate goal.

In this paper we work with DF0L systems, a generalization of D0L systems which allows for multiple axioms. The need for this added generality arises naturally when working within a D0L system, for example when one wants to take powers of the morphism without changing the language (see Example 2.6).

We moreover study systems where the morphism is not necessarily injective. This is a departure from many earlier papers, where injectivity is needed for most results. We focus on two notions of circularity, called weak and strong circularity. While these are equivalent for injective systems (and more generally *eventually injective* systems, defined in Sect. 2), they might differ in general.

The first author was supported by the CTU Global Postdoc Fellowship program.

G. Gamard and J. Leroy (Eds.): WORDS 2025, LNCS 15729, pp. 153–165, 2025.
https://doi.org/10.1007/978-3-031-97548-6_14

Our main result, Theorem 4.2, extends to the non-injective case a characterization from [7] stating that strong circularity is equivalent to unbounded repetitiveness. In the non-injective case, the result is no longer a characterization. A consequence of our main result is that when the system is not unboundedly repetitive (a decidable condition by [5]), then we can safely calculate the weak circularity threshold. Finally, we also show that in a strongly circular system, weak circularity is preserved when taking powers of the morphism, something which is not true in general. The questions of whether strong circularity is decidable or preserved under taking powers of the morphism remain open.

2 DF0L Systems

Let $\mathcal{A}^*$ be the set of all finite words over a finite alphabet $\mathcal{A}$. The empty word is denoted by ε and we let $\mathcal{A}^+ = \mathcal{A}^* \setminus \{\varepsilon\}$.

Definition 2.1. A *DF0L system* is a triplet $S = (\mathcal{A}, \varphi, W)$ where φ is a morphism $\mathcal{A}^* \to \mathcal{A}^*$ and $W \subseteq \mathcal{A}^+$ is a non-empty finite set of non-empty words called *axioms*. If φ is non-erasing, i.e. $\varphi(a) \neq \varepsilon$ for all $a \in A$, then S is called a *PDF0L system*.

We refer to [11] for more details on the terminology of L systems. When W consists of a single element, S is also called a D0L system, and a PD0L system if φ is non-erasing. We note that DF0L systems can arise naturally even when working in the setting of D0L systems, as shown in Example 2.6.

We define the language of a DF0L system $S = (\mathcal{A}, \varphi, W)$ by

$$\mathcal{L}(S) = \{u \in A^* \mid \exists k \geq 0, \exists w \in W, \varphi^k(w) \in A^*uA^*\}.$$

Note that membership in $\mathcal{L}(S)$ is decidable by [12, Lemma 3].

Example 2.2. The *Thue–Morse system* is the D0L system $S = (\{a, b\}, \varphi, \{a\})$ where $\varphi\colon a \mapsto ab, b \mapsto ba$. In this system

$$\mathcal{L}(S) = \{\varepsilon, a, b, aa, ab, ba, bb, aab, aba, abb, baa, bab, bba, aaba, aabb, \dots\}.$$

Let us say that a DF0L system $S = (\mathcal{A}, \varphi, W)$ is *injective* if φ is injective on $\mathcal{L}(S)$. Many well-known examples of D0L systems are injective, like the Thue–Morse system above. We next introduce a weaker form of injectivity, under which the equivalence between the two kinds of circularity still holds (Proposition 3.11).

Definition 2.3. A DF0L system $S = (\mathcal{A}, \varphi, W)$ is called *eventually injective* if the following set is finite:

$$\Delta_S = \{\{u, v\} \subseteq \mathcal{L}(S) \mid u \neq v, \varphi(u) = \varphi(v)\}$$

Failure of injectivity of a PDF0L system $S = (\mathcal{A}, \varphi, W)$ can be measured by

$$\delta_S = \max\{|\varphi(u)| : \exists v, \{u, v\} \in \Delta_S\},$$

with $\delta_S = 0$ when $\Delta_S = \emptyset$. Eventual injectivity is equivalent to $\delta_S < \infty$. We do not know whether eventual injectivity, or even injectivity, is decidable in DF0L systems. We next give two examples of non-injective systems, one of which is eventually injective.

Example 2.4. Let $S = (\{a, b, c\}, \varphi, \{a\})$ where $\varphi\colon a \mapsto abacc, b \mapsto aba, c \mapsto aba$. This system is eventually injective, with

$$\Delta_S = \{\{b, c\}, \{ba, ca\}, \{ab, ac\}, \{bac, cab\}\},$$

and thus $\delta_S = 11$.

Example 2.5. Let $S = (\{a, b, c\}, \varphi, \{a\})$ where $\varphi\colon a \mapsto abaca, b \mapsto aba, c \mapsto aba$. Unlike the previous example, this system is not eventually injective. We can obtain an infinite sequence of elements $\{u_n, v_n\}_{n\in\mathbb{N}}$ in Δ_S as follows: let

$$u_1 = aca, \;\; u_{n+1} = u_1\varphi(u_n), \quad v_1 = aba, \;\; v_{n+1} = v_1\varphi(v_n).$$

It is clear that $u_n \neq v_n$ and $\varphi(u_n) = \varphi(v_n)$. Observing that $bu_n, av_n \in \mathcal{L}(S)$, one can show by induction on n that $u_n, v_n \in \mathcal{L}(S)$ for all $n \in \mathbb{N}$.

A common technique in the study of D0L systems is to pass to a power of the morphism, for instance to ensure certain growth conditions (as in the proof of Theorem 4.2), while preserving the language of the system. Given a system $S = (\mathcal{A}, \varphi, W)$, let us define the k-th power of S ($k \geq 1$) as the system

$$S^k = (\mathcal{A}, \varphi^k, \{\varphi^i(w) \mid w \in W, 0 \leq i < k\}),$$

which satisfies $\mathcal{L}(S^k) = \mathcal{L}(S)$. The next example shows the necessity of using multiple axioms when defining S^k.

Example 2.6. Let $S = (\mathcal{A}, \varphi, b)$ where $\varphi\colon a \mapsto cb, b \mapsto ad, c \mapsto c, d \mapsto d$. Let $S' = (\mathcal{A}, \varphi^2, b)$. The language $\mathcal{L}(S')$ is contained in $\mathcal{L}(S)$, but the inclusion is strict, since for instance $ad \notin \mathcal{L}(S')$.

3 Circularity

In this section, we recall two notions of circularity for DF0L systems called *weak* and *strong* circularity. We follow in part the exposition from [7]. We also sketch algorithms for computing the weak and strong circularity thresholds. We start with the definition of interpretations.

Definition 3.1. Let $S = (\mathcal{A}, \varphi, W)$ be a DF0L system. An *interpretation* of $u \in A^+$ in S is a triplet (s, w, t) such that $\varphi(w) = sut$ and $w \in \mathcal{L}(S)$. An interpretation is *minimal* if $|s| < |\varphi(a)|$ and $|t| < |\varphi(b)|$, where a and b are respectively the first and last letters of w.

Given a morphism φ, we let

$$\lfloor\varphi\rfloor = \min(|\varphi(a)| : a \in \mathcal{A}) \quad \text{and} \quad \lceil\varphi\rceil = \max(|\varphi(a)| : a \in \mathcal{A}).$$

Lemma 3.2. Let $S = (\mathcal{A}, \varphi, W)$ be a PDF0L system and $u \in \mathcal{L}(S)$. If (s, w, t) is a minimal interpretation of u in S, then

$$|u|/\lceil\varphi\rceil \leq |w| \leq 2 + (|u| - 2)/\lfloor\varphi\rfloor.$$

Proof. The leftmost inequality follows from $|u| \leq |\varphi(w)| \leq \lceil\varphi\rceil \cdot |w|$. For the rightmost inequality, first observe that if $|w| = 1$ or $|w| = 0$, then the statement becomes trivial. Thus we assume that $|w| \geq 2$. Let a and b be respectively the first and last letters of w, and let $w = aw'b$. Since the interpretation is minimal, there exist $s', t' \in \mathcal{A}^+$ and $s'', t'' \in \mathcal{A}^*$ such that $\varphi(a) = s''s'$, $\varphi(b) = t't''$ and $u = s'\varphi(w')t'$. It follows that

$$|u| = |s'| + |t'| + |\varphi(w')| \geq 2 + |w'|\lfloor\varphi\rfloor = 2 + (|w| - 2)\lfloor\varphi\rfloor.$$

Since φ is non-erasing we have $\lfloor\varphi\rfloor > 0$, and we obtain the desired inequality by solving for $|w|$. □

Thus in a PDF0L system, the set of minimal interpretations of a given word is finite and computable. We may write a simple algorithm which computes all minimal interpretations of the words of length n by looking at the images of all words of length $\lfloor 2 + (n - 2)/\lfloor\varphi\rfloor\rfloor$.

Example 3.3. In the Thue–Morse system (Example 2.2), the word aba admits exactly two minimal interpretations, namely (b, bb, ε) and (ε, aa, b). On the other hand the word aa admits only one minimal interpretation, namely (b, ba, a).

Next we turn to a finer analysis of how interpretations behave in a DF0L system $S = (\mathcal{A}, \varphi, W)$. A pair (u', u'') is said to be *compatible* with an interpretation (s, w, t) of $u'u''$ in S if there is a pair (w', w'') such that $w'w'' = w$, $\varphi(w') = su'$, $\varphi(w'') = u''t$. A pair (u', u'') is called *admissible* if it is compatible with at least one interpretation of $u'u''$.

Definition 3.4. Let $S = (\mathcal{A}, \varphi, W)$ be a DF0L system.

1. A pair (u', u'') is called *weakly synchronizing* if it is compatible with all interpretations of $u'u''$. We say that the word $u = u'u''$ is *weakly synchronized.*
2. A pair (u', u'') where $u' \neq \varepsilon$ is called *strongly synchronizing* if there is $a \in \mathcal{A}$ such that for all interpretations (s, w, t) of $u'u''$, there is a pair (w', w'') such that $w = w'w''$, $\varphi(w') = su'$, $\varphi(w'') = u''t$, where w' ends with a.

These notions are introduced mainly to break down the definitions of weak and strong circularity (Definition 3.7). The choice to consider the last letter of w' instead of the first letter of w'' in the definition of strong circularity is a matter of preference; both options lead to equivalent definitions of strong circularity.

Remark 3.5. To test whether a pair (u', u'') is weakly or strongly synchronizing, it suffices look at minimal interpretations of $u'u''$. Thus those properties are decidable since there is a finite number of minimal interpretation by Lemma 3.2.

Lemma 3.6. Let S be a DF0L system and let (u_1, u_2) be such that $u_1u_2 \in \mathcal{L}(S)$. If there is a pair (v_1, v_2) such that $u_1 \in \mathcal{A}^*v_1$, $u_2 \in v_2\mathcal{A}^*$, and (v_1, v_2) is weakly (strongly) synchronizing, then (u_1, u_2) is also weakly (strongly) synchronizing.

Proof. Let $S = (\mathcal{A}, \varphi, W)$. Write $u_1 = pv_1, u_2 = v_2q$ and let (s, w, t) be an interpretation of u_1u_2. As (sp, w, qt) is an interpretation of v_1v_2 and (v_1, v_2) is synchronizing, there is a pair (w_1, w_2) such that $w_1w_2 = w$, $\varphi(w_1) = spv_1 = su_1$, $\varphi(w_2) = v_2qt = u_2t$. Thus (s, w, t) is compatible with (u_1, u_2), hence (u_1, u_2) is weakly synchronizing. If (v_1, v_2) is strongly synchronizing, the last letter of w_1 does not depend on w, and so (u_1, u_2) is strongly synchronizing as well. □

In particular a word $u \in \mathcal{L}(S)$ which has a weakly synchronized factor is also weakly synchronized. For example, in the Thue–Morse system, any word of $\mathcal{L}(S)$ which admits aa as a factor is weakly synchronized (see Example 3.3).

Next we define two notions of circularity. The first one is due to Cassaigne [1], while the second one is a variation found in [7] of a notion introduced by Mignosi and Séébold [9].

Definition 3.7. Let $S = (\mathcal{A}, \varphi, W)$ be a DF0L system.

1. S is *weakly circular* if there exists an integer $D \geq 0$ such that all words $u \in \mathcal{L}(S)$ where $|u| > D$ are weakly synchronized.
2. S is *strongly circular* if there exists an integer $D' \geq 0$ such that all admissible pairs (u', u'') where $|u'| > D'$ and $|u''| > D'$ are strongly synchronizing.

The smallest value of D and D' (if they exist) are called the *thresholds* for weak and strong circularity. When S is clear from context, they are denoted by D_w and D_s. By Lemma 3.6, we can check whether a given value D exceeds D_w or D_s by checking respectively all words of length $D+1$ and all pairs (u', u'') where $|u'|, |u''| > D+1$. Since the property of being weakly or strongly synchronizing is decidable (Remark 3.5), this can be turned into algorithms for computing D_w and D_s, if they exist. For instance: starting from $D = \lceil \varphi \rceil - 2$ (an obvious lower bound for D_w), one can test whether all words of length D are weakly synchronized, and else increment the value of D and start over. We can improve on the brute force approach by reusing calculations from earlier steps, thanks to Lemma 3.6.

Example 3.8. In the Thue–Morse system (Example 2.2), the word aba is not weakly synchronized (Example 3.3). But all words of length 4 in $\mathcal{L}(S)$ admit only one minimal interpretation, thus $D_\mathsf{w} = 3$. Similarly, $D_\mathsf{s} = 1$.

Example 3.9. For the system of Example 2.4, one can check that $D_\mathsf{w} = D_\mathsf{s} = 3$. We used this fact to calculate the set Δ_S for this system.

Example 3.10. The system from Example 2.5 has $D_\mathsf{w} = 3$ and $D_\mathsf{s} = 9$.

Thus the previous example is a strongly circular system which is not eventually injective. An example which is weakly but not strongly circular is given in [7, Example 5]. The next proposition extends an observation from [7] which can also be found in [8, Observation 3.6.14, Proposition 3.6.17].

Proposition 3.11. Let $S = (\mathcal{A}, \varphi, W)$ be a DF0L system.

1. If S is strongly circular, then it is weakly circular with $D_{\mathsf{w}} \leq 2D_{\mathsf{s}} + \lceil\varphi\rceil$.
2. If S is eventually injective and weakly circular, then it is strongly circular with $D_{\mathsf{s}} \leq D_{\mathsf{w}} + \delta_S + 1$.

Proof. Let u be a word such that $|u| > 2D_{\mathsf{s}} + \lceil\varphi\rceil$. Let (s, w, t) be an interpretation of u and $w = w_1 w_2$ where w_2 is the shortest suffix of w such that $|\varphi(w_2)| - |t| > D_{\mathsf{s}}$. Write $\varphi(w_1) = su_1$, $\varphi(w_2) = u_2 t$, and $w_2 = aw_3$ where $u_1, u_2 \in \mathcal{A}^*$ and $a \in \mathcal{A}$. By minimality of w_2, we have $|\varphi(w_3)| - |t| \leq D_{\mathsf{s}}$ and

$$|u_1| = |u| - |\varphi(a)| - (|\varphi(w_3)| - |t|) \geq |u| - \lceil\varphi\rceil - D_{\mathsf{s}} > D_{\mathsf{s}}.$$

Thus (u_1, u_2) is strongly synchronizing, and in particular weakly synchronizing.

Let (u_1, u_2) be an admissible pair such that $|u_1| = |u_2| = D_{\mathsf{w}} + \delta_S + 2$. Let (s, w, t) be an interpretation of $u_1 u_2$ compatible with (u_1, u_2) and $(\bar{s}, \bar{w}, \bar{t})$ be any other interpretation of $u_1 u_2$. Write $u_1 = u_3 r$ where $u_3, r \in \mathcal{A}^*$ and $|r| = \delta_S + 1$. By assumption, u_3 and u_2 are weakly synchronized, so we may find factorizations $u_1 = p_1 q_1 r$, $u_2 = p_2 q_2$, $w = w_1 w_3 w_2$, and $\bar{w} = \bar{w}_1 \bar{w}_3 \bar{w}_2$ such that

$$s p_1 = \varphi(w_1), \; \bar{s} p_1 = \varphi(\bar{w}_1), \quad q_2 t = \varphi(w_2), \; q_2 \bar{t} = \varphi(\bar{w}_2),$$
$$q_1 r p_2 = \varphi(w_3) = \varphi(\bar{w}_3).$$

Since $|q_1 r p_2| > \delta_S$, it follows that $w_3 = \bar{w}_3$. Moreover since (s, w, t) is compatible with (u_1, u_2), there must be a factorization $w_3 = xy$ with $x \neq \varepsilon$ such that $\varphi(w_1 x) = su_1$. It follows that $\bar{w} = \bar{w}_1 x y \bar{w}_2$ with $\varphi(\bar{w}_1 x) = \bar{s} u_1$. This shows that (u_1, u_2) is strongly synchronizing. □

Proposition 3.12. Let $S = (\mathcal{A}, \varphi, W)$ be a PDF0L system and $u \in \mathcal{L}(S)$. If u is weakly synchronized in S^k where $|u| > \lceil\varphi\rceil \cdot \max\{|\varphi^{k-2}(x)| : x \in W\}$, then u is also weakly synchronized in S.

Proof. Let $C = \lceil\varphi\rceil \cdot \max\{|\varphi^{k-2}(x)| : x \in W\}$. Take a word $u \in \mathcal{L}(S)$ such that $|u| > C$ which is weakly synchronized in S^k. Thus there is a pair (u_1, u_2) such that $u_1 u_2 = u$ and which is weakly synchronizing in S^k. Let us show that (u_1, u_2) is also weakly synchronizing in S. For this, fix an interpretation (s, w, t) of u. By Lemma 3.2 we have $|w| > |\varphi^{k-2}(x)|$ for every $x \in W$, so w is a factor of $\varphi^{k-1}(z)$ for some $z \in \mathcal{L}(S)$. Let $\varphi^{k-1}(z) = pwq$. Note that $\varphi^k(z) = \varphi(p) s u t \varphi(q)$, so $(\varphi(p)s, z, t\varphi(q))$ is an interpretation of u in S^k. Since (u_1, u_2) is synchronizing in S^k, there is a pair (z_1, z_2) such that

$$z = z_1 z_2, \quad \varphi^k(z_1) = \varphi(p) s u_1, \quad \varphi^k(z_2) = u_2 t \varphi(q).$$

Since $\varphi^{k-1}(z_1)$ is a prefix of $\varphi^{k-1}(z) = pwq$, it follows that $\varphi^{k-1}(z_1)$ and pw are prefix comparable. If pw would be a proper prefix of $\varphi^{k-1}(z_1)$, then $\varphi(pw) = \varphi(p)sut$ would be a proper prefix of $\varphi^k(z_1) = \varphi(p)su_1$, which is a contradiction (where we used the fact that φ is non-erasing). Thus we conclude that $\varphi^{k-1}(z_1)$ is a prefix of pw. Likewise it is clear that p must be a prefix of $\varphi^{k-1}(z_1)$. Therefore there is a pair (w_1, w_2) such that $w_1 w_2 = w$ and $\varphi^{k-1}(z_1) = pw_1$, and as a result $\varphi^{k-1}(z_2) = w_2 q$. Finally we have

$$\varphi(p)\varphi(w_1) = \varphi(pw_1) = \varphi^k(z_1) = \varphi(p)su_1,$$

and therefore $\varphi(w_1) = su_1$. Likewise we conclude that $\varphi(w_2) = u_2 t$, which shows that (u_1, u_2) is compatible with (s, w, t). □

In particular, if S^k is weakly circular for some $k \geq 1$, then so is S. The next example shows that the converse is false.

Example 3.13. Let $S = (\{a, b, c\}, \varphi, \{a\})$ where $\varphi\colon a \mapsto aac, b \mapsto bc, c \mapsto bc$. The system S is weakly circular with weak threshold 1, but the power $S^2 = (\{a, b, c\}, \varphi^2, \{a, aac\})$ is not. For instance, the words $(bc)^{4k}$ have interpretations which never synchronize, such as $(\varepsilon, (bc)^k, \varepsilon)$ and $(bc, (bc)^k b, bc)$.

4 Repetitiveness

In this section, we relate strong circularity with repetitiveness of the system. Let $S = (\mathcal{A}, \varphi, W)$ be a DF0L system. We say that a word $w \in \mathcal{A}^*$ is *bounded* if $\lim_{k\to\infty} |\varphi^k(w)| < \infty$, and *unbounded* otherwise.

Definition 4.1. A DF0L system is called *unboundedly repetitive* if there exists an unbounded word $v \in \mathcal{L}(S)$ such that $v^k \in \mathcal{L}(S)$ for all $k \in \mathbb{N}$.

Recall that a word w is *primitive* if $u^n = w \implies n = 1$. For every word $u \neq \varepsilon$, there exists a unique primitive word $\rho(u)$, called the *primitive root* of u, such that $u = \rho(u)^n$. It is well known that $\rho(x) = \rho(y)$ if and only if $xy = yx$. When studying repetitiveness of DF0L systems, it is natural to consider the set

$$\Omega(S) = \left\{v \in \mathcal{L}(S) \mid v \text{ is primitive and } \forall k \in \mathbb{N}, v^k \in \mathcal{L}(S)\right\}.$$

The second and third author proved in [7] that in an injective D0L system, strong circularity is equivalent to unbounded repetitiveness. In this section, we will prove the following generalization of this result.

Theorem 4.2. *Let $S = (\mathcal{A}, \varphi, W)$ be a PDF0L system.*

1. *If S is not weakly circular, then it is unboundedly repetitive.*
2. *If S is unboundedly repetitive, then it is not strongly circular.*

Thanks to Proposition 3.11 it follows that, when the system is eventually injective, both forms of circularity are equivalent to the absence of unbounded elements in $\Omega(S)$, which is a decidable property by [5].

Observe that unbounded repetitiveness is well-behaved with respect to taking powers, since this does not change the language or the set of unbounded words. Thus we obtain the following corollary of Theorem 4.2.

Corollary 4.3. *If a PDF0L system S is strongly circular, then all of the powers S^n for $n \geq 1$ are weakly circular.*

As shown in Example 3.13 this may fail when the system is not strongly circular. We do not know if strong circularity is stable under taking powers.

The rest of the section will be devoted to the proof of Theorem 4.2, which follows a similar structure as the injective case from [7]. Several new steps are needed to account for the lack of injectivity, but some arguments have also been streamlined, in particular in the proof of the first part.

4.1 Preparatory Results for Part 1

This subsection prepares the proof of part 1 of Theorem 4.2. We start with a simple technical lemma which extends [7, Lemma 8].

Lemma 4.4. If S is a PDF0L system, then every long enough bounded word in $\mathcal{L}(S)$ is weakly synchronized.

Proof. By [7, Lemma 8], the result holds for PD0L systems. Applying it to $T = (\mathcal{A}, \varphi, \prod_{w \in W} w)$ and noting that $\mathcal{L}(S) \subseteq \mathcal{L}(T)$, it follows that a word in $\mathcal{L}(S)$ weakly synchronized in T is also weakly synchronized in S. □

Next we recall some results from [6]. Let $\varphi : \mathcal{A}^* \to \mathcal{A}^*$ be a morphism and denote by $\mathrm{alph}(w)$ the set of all letters which occur in w. A subalphabet $\mathcal{B} \subseteq \mathcal{A}$ is called *p-invariant* if $\mathcal{B}$ contains an unbounded letter and $\mathcal{B} = \bigcup_{a \in \mathcal{B}} \mathrm{alph}(\varphi^p(a))$ [6, Definition 4]. For p fixed, we say that $\mathcal{B}$ is *minimal* if it is minimal for inclusion among all p-invariant subalphabets.

Lemma 4.5 ([6, Lemmas 5 and 6])**.** Let φ be a morphism on $\mathcal{A}$. There exists $p \geq 1$ such that $\mathrm{alph}(\varphi^p(a)) = \mathrm{alph}(\varphi^{pk}(a))$ for any $a \in \mathcal{A}$ and positive k. In particular, if a is unbounded, $\mathcal{B} = \mathrm{alph}(\varphi^p(a))$ is p-invariant. Moreover, if $\mathcal{B}$ is minimal, then $\mathcal{B} = \mathrm{alph}(\varphi^p(g))$ for every unbounded letter $g \in \mathcal{B}$.

Given a morphism φ, we say that a sequence of words $(w_i)_{i \in \mathbb{N}}$ is *pushy* if the set of bounded words which appear as factors in the words w_i, $i \in \mathbb{N}$, is finite, and otherwise we say that $(w_i)_{i \in \mathbb{N}}$ is *non-pushy* [6,8].

Lemma 4.6 ([6, Corollary 10])**.** Let φ be a morphism and $(w_i)_{i \in \mathbb{N}}$ be a non-pushy sequence of words such that $\lim_{i \to \infty} |w_i| \to \infty$. Let p be the integer of Lemma 4.5. There exists a minimal p-invariant subalphabet $\mathcal{B}$ and an unbounded letter $g \in \mathcal{B}$ with the following property: for all $n \in \mathbb{N}$, there exists $k \in \mathbb{N}$ such that $|\varphi^{pk}(g)| > n$ and $\varphi^{pk}(g)$ occurs as a factor in some w_i, $i \in \mathbb{N}$.

4.2 Preparatory Results for Part 2

We now move on to results used in the proof of part 2 of Theorem 4.2. We first recall the following notion due to Ehrenfeucht and Rozenberg [3].

Definition 4.7. Let $\mathcal{A}$ and $\mathcal{B}$ be finite alphabets and $\varphi\colon \mathcal{A}^* \to \mathcal{A}^*$, $\psi\colon \mathcal{B}^* \to \mathcal{B}^*$, $\alpha\colon \mathcal{A}^* \to \mathcal{B}^*$, and $\beta\colon \mathcal{B}^* \to \mathcal{A}^*$ be morphisms. We say that φ and ψ are *twinned* with respect to (α, β) if $\beta \circ \alpha = \varphi$ and $\alpha \circ \beta = \psi$. If $\#\mathcal{B} < \#\mathcal{A}$ then we say that ψ is a *simplification* of φ with respect to (α, β).

Every non-injective morphism has an injective simplification [3, Corollary 1]. Moreover if φ and ψ are twinned with respect to (α, β), then for all $k \in \mathbb{N}$

$$\alpha \circ \varphi^k = \psi^k \circ \alpha, \qquad \varphi^k \circ \beta = \beta \circ \psi^k. \tag{1}$$

Lemma 4.8. Let $S = (\mathcal{A}, \varphi, W)$ be a PDF0L system and let $\psi\colon \mathcal{B}^* \to \mathcal{B}^*$ be a morphism. Assume that φ and ψ are twinned with respect to (α, β). The system $T = (\mathcal{B}, \psi, \alpha(W))$ satisfies $\alpha(\mathcal{L}(S)) \subseteq \mathcal{L}(T)$ and $\beta(\mathcal{L}(T)) \subseteq \mathcal{L}(S)$.

Proof. First let $v \in \mathcal{L}(S)$. Then v is a factor of $\varphi^n(w)$ for some $n \in \mathbb{N}$ and $w \in W$, hence $\alpha(v)$ is a factor of $\alpha \circ \varphi^n(w) = \psi^n \circ \alpha(w)$, where the last equality follows from (1). By definition of T we have $\alpha(v) \in \mathcal{L}(T)$. Next let $S' = (\mathcal{A}, \varphi, \varphi(W))$. Then $\mathcal{L}(S') \subseteq \mathcal{L}(S)$ and by the first inclusion $\beta(\mathcal{L}(T)) \subseteq \mathcal{L}(S') \subseteq \mathcal{L}(S)$. □

We say that two words u, v are conjugate, denoted $u \sim v$, if $u = xy$ and $v = yx$ for some $x, y \in \mathcal{A}^*$. Equivalently, $u \sim v$ when $|u| = |v|$ and $\rho(u) \sim \rho(v)$. Notice that if $u \sim v$ and $u \in \Omega(S)$, then $v \in \Omega(S)$. Next we give a series of two lemmas. The first is a rephrasing of [5, Lemma 4].

Lemma 4.9. Let $\psi\colon \mathcal{B}^* \to \mathcal{A}^*$ be an injective morphism and $z \in \mathcal{B}^*, v \in \mathcal{A}^*$ such that v is primitive and $\psi(z)$ is a factor of v^k, $k \geq 1$. If $|z| \geq (\ell + 1)|v|$ and $\ell \geq 2$, then there is a primitive word u such that u^ℓ is a factor of z and $\rho(\psi(u)) \sim v$.

Proof. Take v_1 such that $v_1 \sim v$ and $\psi(z)$ is a prefix of v_1^k. Let $v_1 = p_i s_i$ with $|p_i| = i$ for $0 \leq i < |v|$. For each j, let q_j be the prefix of length j of z, and let $k(j)$ and $i(j)$ be the integers such that $\psi(q_j) = {v_1}^{k(j)} p_{i(j)}$.

As $|z| \geq (\ell + 1)|v|$, by the pigeonhole principle, there is a strictly increasing sequence $j_1 < j_2 < \cdots < j_s$ such that $s > \ell$ and $i(j_t) = i(j_1)$ for all $1 \leq t \leq s$. Observe that $k(j_t) < k(j_{t+1})$, otherwise we would have $\psi(q_{j_t}) = \psi(q_{j_{t+1}})$, which would contradict injectivity. If we let $v_2 = s_{i(j_1)} p_{i(j_1)}$, then

$$\begin{aligned} \psi(q_{j_{t+1}}) &= v_1^{k(j_{t+1})} p_{i(j_{t+1})} = v_1^{k(j_t)} v_1^{k(j_{t+1}) - k(j_t)} p_{i(j_t)} \\ &= v_1^{k(j_t)} p_{i(j_t)} v_2^{k(j_{t+1}) - k(j_t)} = \psi(q_{j_t}) v_2^{k(j_t) - k(j_1)}. \end{aligned}$$

Let r_t be the suffix of length $j_{t+1} - j_t$ of $q_{j_{t+1}}$, for $t \in \{1, \ldots, s-1\}$. Then by the above equality we have $\psi(r_t) = v_2^{k(j_{t+1}) - k(j_t)}$. Let $u = \rho(r_1)$. Note that $\rho(\psi(r_t)) = \rho(\psi(r_1)) = v_2$, which implies that $\rho(r_t) = \rho(r_1) = u$ since ψ is injective. Thus z has a factor $r_1 \cdots r_{s-1} = u^n$ for some $n \geq \ell$. Since $\rho(\psi(u)) = \rho(\psi(r_1)) = v_2 \sim v$, this concludes the proof. □

Lemma 4.10. Let $S = (\mathcal{A}, \varphi, W)$ be a PDF0L system. For all $v \in \Omega(S)$, there exists $u \in \Omega(S)$ such that $\rho(\varphi(u)) \sim v$.

Proof. First we assume that φ is injective. Let $v \in \Omega(S)$. For each $\ell \in \mathbb{N}$, fix a word $z_\ell \in \mathcal{L}(S)$ such that $|z_\ell| \geq (\ell+1)|v|$ and $\varphi(z_\ell)$ is a factor of some v^k. By Lemma 4.9 there exists a primitive word u_ℓ such that u_ℓ^ℓ is a factor of z_ℓ and $\rho(\varphi(u_\ell)) \sim v$. Since φ is injective, it follows that the words u_ℓ, $\ell \in \mathbb{N}$ are all conjugate, and so there are only finitely many of them. By the pigeonhole principle, there is $u \in \mathcal{L}(S)$ such that $u = u_\ell$ for infinitely many ℓ. In particular, $u \in \Omega(S)$ and $\rho(\varphi(u)) \sim v$.

Now let us prove the general case. Let ψ be an injective simplification of φ with respect to (α, β) and consider the system $T = (\mathcal{B}, \psi, \alpha(W))$. Observe that β must be injective since $\psi = \alpha \circ \beta$ is, and likewise α must be non-erasing since $\varphi = \beta \circ \alpha$ is. Let $v \in \Omega(S)$.

First, we claim that there exists $\bar{v} \in \Omega(T)$ such that $\rho(\beta(\bar{v})) \sim v$. Fix $\ell \in \mathbb{N}$ and choose $z_\ell \in \mathcal{L}(S)$ such that $\varphi(z_\ell)$ is a factor of v^k for some k and $|z_\ell| \geq (\ell+1)|v|$. Let $\bar{z}_\ell = \alpha(z_\ell)$. As α is non-erasing we have $|\bar{z}_\ell| \geq (\ell+1)|v|$. Since β is injective, by Lemma 4.9, there is a primitive word $\bar{v}_\ell$ such that $\bar{v}_\ell^\ell$ is a factor of $\bar{z}_\ell$ and $\rho(\beta(\bar{v}_\ell)) \sim v$. As in the first part of the proof, we observe that elements in the set $\{\bar{v}_\ell \mid \ell \in \mathbb{N}\}$ are all conjugate, and thus by the pigeonhole principle there exists $\bar{v}$ such that $\bar{v} = \bar{v}_\ell$ for infinitely many ℓ. Since $\bar{z}_\ell \in \mathcal{L}(T)$ by Lemma 4.8, this concludes the proof of the claim.

Using the first part of the proof, there exists $\bar{u} \in \Omega(T)$ such that $\rho(\psi(\bar{u})) \sim \bar{v}$, and hence, $\rho(\beta \circ \psi(\bar{u})) \sim \rho(\beta(\bar{v})) \sim v$. Finally if we let $u = \rho(\beta(\bar{u}))$, then

$$\rho(\varphi(u)) = \rho(\varphi \circ \beta(\bar{u})) = \rho(\beta \circ \psi(\bar{u})) \sim v$$

where we used that $\rho(\varphi(x)) = \rho(\varphi(\rho(x))$. Since $\beta(\bar{u}) \in \mathcal{L}(S)$ by Lemma 4.8, this concludes the proof. □

Finally we extend [5, Theorem 15] to non-injective systems. Given $u \in \mathcal{A}^*$, let u^ω denote the infinite periodic word $uuu \cdots \in \mathcal{A}^{\mathbb{N}}$.

Proposition 4.11. Let $S = (\mathcal{A}, \varphi, W)$ be a PDF0L system. Let $v \in \Omega(S)$ be unbounded. There exists an unbounded letter $a \in \mathcal{A}$, an integer ℓ, $1 \leq \ell \leq \#\mathcal{A}$, and a word $u \in a\mathcal{A}^* \cap \mathcal{L}(S)$ such that $v \sim u$ and $\lim_{k\to\infty} \varphi^{k\ell}(a) = u^\omega$.

Proof. Take $j > \#\mathcal{A}(|v|+2)$ such that $|\varphi^j(a)| \geq |v|$ for all unbounded letters $a \in \mathcal{A}$. Applying Lemma 4.10 j times, there exists $z \in \Omega(S)$ such that $\rho(\varphi^j(z)) \sim v$. As v contains an unbounded letter, so does z. Set $z = p_0 a_0 w_0$ where p_0 is a bounded word, a_0 is an unbounded letter and $w_0 \in \mathcal{A}^*$. For $i \geq 1$ set $\varphi^i(a_0) = p_i a_i w_i$ where p_i is a bounded word, a_i is an unbounded letter and $w_i \in \mathcal{A}^*$. As $\mathcal{A}$ is finite, there exist integers n and ℓ with $n < \#\mathcal{A}$ and $\ell \leq \#\mathcal{A}$ such that $a_n = a_{n+\ell}$. It follows that $a_i = a_{i+\ell}$ for all $i \geq n$.

Assume $p_i \neq \varepsilon$ for some i with $n < i \leq n+\ell$. Since $j > \#\mathcal{A}(|v|+2)$, φ is non-erasing, and $\ell \leq \#\mathcal{A}$, it follows that $\varphi^j(z)$ contains a bounded prefix of length greater than $|v|$. Since $|\rho(\varphi^j(z))| = |v|$ it follows that $\varphi^j(z)$ consists only of bounded letters, a contradiction. Hence, $p_i = \varepsilon$ for all i with $n < i \leq n+\ell$. We conclude that $p_i = \varepsilon$ for all i with $n < i$. Finally, since $\varphi^\ell(a_{j-\ell}) = a_{j-\ell}w$ for some $w \in \mathcal{A}^*$ and we can select j arbitrarily large, we conclude that $\rho(\varphi^{\ell k}(a_{j-\ell})) \sim v$ for any k such that $|\varphi^{\ell k}(a_{j-\ell})| \geq |v|$. □

Lemma 4.12. Let $\varphi\colon \mathcal{A}^* \to \mathcal{A}^*$ be a morphism, $a \in \mathcal{A}$, and $w \in \mathcal{A}^*$ such that $\lim_{k\to\infty} \varphi^k(a) = w^\omega$ and w is primitive. Then $\varphi(w) = w^n$ for some $n > 1$.

Proof. Let p be a proper prefix of w and $n \in \mathbb{N}$ such that $\varphi(w) = w^n p$. Since $\lim_{k\to\infty} \varphi^k(a) = w^\omega$, we have $|\varphi(w)| > |w|$ and thus $n \geq 1$. If follows that $\varphi(ww) = w^n p w^n p$ is a prefix of w^{2n+2}. By [10, Property 2.3], this implies $p = \varepsilon$. Thus $\varphi(w) = w^n$, where $n > 1$ since $|\varphi(w)| > |w|$. □

4.3 Proof of Theorem 4.2

Part 1. Assume that S is not weakly circular. It follows that there exists a sequence $(w_i)_{i\in\mathbb{N}}$ such that $w_i \in \mathcal{L}(S)$, w_i is not weakly synchronized, and $|w_i| \to +\infty$ as $i \to \infty$. Since the words w_i cannot have weakly synchronized factors (Lemma 3.6), it follows from Lemma 4.4 that $(w_i)_{i\in\mathbb{N}}$ is a non-pushy sequence. Let p be the integer from Lemma 4.5 and let $\psi = \varphi^p$. By Lemma 4.6, we may fix a minimal invariant subalphabet $\mathcal{B}$ and an unbounded letter $g \in \mathcal{B}$ such that for every $n \in \mathbb{N}$, there is $k \in \mathbb{N}$ such that $|\psi^k(g)| > n$ and $\psi^k(g)$ occurs as a factor in $(w_i)_{i\in\mathbb{N}}$. As $\psi^k(g)$ is a factor of w_j for some j and k, it cannot be weakly synchronized in S. By Proposition 3.12, it follows that for n large enough, $\psi^k(g)$ is not weakly synchronized in S^p. Moreover, as $\mathcal{B}$ is a minimal invariant subalphabet, it follows that $\psi^\ell(g)$ is a factor of $\psi^k(g)$ for all ℓ with $\ell \leq k$, thus $\psi^\ell(g)$ is not weakly synchronized in S^p for all $\ell \leq k$. Since this is true for infinitely many integers k, we conclude that in fact $\psi^\ell(g)$ is not weakly synchronized in S^p for all $\ell \in \mathbb{N}$.

Fix integers $t, \ell \in \mathbb{N}$ and put $u = \psi^t(g)$. Observe that $(\varepsilon, \psi^{\ell-1}(u), \varepsilon)$ is an interpretation of $\psi^\ell(u)$ in S^p which is compatible with $\big(\varepsilon, \psi^\ell(u)\big)$. As $\psi^\ell(u)$ is not weakly synchronized, we can find an interpretation which is not compatible with the pair $(\varepsilon, \psi^\ell(u))$. Moreover, observe that $\mathcal{B} \cap \mathcal{L}(S)$ is extendable (every element has left and right extensions in the language) and thus we may take this interpretation to be as long as we need. In particular, we can assume it has the form $(s_\ell, \psi^{\ell-1}(v_\ell), t_\ell)$, for some $v_\ell \in \mathcal{L}(S)$ such that s_ℓ is a prefix of $\psi^\ell(a)$ where a is the first letter of v_ℓ, and t_ℓ is a suffix of $\psi^\ell(b)$ for the last letter b of v_ℓ. Since the interpretation is not compatible with $(\varepsilon, \psi^\ell(u))$, we have $s_\ell \neq \varepsilon$.

We claim that the words v_ℓ are bounded in length. Write $v_\ell = a\bar{v}_\ell b$ where $a, b \in \mathcal{A}$, $\bar{v}_\ell \in \mathcal{A}^*$. Since every letter in $\mathrm{alph}(\bar{v}_\ell)$ is mapped by ψ^ℓ to a factor of $\psi(u_\ell) \in \mathcal{B}^*$, Lemma 4.5 implies that $\bar{v}_\ell \in \mathcal{B}^*$. Moreover, the sequence $(\psi^\ell(u))_{\ell\in\mathbb{N}}$ is non-pushy, and ψ^ℓ maps bounded factors of $\bar{v}_\ell$ to bounded factors of $\psi^\ell(u)$, thus $(\bar{v}_\ell)_{\ell\in\mathbb{N}}$ is also non-pushy. Given $x \in \mathcal{A}^*$, let $\#_{\mathsf{u}}x$ be the number of unbounded letters in x. We are reduced to show that $\#_{\mathsf{u}}\bar{v}_\ell$ is bounded independently of ℓ. Since $\mathcal{B}$ is minimal, Lemma 4.5 implies that for all unbounded letters $c \in \mathcal{B}$:

$$|\psi^\ell(c)|/\lceil\psi\rceil \leq |\psi^{\ell-1}(g)| \leq |\psi^\ell(c)| \cdot \lceil\psi\rceil.$$

and therefore

$$\#_{\mathsf{u}}\bar{v}_\ell \cdot |\psi^{\ell-1}(g)| \le |\psi^\ell(\bar{v}_\ell)| \cdot \lceil\psi\rceil \le |\psi^\ell(u)| \cdot \lceil\psi\rceil \le \#_{\mathsf{u}}u \cdot |\psi^{\ell-1}(g)| \cdot \lceil\psi\rceil^2.$$

Thus $\#_{\mathsf{u}}v_\ell$ is bounded independently of ℓ, concluding the proof of the claim.

By the pigeonhole principle, there exists m and n such that $n < m$ and $v_m = v_n = v$, which yields

$$\psi^{m-n}(s_n)\psi^m(u)\psi^{m-n}(t_n) = \psi^m(v) = s_m\psi^m(u)t_m, \tag{2}$$

$$\psi^{m-1}(v) = \psi^{m-n-1}(s_n)\psi^{m-1}(u)\psi^{m-n-1}(t_n). \tag{3}$$

Since, by assumption, $(\varepsilon, \psi^m(u))$ is not compatible with $(s_m, \psi^{m-1}(v), t_m)$, it follows from (3) that $\psi^{m-n}(s_n) \ne s_m$. But by (2), $\psi^{m-n}(s_n)$ and s_m are prefix comparable, so there is $z \in \mathcal{A}^+$ such that $s_m = \psi^{m-n}(s_n)z$ or $\psi^{m-n}(s_n) = s_m z$. In both cases we have from (2) that $\psi^m(u)$ is prefix of $z\psi^m(u)$. As $\psi^{m-n}(s_n)$ and s_m are prefixes of $\psi^\ell(a)$ where a is the first letter of v, we have $|z| < |\psi^m(a)|$.

Let us write $u = u_t$, $z = z_t$, $a = a_t$ and $m = m_t$, as all of these objects depend on t. As $\mathcal{B}$ is p-invariant and $(\psi^k(g))_{k\in\mathbb{N}}$ is a non-pushy sequence, $\psi(g)$ contains at least two unbounded letters. By Lemma 4.5, the letter a_t occurs at least $t-1$ times in $\psi^t(g)$. Therefore,

$$|\psi^{m_t}(u_t)| = |\psi^{m_t}(\psi^t(g))| \ge (t-1)|\psi^{m_t}(a_t)| = (t-1)|z_t|.$$

Since $\psi^{m_t}(u_t)$ is a prefix of $z_t\psi^{m_t}(u_t)$, this implies that z_t^{t-1} is a prefix of $\psi^{m_t}(u_t)$ for all t, and thus $z_t^{t-1} \in \mathcal{L}(S)$. As the sequence $(\psi^\ell(u_t))_{t\in\mathbb{N}}$ is non-pushy, z_t^{t-1} must be unbounded, and therefore S is unboundedly repetitive.

Part 2. Assume that S is unboundedly repetitive, and fix an unbounded element $v \in \Omega(S)$. By Proposition 4.11, there is an unbounded letter $a \in \mathcal{A}$, a positive integer $\ell \le \#\mathcal{A}$, and a word $u \in a\mathcal{A}^* \cap \mathcal{L}(S)$ such that $v \sim u$ and $\lim_{k\to\infty} \varphi^{\ell k}(a) = u^\omega$. By Lemma 4.12, we must have $\varphi^\ell(u) = u^n$ for some $n > 1$. Let $j = \min\{i \in \mathbb{N} \mid \varphi^i(u) \text{ is not primitive}\}$. Note that $1 \le j \le \ell$. Let $z = \varphi^{j-1}(u)$. Write $\varphi(z) = x^m$ where $x = \rho(\varphi(z))$. It follows from the definition of j that $z \in \Omega(S)$ and $m > 1$.

By Proposition 3.11, if S is not weakly circular, it is not strongly circular. Thus, we assume that S is weakly circular. For all $i \in \mathbb{N}$ large enough, the word $x^{im} = \varphi(z^i)$ must be weakly synchronized. It follows that there are words $s, t \in \mathcal{A}^*$ such that $x = st$ and $(s(ts)^{j_1}, (ts)^{j_2}t)$ is a weakly synchronizing pair with $j_1 + j_2 = im - 1$. Let $\bar{x} = ts$. It follows from Lemma 3.6 that $(s\bar{x}^{j_1+k_1}, \bar{x}^{j_1+k_2}t)$ is also weakly synchronizing for all $k_1, k_2 \in \mathbb{N}$. Thus, we may assume that j_1, j_2 are both arbitrarily large and, in particular, $j_1 \ge m$.

Since $(\varepsilon, z^i, \varepsilon)$ and (x, z^{i+1}, x^{m-1}) are two interpretations of x^{im}, there are pairs (z_1, z_2) and (z_3, z_4) such that $z = z_1z_2 = z_3z_4$ and

$$\varphi\left(z_1(z_2z_1)^{\ell_1}\right) = s\bar{x}^{j_1}, \qquad \varphi\left((z_2z_1)^{\ell_2}z_1\right) = \bar{x}^{j_2}t,$$

$$\varphi\left(z_3(z_4z_3)^{\ell_1}\right) = xs\bar{x}^{j_1}, \qquad \varphi\left((z_4z_3)^{\ell_2+1}z_4\right) = \bar{x}^{j_2}tx^{m-1}$$

for some $\ell_1, \ell_2 \in \mathbb{N}$ such that $\ell_1 + \ell_2 = i - 1$. As $|s\bar{x}^{j_1}| = \left|\varphi\left(z_1(z_2z_1)^{\ell_1}\right)\right| = \left|\varphi\left(z_3(z_4z_3)^{\ell_1}\right)\right| - |\bar{x}|$ and $z_1z_2 = z_3z_4$, we have $|z_1| < |z_3|$. Since z is primitive, it follows that $z_2z_1 \neq z_4z_3$. Let q be the longest common suffix of z_2z_1 and z_4z_3. Then $s\bar{x}^{j_1}\varphi(q)^{-1}$ is well-defined due to $j_1 \geq m$, and $(s\bar{x}^{j_1}\varphi(q)^{-1}, \varphi(q)\bar{x}^{j_2}t)$ is compatible with both interpretations $(\varepsilon, z^i, \varepsilon)$ and (x, z^{i+1}, x^{m-1}) with

$$\varphi(z_1(z_2z_1)^{\ell_1}q^{-1}) = s\bar{x}^{j_1}\varphi(q)^{-1}, \quad \varphi(z_3(z_4z_3)^{\ell_1}q^{-1}) = xs\bar{x}^{j_1}\varphi(q)^{-1}.$$

By the definition of q, the last letters of $z_1(z_2z_1)^{\ell_1}q^{-1}$ and $z_3(z_4z_3)^{\ell_1}q^{-1}$ are distinct. This shows that $(s\bar{x}^{j_1}\varphi(q)^{-1}, \varphi(q)\bar{x}^{j_2}t)$ is admissible but not strongly synchronizing. Since j_1, j_2 can be arbitrarily large, we conclude that S is not strongly circular. □

References

1. Cassaigne, J.: An algorithm to test if a given circular HD0L-language avoids a pattern. In: Pehrson, B., Simon, I. (eds.) Proceedings of the IFIP 13th World Computer Congress. pp. 459–464. Elsevier (1994)
2. Cassaigne, J.: Complexité et facteurs spéciaux. Bull. Bel. Math. Soc. Simon Stevin **4**(1), 67–88 (1997). https://doi.org/10.36045/bbms/1105730624
3. Ehrenfeucht, A., Rozenberg, G.: Simplifications of homomorphisms. Inform. Control **38**(3), 298–309 (1978). https://doi.org/10.1016/s0019-9958(78)90095-5
4. Frid, A., Avgustinovich, S.V.: On bispecial words and subword complexity of DOL sequences. In: Ding, C., Helleseth, T., Niederreiter, H. (eds.) Sequences and their Applications. Discrete Mathematics and Theoretical Computer Science. Springer, London (1999). https://doi.org/10.1007/978-1-4471-0551-0_13
5. Klouda, K., Starosta, Š: An algorithm for enumerating all infinite repetitions in a D0L-system. J. Discrete Algorithms **33**, 130–138 (2015). https://doi.org/10.1016/j.jda.2015.03.006
6. Klouda, K., Starosta, Š: The number of primitive words of unbounded exponent in the language of an HD0L-system is finite. J. Comb. Theory, Ser. A **206**, 105904 (2024). https://doi.org/10.1016/j.jcta.2024.105904
7. Klouda, K., Starosta, Š: Characterization of circular D0L-systems. Theoret. Comput. Sci. **790**, 131–137 (2019). https://doi.org/10.1016/j.tcs.2019.04.021
8. Kyriakoglou, R.: Iterated morphisms, combinatorics on words and symbolic dynamical systems. Ph.D. thesis, Université Paris-Est (2019)
9. Mignosi, F., Séébold, P.: If a DOL language is k-power free then it is circular. In: Lingas, A., Karlsson, R., Carlsson, S. (eds.) ICALP 1993. LNCS, vol. 700, pp. 507–518. Springer, Heidelberg (1993). https://doi.org/10.1007/3-540-56939-1_98
10. Mossé, B.: Puissance de mots et reconnaissabilité des points fixes d'une substitution. Theor. Comput. Sci. **99**(2), 327–334 (1992). https://doi.org/10.1016/0304-3975(92)90357-L
11. Rozenberg, G., Salomaa, A.: The Mathematical Theory of L Systems. Academic Press (1980)
12. Salo, V.: Decidability and universality of quasiminimal subshifts. J. Comput. Syst. Sci. **89**, 288–314 (2017). https://doi.org/10.1016/j.jcss.2017.05.017

Shuffle Squares and Nest-Free Graphs

Jarosław Grytczuk[1], Bartłomiej Pawlik[2(✉)], and Andrzej Ruciński[3]

[1] Faculty of Mathematics and Information Science, Warsaw University of Technology, 00-662 Warsaw, Poland
jaroslaw.grytczuk@pw.edu.pl

[2] Institute of Mathematics, Silesian University of Technology, 44-100 Gliwice, Poland
bpawlik@polsl.pl

[3] Faculty of Mathematics and Computer Science, Adam Mickiewicz University, 61-614 Poznań, Poland
rucinski@amu.edu.pl

Abstract. A *shuffle square* is a word consisting of two shuffled copies of the same word. For instance, the French word tuteurer is a shuffle square, as it can be split into two copies of the word tuer. An *ordered graph* is a graph with a fixed linear order of vertices.

We propose a representation of shuffle squares in terms of special nest-free ordered graphs and demonstrate the usefulness of this approach by applying it to several problems. Among others, we prove that binary words of the type $(\mathtt{1001})^n$, n odd, are not shuffle squares and, moreover, they are the only such words among all binary words whose every 1-run has length one or two, while every 0-run has length two. We also provide a counterexample to a believable stipulation that binary words of the form $\mathtt{1}^n\mathtt{0}^{n-2}\mathtt{1}^{n-4}\cdots$, n odd, are far from being shuffle squares (the distance measured by the minimum number of letters one has to delete in order to turn a word into a shuffle square).

Keywords: combinatorics on words · shuffle squares · ordered graphs

1 Introduction

1.1 Definitions

Let $k \geqslant 1$ be an integer and let A be an alphabet of letters, with $|A| = k$. By a *k-ary word* of *length* n we mean any sequence $W = w_1 \cdots w_n$ with $w_i \in A$ for all $1 \leqslant i \leqslant n$. For *binary* words we typically take the alphabet $\{\mathtt{0}, \mathtt{1}\}$ or $\{\mathtt{A}, \mathtt{B}\}$, while for bigger alphabets we rather use the set $[k] := \{\mathtt{1}, \mathtt{2}, \ldots, \mathtt{k}\}$. The length of a word W will be denoted by $|W|$.

Any subsequence $W' = w_{i_1} \cdots w_{i_t}$ of a word $W = w_1 \cdots w_n$ is called a *subword* of W and its set of indices, $\mathrm{supp}(W') = \{i_1, \ldots, i_t\}$, is called a *support* of W'. If the support of a subword W' consists of consecutive integers, then

J. Grytczuk—Supported by Narodowe Centrum Nauki, grant 2020/37/B/ST1/03298.
A. Ruciński—Supported by Narodowe Centrum Nauki, grant 2024/53/B/ST1/00164.

G. Gamard and J. Leroy (Eds.): WORDS 2025, LNCS 15729, pp. 166–178, 2025.
https://doi.org/10.1007/978-3-031-97548-6_15

we call such W' a *block* of W. A pair of subwords X, Y of W with $X = Y$ and disjoint supports will be called *twins*. We will be using the shorthand $\mathrm{supp}(X, Y) = \mathrm{supp}(X) \cup \mathrm{supp}(Y)$ wherever convenient. The length of twins is defined as the length of either one of them, $|X| = |Y|$. We will sometime refer also to the *double length* of twins which equals twice the length and measures how much of the word is taken by the twins. The elements of W which do not belong to either of the twins are called *gaps* (with respect to $\{X, Y\}$).

Example 1. Let $W = \mathtt{1\ 1\ 1\ 0\ 0\ 1\ 0\ 0\ 0\ 1\ 1\ 0}$. One pair of twins (X, Y) in W of length 5 (so of double length 10) is given by $\mathrm{supp}(X) = \{1, 6, 8, 9, 12\}$ and $\mathrm{supp}(Y) = \{2, 3, 4, 5, 7\}$, as indicated by colors and under/overlines: $W = \underline{1}\ \overline{1}\ \overline{1}\ \overline{0}\ \overline{0}\ \underline{1}\ \overline{0}\ \underline{0}\ \underline{0}\ 1\ 1\ \underline{0}$. Indeed, we have $X = Y = \mathtt{11000}$. Note that w_{10} and w_{11} are gaps here.

Given two words $U = u_1 \cdots u_n$ and $W = w_1 \cdots w_m$ their *concatenation* is defined as the word $UW = u_1 \cdots u_n w_1 \cdots w_m$. This definition extends naturally to any finite number of words. In particular, W^r denotes the concatenation of r copies of the word W. For $w \in A$, we may write w^r instead of $ww \cdots w$. For instance, $W = \mathtt{111001000110} = \mathtt{1}^3\mathtt{0}^2\mathtt{10}^3\mathtt{1}^2\mathtt{0}$. This notation comes in handy when there are long blocks of identical letters. A *w-run* in a word W is a maximal block F of W of the form $F = w^r$ for some $r \geqslant 1$. Here "maximal" means that no other block F' of W which contains F is of the form $F' = w^{r'}$, with $r' > r$. By a *run* we mean a w-run for any $w \in A$. Every word can be uniquely decomposed into a concatenation of its runs. For instance, in 011011100 there are two 1-runs (11, 111) and three 0-runs (0, 0, 00), and the corresponding decomposition is $\mathtt{0|11|0|111|00} = \mathtt{01}^2\mathtt{01}^3\mathtt{0}^2$.

For every non-empty word X we call the word $X^2 = XX$ a *square*. For example couscous is a square for $X =$ cous, as is 01100110 for $X = \mathtt{0110}$. In this note, however, we are interested in a (much) broader notion. A *shuffle square* is any word W which contains twins of double length $|W|$; such twins will be called *perfect*. For instance, $\underline{0}\overline{0001}\underline{001}$ is a shuffle square with the twins equal to 0001.

Shuffle squares were introduced in 2012 by Henshall, Rampersad, and Shallit [12], and since then their various aspects have been intensively studied (see [4–6,10,11]. In particular, Bulteau and Vialette [4] proved that determining if a given binary word is a shuffle square is NP-complete. Also, Bulteau, Jugé, and Vialette [5] proved that shuffle squares are *avoidable* over an alphabet of size six, which is currently the best estimate.

Obviously, a necessary condition for a word W to be a shuffle square is that each letter appears in W an even number of times. We will sometimes refer to such a word as *even*. For example, words reappear and tattletale are even. (Elsewhere, cf. [7,14], such words are call *tangrams*.)

Note that if there are twins of length t in a word W, then there are also *monotone* twins X, Y of length t in W, that is, $i_h < j_h$ for all $1 \leqslant h \leqslant t$, where $\mathrm{supp}(X) = \{i_1, \ldots, i_t\}$ and $\mathrm{supp}(Y) = \{j_1, \ldots, j_t\}$. Moreover, there are also *canonical* twins which, in addition to being monotone, partition each run of W

into three consecutive blocks (some might be empty) with the elements of X to the left, followed by the elements of Y, followed by gaps. It is crucial for us that among the longest twins in a word there are always canonical ones. For a formal explanation see [9].

Example 2. In Example 1 the twins are not monotone, but after swapping four pairs of elements between X and Y (all but the first one), the pair of twins indicated here: $W = \underline{1}\ \overline{1}\ \underline{1}\ \underline{0}\ \underline{0}\ \overline{1}\ \underline{0}\ \overline{0}\ \overline{0}\ 1\ 1\ \overline{0}$, is monotone, though not yet canonical due to the (dis)order in the first 1-run. A single "rewiring" operation makes it canonical: $W = \underline{1}\ \underline{1}\ \overline{1}\ \underline{0}\ \underline{0}\ \overline{1}\ \underline{0}\ \overline{0}\ \overline{0}\ 1\ 1\ \overline{0}$ (see Fig. 1).

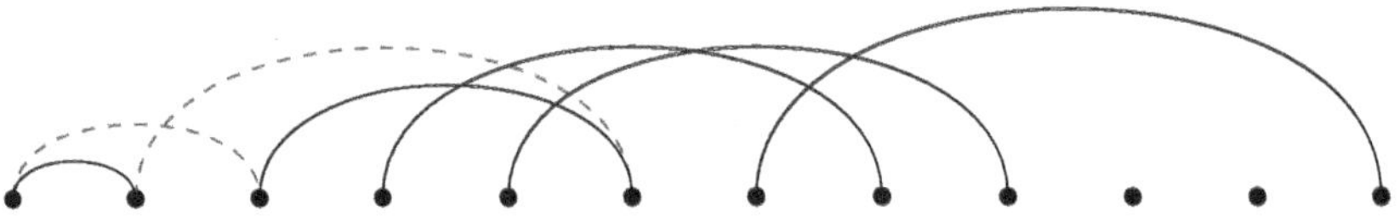

Fig. 1. Non-canonical and canonical (red dashed) twins in the word W = `111001000110`. The arcs join the i-th element of X with the i-th element of Y, $i = 1, \ldots, 5$. (Color figure online)

1.2 Results and Method

In this paper we establish conditions under which binary words belonging to a specified class are shuffle squares. Among others, in Sect. 3 we consider words with four runs and words with all 1-runs of length 1 (see Propositions 2 and 3). In Sect. 4, we give our main result. It states (see Proposition 4 and Theorem 1) that among the binary words with all 0-runs of length two and all 1-runs of length at most two, only the words $(1001)^n$ for n odd, are *not* shuffle squares. That any odd power of the word 1001 is not a shuffle square was conjectured by Komisarski [13], who used to call it the *odd ABBA problem*. Let us also mention that an alleged proof was presented by Fortuna [8], though, up to our knowledge, it has not yet been written down. Actually, Theorem 1 is much more general as it covers not only "odd ABBAs", but many other binary words with an odd number of 0-runs of the same length.

In the last section, we tackle the question of how far a binary word may be from a shuffle square. Given a word W of length n, let $f(W)$ be the length of the longest twins in W and set $g(W) = n - 2f(W)$. Further, let $g(n) = \max g(W)$ over all W, with $|W| = n$. By a sophisticated approach via a regularity lemma for words, Axenovich, Person, and Puzynina [1] showed that $g(n) = o(n)$ and, by providing a construction, that $g(n) = \Omega(\log n)$. The lower bound was improved by Bukh [3] to $g(n) = \Omega(n^{1/3})$ by refining the construction from [1] (see [2] for a proof). In a personal communication Bukh has expressed his belief that, in fact, $g(n) = \Omega(\sqrt{n})$ and that the real challenge would be to improve the upper bound, where one would need to refine the regularity approach from [1]. Focusing on the lower bound, in [2] we shared that belief and implicitly stated the following conjecture.

Conjecture 1. There is a constant c such that for every n there is a binary word of length n with largest twins leaving out at least $c\sqrt{n}$ elements. In short, $g(n) = \Omega(\sqrt{n})$.

In view of the constructions from [1,3], we had believed for long that a natural candidate to facilitate Conjecture 1 is the binary word $O_{m,r} = 1^m 0^{m-2} 1^{m-4} \dots$ consisting of r runs whose lengths form a decreasing sequence of *consecutive* odd numbers, for some odd m and $r \leqslant (m+1)/2$. Here we rebuke this belief by finding twins in $O_{m,r}$ of the double length at least $|O_{m,r}| - 23$, for all m and r. This result does not mean that Conjecture 1 is false. In search for better constructions, one should perhaps allow varying differences between consecutive, still odd, terms.

But at least as important as the obtained results, we believe, is the method we propose to prove them with. To each pair of canonical twins we uniquely assign an *ordered graph* (see the formal definition in he next section). Thus, instead of looking for twins in a word or deciding if it is a shuffle square, we construct a corresponding graph or show that such a graph does not exist. We believe that this approach is more transparent, insightful, and ultimately has some potential in this area of research.

2 Graphic Representations

Here we introduce our main tool – a representation of shuffle squares and, more generally, canonical twins in terms of ordered graphs with no nests. An *ordered graph* is just a graph with its vertex set linearly ordered. We allow loops and parallel edges. A *nest* in an ordered graph consists of a pair of disjoint edges (including loops) e and f with $\min e < \min f$ and $\max e > \max f$. An ordered graph is *nest-free* if no pair of its edges forms a nest.

Let a k-ary word $W = w_1 \cdots w_n$ have canonical twins $X = w_{i_1} \cdots w_{i_t}$ and $Y = w_{j_1} \cdots w_{j_t}$. In particular, $\{i_1, \dots, i_t\} \cap \{j_1, \dots, j_t\} = \emptyset$, and for all $h = 1, \dots, t$, we have $i_h < j_h$ and $w_{i_h} = w_{j_h}$. As an intermediate step, define an ordered graph $M = M_{X,Y}$ on the set of vertices $\{1, \dots, n\}$ with edges $\{i_1, j_1\}, \dots, \{i_t, j_t\}$. So, M consists of a *matching* of size t and $n - 2t$ isolated vertices. Clearly, there are no nests in M.

Let $W = U_1 \cdots U_m$, where each block U_h is a run in W. Now, contract each set $\operatorname{supp}(U_h)$ into a single vertex u_h, keeping all the edges of M in place and keeping the order $u_1 < \cdots < u_m$. We denote the obtained ordered multigraph by $G = G_{X,Y}$. Observe that G is still nest-free and, moreover, consists of k separated graphs on vertex sets $N_w = \{h : U_h \text{ is a } w\text{-run}\}$, for each letter $w \in [k]$. The degrees of vertices in G are $\deg_G(u_h) = |\operatorname{supp}(U_h) \cap \operatorname{supp}(X, Y)|$, $h = 1, \dots, m$.

From G it is straightforward to uniquely reconstruct (X, Y). The uniqueness is a consequence of the fact that we restrict ourselves to canonical twins. We simply scan the vertices u of G from left to right and record their *right degrees* $\deg_G^{(\text{right})}(u)$ defined as the number of edges going from u to the right and *left degrees* $\deg_G^{(\text{left})}(u)$ defined analogously. Then from each run U_h, assign the first

$\deg_G^{(\text{right})}(u)$ elements to X and the next $\deg_G^{(\text{left})}(u)$ to Y, leaving the remaining vertices of U_h, if there are any, unassigned (gaps). But if we are only interested in the length of the twins, then just by looking at G we learn that there is a pair of twins in W of double length equal to $\sum \deg_G(u_h)$, where loops count as two.

This allows us to apply this "graphic" approach also in situations when we do not know any twins in W and our goal is to find them as large as possible. Then we simply attempt to construct a nest-free ordered graph G on vertices u_h, $h = 1, \ldots, m$, with no edges joining vertices from different sets N_w, with degrees as close as possible to the lengths of the corresponding runs U_h. In particular, if we succeed to create such a graph G with all $\deg_G(u_h) = |U_h|$, then we know that W is a shuffle square and we can produce a pair of perfect twins witnessing this property.

Let us summarize the above discussion in the following statement characterizing shuffle squares in terms of their ordered graph representations.

Proposition 1. *Let W be a k-ary word and let $W = U_1 \cdots U_m$ be a decomposition of W into runs U_i. Then W is a shuffle square if and only if there exists an ordered graph G on the set $\{u_1, \ldots, u_m\}$ satisfying the following conditions.*

1. *If u_i is joined to u_j by an edge, then the corresponding runs U_i and U_j are w-runs for the same letter w.*
2. *The degree of each vertex u_i satisfies $\deg_G(u_i) = |U_i|$.*
3. *Graph G is nest-free.*

Example 3. Let $W = \underline{1}\ \underline{1}\ \overline{1}\ \underline{2}\ \underline{2}\ \underline{3}\ 3\ 3\ \overline{1}\ \overline{2}\ \overline{2}\ \overline{3}$ with canonical twins as indicated. The corresponding matching M and graph G are given in Fig. 2. Note that there is no way to change G so that $\deg_G(u_3) = 3$ while keeping all other degrees intact. Indeed, u_6 has to be adjacent to u_3, so there cannot be a loop at u_5 which, in turn, implies that there is no loop at u_3 either.

Example 4. Let $W = \underline{1}\ 1\ 1\ \underline{0}\ \underline{0}\ \overline{1}\ \overline{0}\ \overline{0}\ 0\ \underline{1}\ \overline{1}\ 0$ and (X, Y) be monotone twins in W of length 4 given by $\mathrm{supp}(X) = \{1, 4, 5, 10\}$ and $\mathrm{supp}(Y) = \{6, 7, 8, 11\}$ (so, both equal 1001) as indicated. The corresponding matching M and graph G are given in Fig. 3. By adding a single loop at vertex u_1, we immediately obtain a pair of twins of length 5 in W, $X' = Y' = 11001$, a bit different from those in Example 2.

As W is even, we might wonder if it is a shuffle square, that is, if it possesses a pair of twins of length 6. This can be quickly answered in negative, using the graph representation as a tool. If W were a shuffle square, there would exist an ordered nest-free graph G on vertex set $\{u_1, \ldots, u_6\}$ with no edges between $\{u_1, u_3, u_5\}$ and $\{u_2, u_4, u_6\}$ and degrees $\deg_G(u_h) = |U_h|$, $h = 1, \ldots, 6$. However, in Fig. 3 we see that such a graph does not exist. Indeed, vertex u_1 cannot be connected to u_5 as then all edges at the bottom part would need to be adjacent to u_6 which is impossible. So, there must be a loop at u_5 which cuts off vertex u_6 at the bottom.

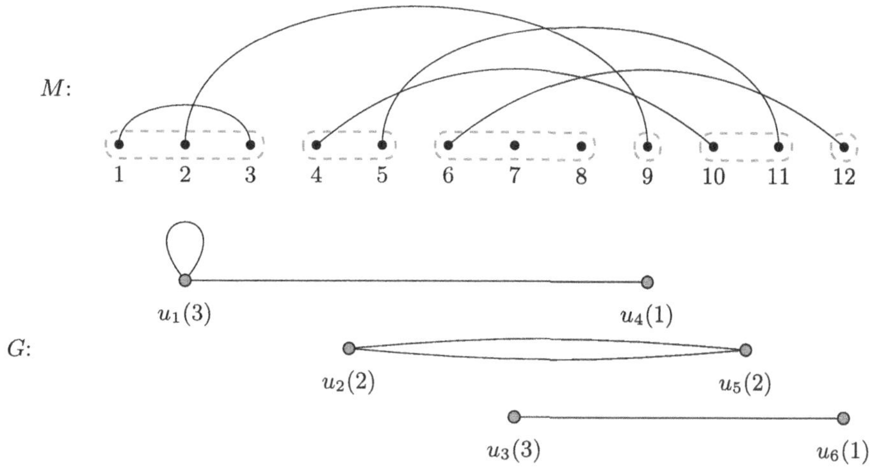

Fig. 2. Matching M and graph G for $W = \underline{1}\ \underline{1}\ \overline{1}\ \underline{2}\ \underline{2}\ \underline{3}\ 3\ 3\ \overline{1}\ \overline{2}\ \overline{2}\ \overline{3}$.

Example 5. Is $W = 110^5 10^6 1110^4 110^7$ a shuffle square? One may try a tedious trial-and-error approach, but instead we draw a nest-free graph G, see Fig. 4, with degrees equal to the sizes of the blocks in W. This yields the answer "yes" to the above question, and we may still construct a pair of perfect twins in W based on G. It is $W = \underline{1}\ \overline{1}\ \underline{0^5}\ \underline{1}\ \underline{0}\ \overline{0^5}\ \underline{1^2}\ \overline{1}\ \underline{0^3}\ \overline{0}\ \overline{1^2}\ \underline{0^2}\ \overline{0^5}$.

3 Two Easy Pieces

In this section we prove two relatively simple results on shuffle squares which demonstrate the elegance and simplicity of the graphic representation. An even word is *dull* if all its runs have even length. It is called dull, because it is trivially a shuffle square with the perfect twins just alternating.

Note that no even binary word with the total of *three* runs is a shuffle square unless it is dull. Indeed, the only way to construct a graph G with vertices $u_1 < u_2 < u_3$ of degrees $|U_1|, |U_2|, |U_3|$, respectively, and with no edges between $\{u_1, u_3\}$ and u_2, is to put $|U_i|/2$ loops at each vertex u_i, $i = 1, 2, 3$, as otherwise nests would be created.

We now settle the case of *four* runs. The result appeared already in [10].

Proposition 2. *Let a, b, c, d be positive integers, not all even, but with $a+c$ and $b+d$ both even. Then the (even but not dull) word $W = 1^a 0^b 1^c 0^d$ is a shuffle square if and only if $a \geqslant c$ and $b \leqslant d$.*

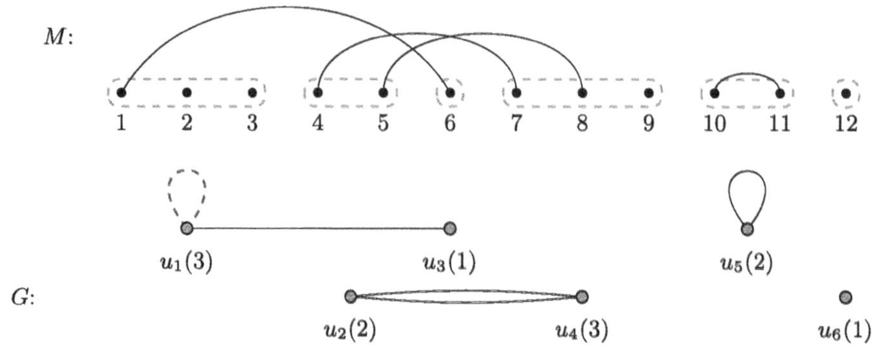

Fig. 3. Matching M and graph G for $W = \underline{1}\,1\,1\,\underline{0}\,\underline{0}\,\overline{1}\,\overline{0}\,\overline{0}\,0\,\underline{1}\,\overline{1}\,0$.

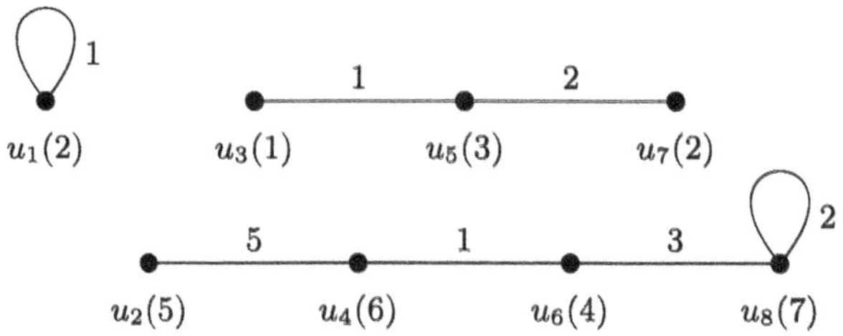

Fig. 4. The word $W = 1\,1\,0^5\,1\,0\,0^5\,1^2\,1\,0^3\,0\,1^2\,0^2\,0^5$ is a shuffle square.

Proof. Consider a nest-free ordered graph on vertex set $u_1 < u_2 < u_3 < u_4$ with capacities a, b, c, d, respectively, and edges permissible only between pairs u_1, u_3 and u_2, u_4 (in addition to loops). If there is a shuffle square in W, then both edges, u_1, u_3 and u_2, u_4, must be present (with some multiplicities), as loops can only accommodate even degrees. But then there cannot be any loops either at u_2 or at u_3, implying that $a \geqslant c$ and $b \leqslant d$ (see Fig. 5).

On the other hand, if $a \geqslant c$ and $b \leqslant d$, then we draw edge u_1u_3 with multiplicity c, edge u_2u_4 with multiplicity b and add $(a-c)/2$ loops at u_1 and $(d-b)/2$ loops at u_4 (see Fig. 5). The obtained ordered graph is nest-free and thus corresponds to a pair of perfect twins in W of the form

$$\underline{1^{(a-c)/2}}\;\overline{1^{(a-c)/2}}\;\underline{1^c}\;\underline{0^b}\;\overline{1^c}\;\overline{0^b}\;\underline{0^{(d-b)/2}}\;\overline{0^{(d-b)/2}}.$$

Note that both twins form the word $1^{(a+c)/2}0^{(b+d)/2}$. □

As a corollary of Proposition 2 we infer that every even word $w = 1^a0^b1^c0^d$ with all a, b, c, d positive, can be rotated (cyclically permuted) so that the outcome is a shuffle square. Indeed, the orderings $abcd, bcda, cdab, dabc$ represent all four possible pairwise relations between a, c and between b, d, so at least one of them will satisfy the conditions in Proposition 2 (with a, b, c, d renamed if needed).

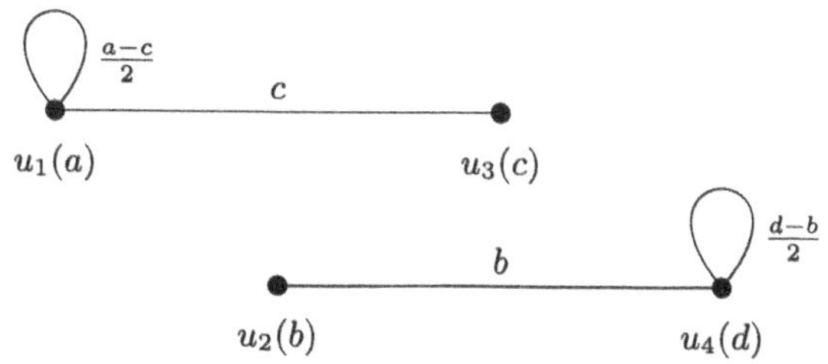

Fig. 5. Graph G for the shuffle square $1^a 0^b 1^c 0^d$.

Finally in this section, we consider even binary words with all 1's separated from each other, that is with all 1-runs of length 1, but with no restrictions on the 0's. That is, we are looking at words of the form

$$W = 1\,0^{a_1}1\,0^{a_2}\cdots1\,0^{a_{2m}}, \tag{1}$$

for positive integers $m, a_1, \ldots, a_{2m}$. Our goal is to define conditions for the values of a_i that make W a shuffle square with the 1's *belonging alternately to perfect twins*.

This could be solved without referring to the ordered graphs, but we prefer the latter approach as it seems to be more insightful. Any graph representing perfect twins in word W must attain a structure presented in Fig. 6, with the additional constraint that for each $i = 1, 3, \ldots, 2m - 3$, either y_i or z_{i+1} must be 0. The variables x_i, y_i, z_i represent multiplicities of the respective edges, so we have altogether $4m - 2$ unknowns to find. Let us denote the top vertices by $u_1, \ldots, u_{2m}$ and the bottom vertices by $v_1, \ldots, v_{2m}$.

$$r_i = \deg^{(right)}(v_i), \quad \ell_i = \deg^{(left)}(v_i), \quad r_i + r_{i+1} = \ell_{i+1} + \ell_{i+2}.$$

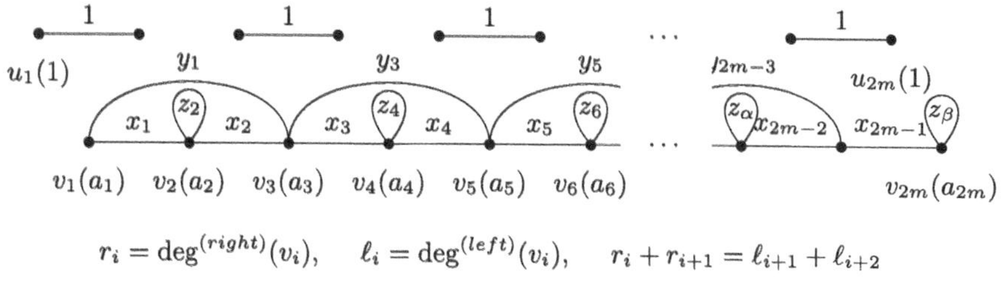

Fig. 6. The structure of the graph representing perfect twins in the shuffle square $1\,0^{a_1}1\,0^{a_2}\cdots1\,0^{a_{2m}}$.

One way to proceed from this would be to set up a system of $2m$ equations guaranteeing that for each $i = 1, \ldots, 2m$, the degree of vertex v_i must be equal to a_i, supplemented by $m-1$ equations of the form $y_i z_{i+1} = 0$, $i = 1, 3, \ldots, 2m-3$. However, it seems simpler to apply more coarse setting in terms of the right degrees $r_i = \deg_G^{(right)}(v_i)$ and left degrees $\ell_i = \deg_G^{(left)}(v_i)$ and utilize the observation that $r_1 = a_1$ and, for each $i = 1, 3, \ldots, 2m-1$, $r_i + r_{i+1} = \ell_{i+1} + \ell_{i+2}$.

Thus, after getting rid of the ℓ_i's (as $r_i + \ell_i = a_i$ for all i), we have just $2m - 1$ variables and m equations. Moreover, we avoid the annoying system of $m - 1$ alternatives: $y_i = 0$ or $z_{i+1} = 0$. But the biggest benefit we earn from this substitution is that the m equations each involve a separate set of variables.

So, it remains to solve this system of m equations. We arbitrarily choose the r_i's with odd i as free variables which determine the remaining ones. Then, the system becomes

$$\begin{cases} r_2 = \frac{1}{2}(-a_1 + a_2 + a_3 - r_3) \\ r_{2i+2} = \frac{1}{2}(a_{2i+2} + a_{2i+3} - r_{2i+1} - r_{2i+3}), 1 \leqslant i \leqslant m-2 \\ r_{2m} = \frac{1}{2}(a_{2m} - r_{2m-1}) \end{cases} \tag{2}$$

As for all i we have the natural constraints $0 \leqslant r_i \leqslant a_i$, we need to show that the system of inequalities

$$\begin{cases} (a_3 - a_1) - u_2 \leqslant r_3 \leqslant (a_3 - a_1) + a_2 \\ a_{2i+3} - a_{2i+2} \leqslant r_{2i+1} + r_{2i+3} \leqslant a_{2i+3} + a_{2i+2}, 1 \leqslant i \leqslant m-2 \\ r_{2m-1} \leqslant a_{2m} \end{cases} \tag{3}$$

has integer-valued solutions $r_3, \ldots, r_{2m-1}$ such that for all $i = 3, \ldots, 2m-1$ we have $0 \leqslant r_i \leqslant a_i$ and the quantities in the parenthesis in (2) are all even (so that the values of $r_2, r_4 \ldots, r_{2m}$ are integers). If for a given sequence of a_i's there exist r_i's satisfying the above conditions, then W has perfect twins of the form

$$1\,0^{a_1+r_2}1\,0^{r_3+r_4}\ldots 1\,0^{r_{2m-1}+r_{2m}} = 1\,0^{\ell_2+\ell_3}1\,0^{\ell_4+\ell_5}\ldots 1\,0^{\ell_{2m}}.$$

Thus we proved the following result.

Proposition 3. *A binary word given by* (1) *is a shuffle square with alternating* 1*'s if and only if the combined systems of inequalities* (3) *and equations* (2) *have integer solutions with* $0 \leqslant r_i \leqslant a_i$ *for all* i. □

In particular, for $m = 1$, the sole condition for $W = 1\,0^{a_1}\,1\,0^{a_2}$ to be a shuffle square is, trivially, that $a_1 \leq a_2$. For $m = 2$, the above system of inequalities is reduced to just

$$\begin{cases} (a_3 - a_1) - a_2 \leqslant r_3 \leqslant (a_3 - a_1) + a_2 \\ r_3 \leqslant a_4 \end{cases}$$

So, whether W is a shuffle square is determined by the validity of the single inequality $a_3 + a_2 - a_1 \geq 0$. Then any choice of

$$r_3 \in \{\max(0, a_3 - a_2 - a_1), \min(a_3 + a_2 - a_1, a_3, a_4)\}$$

which has the same parity as $a_3 + a_2 - a_1$ (and so, the same parity as a_4) yields a feasible solution (r_2, r_3, r_4), where

$$\begin{cases} r_2 = \frac{1}{2}(-a_1 + a_2 + a_3 - r_3) \\ r_4 = \frac{1}{2}(a_4 - r_3) \end{cases}$$

Interestingly, if a word $W = 1\,0^{a_1}1\,0^{a_2}1\,0^{a_3}1\,0^{a_3}$ does not satisfy the condition $a_3 + a_2 - a_1 \geqslant 0$, then there is always a rotation so that the new word still begins with a 1, but now is a shuffle square with the 1's alternating. Indeed, it suffices to redefine a_1 as the minimum of all four parameters and start the word there. For instance, let $W = 1\,0^{16}1\,0^{9}1\,0^{5}1\,0^{4}$. Here $a_3+a_2-a_1 = 5+9-16 < 0$, so this word is not a shuffle square. But beginning at, say, $1\,0^4$ we get a new word $W' = 1\,0^{4}1\,0^{16}1\,0^{9}1\,0^{5}$, a rotation of W, for which the condition $a_3+a_2-a_1 \geqslant 0$ holds. Thus, W' is a desired shuffle square.

This is no longer true for $m \geqslant 3$. For instance, $W = 1\,0\,1\,0\,1\,0^{5}1\,0\,1\,0\,1\,0^{5}$ does not have a shuffle square consisting of twins which get the 1's alternately, no matter at which 1 we begin. (On the other hand, W is a perfect square.)

4 The Odd ABBA Problem

Recall that if every run of a binary word W has an even length, in particular, length two, then, trivially, W is a shuffle square. It is perhaps a bit surprising that a modest relaxation allowing runs of length one, but only for one of the two letters, say letter 1, becomes quite non-trivial.

One can quickly come up with such words which are not shuffle squares, like 1001 or 100110011001. Our main result in this section states that words of that form, more generally, of the form $(1001)^n$, for n odd, are indeed *not* shuffle squares.

As a complementary result, we show in [9] that all other even binary words with all 0-runs of length exactly two and all 1-runs of length at most two, are shuffle squares. So, in fact, the odd repeats of the word 1001 are the sole exception in this class.

Proposition 4. *Every even binary word in which every* 0*-run has length two, while every* 1*-run has length at most two is a shuffle square, unless it is of the form* $(1001)^n$ *for some odd* n. □

The proof of our main result relies on the following lemma.

Lemma 1. *Let* $H = P \cup C$ *be an ordered graph on the set of vertices* $\{1, 2, \ldots, n\}$*, consisting of two vertex disjoint subgraphs: a path* P *with endpoints at* 1 *and* n*, and a cycle* C*. If* H *is nest-free, then* C *is an* even *cycle.*

Proof. Let a nest-free graph H satisfy the assumptions of the lemma and let $E(P)$ denote the set of edges of the path P. Consider an order relation $\prec$ on $E(P)$ defined as follows: for each pair of distinct edges $e, f \in E(P)$, $e \prec f$ if $\min e \leqslant \min f$ and $\max e \leqslant \max f$. Since P is nest-free, this relation is a linear order and one can enumerate the edges of P accordingly to that order as $E(P) = \{e_1 \prec e_2 \prec \cdots \prec e_p\}$. Let $b(e_i) \equiv i \pmod 2$, $i = 1, \ldots, p$, where $p = |E(P)|$, be the parity of edge e_i.

Let $V(C) = \{x_1, \ldots, x_q\}$ be the vertex set of the cycle C, where the vertices are numbered in the order they are traversed by C (we fix one of the two possible directions). We say that a vertex x of C is *hugged* by an edge e of P if $\min e < x < \max e$. For each $i = 1, \ldots, q$, let $X_i = \{e \in P : x_i \text{ is hugged by } e\}$ and set $r_i = |X_i|$.

Observe that for each $i = 1, \ldots, q$ the number r_i is odd, because path P, on the way from 1 to n (possibly zigzaging), "passes" each vertex of C either once, or trice, or five times, etc. Moreover, by the definition of hugging, each set X_i consists of a block of consecutive edges of P under $\prec$. Finally, by the nest-freeness of H, for each $i = 1, \ldots, q-1$, the sets X_i and X_{i+1} are disjoint and positioned "side-to-side", that is, X_{i+1} either directly follows or directly precedes X_i.

Thus, we may define a 2-coloring of $V(C)$ by assigning to $x_i \in V(C)$, $i = 1, \ldots, q$, color $b(e)$, where e is the first (or the last) edge in X_i. Clearly, adjacent vertices on C receive different colors, which implies that q is even. □

Using this lemma we may easily get our main result.

Theorem 1. *Let $m, r, s, t, a_1, \ldots, a_{2m}$ be positive integers such that s, t are odd while $r, a_1, \ldots, a_{2m}$ are even. Then the word $W = \mathtt{1}^s\mathtt{0}^r\mathtt{1}^{a_1}\mathtt{0}^r\mathtt{1}^{a_2}\cdots\mathtt{1}^{a_{2m}}\mathtt{0}^r\mathtt{1}^t$ is not a shuffle square. In particular, the word $(\mathtt{1001})^{2m+1}$ is not a shuffle square.*

Proof. Suppose on the contrary that W is a shuffle square and let $G = G_0 \cup G_1$ be an ordered nest-free graph representing W as in Proposition 1, where G_0 and G_1 are the subgraphs of G corresponding, respectively, to 0-runs and 1-runs. Thus, the degree sequence of G_0 is $(\underbrace{r, r, \ldots, r}_{2m+1})$ and the degree sequence of G_1 is $(s, a_1, a_2, \ldots, a_{2m}, t)$. Since G_0 is an r-regular graph with an odd number of vertices, it cannot be bipartite, so it contains and odd cycle C. On the other hand, G_1 contains a path P whose ends are the only vertices of G_1 with odd degrees, namely the first and the last vertex. It follows by Lemma 1 that $H = C \cup P \subseteq G$ is not nest-free, and so is G, a contradiction. □

5 Largest Twins

We first describe a general class of words whose two particular instances yielded lower bounds on $g(n)$ in [1,3]. For positive integers $n_1, \ldots, n_r$, let $W(n_1, \ldots, n_r) = U_1 \cdots U_r$ be the binary word whose consecutive runs U_i have lengths $|U_i| = n_i$, $i = 1, \ldots, r$. We adopt the convention that U_1 is a 1-run. The main features of the words used in [1,3] were that the runs rapidly decrease in length and that their lengths are odd.

An easy generalization of these two special words led us in [2] to an observation that if $n_1 > n_2 > \cdots > n_r$ are all odd, then $g(W(n_1, \ldots, n_r)) \geqslant \min\{r, \delta\}$, where δ is the smallest difference between two consecutive terms (see [2, Lemma 3.1]). The proof idea is that if twins take equally from each run, then they "lose" at least one element per run. Otherwise, the leading twin is ahead at some point and the follower has to "jump" over the second next run, a loss that can never be made up, due to the run lengths getting smaller and smaller.

The particular words used in [1,3] were $A_r := W(3^{r-1}, 3^{r-2}, \ldots, 3^2, 3, 1)$ and $B_r := W(r^2+1, (r-1)r+1, \ldots, 2r+1, r+1)$, respectively. Simple calculations reveal that $r = \Theta(\log |A_r|)$ and $r = \Theta(|B_r|^{1/3})$, leading to the desired bounds. (For n not expressible as $|A_r|$ or $|B_r|$ for any r an obvious extrapolation has been applied.)

In view of the constructions from [1,3] a natural candidate to facilitate Conjecture 1 seemed to be, for an odd integer m and $r \leq (m+1)/2$, a binary word $O_{m,r} := W(m, m-2, m-4, \ldots, m-2r+2)$ consisting of r *consecutive* odd numbers. Indeed, $n := |O_{m,r}| = r(m-r+1) = \Theta(m^2)$ for $r = \Theta(m)$ and thus, showing that $g(O_{m,r}) = \Omega(m)$ would do the job. Below we show that independently of m and r, $O_{m,r}$, quite surprisingly, is very close to a shuffle square even for large m.

Proposition 5. *For every $m \geqslant 2r-1$, m odd, the word $O_{m,r}$ contains twins of total length at least $|O_{m,r}| - 23$, that is, $g(O_{m,r}) \leqslant 23$.*

Proof. Our aim is to find canonical twins in $O_{m,r}$ of double length at least $|O_{m,r}| - 23$. To this end, we are going to construct a nest-free graph on r vertices whose *deficit* defined as $|O_{m,r}| - \sum_{u \in V(G)} \deg_G(v)$ is at most 23. We may and do assume that $r \geqslant 24$, as otherwise we could draw as many as possible loops at every vertex, getting the deficit of $r \leqslant 23$. For sheer convenience, we also assume that r is divisible by 8. If not, we add loops at up to the first seven vertices on, say, the left end of $O_{m,r}$, procuring a deficit of at most 7, and build G on the remaining vertices. So, let $r = 8k$ for some integer $k \geqslant 3$.

As usual, there are two kinds of vertices, say, red and blue, alternating, and there are no red-blue edges whatsoever. Let us identify the vertices with their capacities, so we have the red set $V_1 = \{m, m-4, m-8, \ldots, m-2r+4\}$ and the blue set $V_2 = \{m-2, m-6, \ldots, m-2r+2\}$.

We begin by describing the "red" subgraph $G[V_1]$. We split V_1 into k consecutive 4-tuples $U_i = \{i, i-4, i-8, i-12\}$, $i = m, m-16, \ldots, m-2r+16$, with edge sets $E_i = \{\{i, i-4\}, \{i, i-8\}, \{i-4, i-12\}\}$ and the edges bearing multiplicities $\mu(i, i-4) = 8$, $\mu(i, i-8) = i-8$, and $\mu(i-4, i-12) = i-12$ (see Fig. 7 - the top part). Note that this way the deficit of the red subgraph $G[V_1]$ is zero. Indeed, for each i, $\deg_G(i) = 8 + (i-8) = i$, $\deg(i-4) = 8 + (i-12) = i-4$, $d(i-8) = i-8$, and $d(i-12) = i-12$.

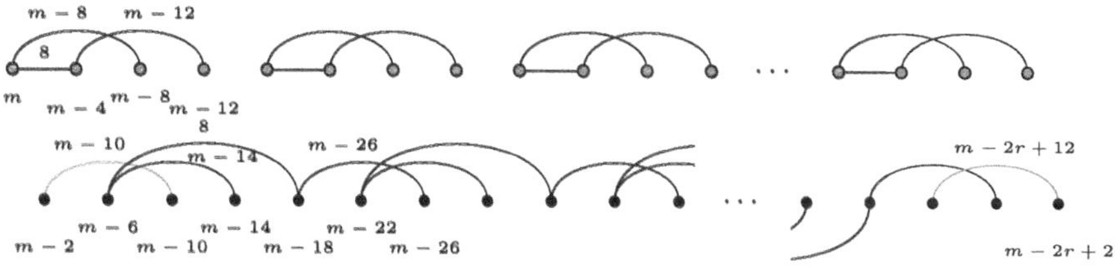

Fig. 7. Construction from the proof of Proposition 5.

The "blue" subgraph is a bit more complicated, as we have to avoid traps set up by the red graph. Nevertheless, its structure is almost the same: it consists of disjoint paths of lengths three, except that now these paths are slightly more stretched out, and two isolated edges, one at the beginning and one at the end. It is these two edges what raises the deficit by 16. Precisely, we have $\mu(m-2, m-10) = m-10$ and $\mu(m-2r+10, m-2r+1) = m-2r+1$, while the remaining vertices are broken up into $k-1$ 4-tuples $W_i = \{i, i-8, i-12, i-20\}$, $i = m-6, m-22, \ldots, m-2r+26$, with edges sets $E_i = \{\{i, i-8\}, \{i, i-12\}, \{i-12, i-20\}\}$ and the edges of multiplicities $\mu(i, i-8) = i-8$, $\mu(i, i-12) = 8$, and $\mu(i-12, i-18) = i-18$ (see Fig. 7 – the bottom part). The subgraphs $G[W_i]$ contribute no deficit, so that the total deficit of G is at most $7+8+8 = 23$. □

Acknowledgment. We would like to thank Andrzej Komisarski for sharing with us his lovely ABBA problem, as well as some results of his impressive computer experiments.

References

1. Axenovich, M., Person, Y., Puzynina, S.: A regularity lemma and twins in words. J. Combin. Theory Ser. A **120**, 733–743 (2012)
2. Basu, A., Ruciński, A.: How far are ternary words from shuffle squares? Ars Math. Contemp. **25**(3), #P3.08 (2024). https://doi.org/10.26493/1855-3974.3211.86d
3. Bukh, B.: unpublished note (2013) (1 page)
4. Bulteau, L., Vialette, S.: Recognizing binary shuffle squares is NP-hard. Theor. Comput. Sci. **806**, 116–132 (2020)
5. Bulteau, L., Jugé, V., Vialette, S.: On shuffled-square-free words. Theor. Comput. Sci. **941**, 91–103 (2023)
6. Buss, S., Soltys, M.: Unshuffling a square is NP-hard. J. Comput. System Sci. **80**, 766–776 (2014)
7. Dębski, M., Grytczuk, J., Pawlik, B., Przybyło, J., Śleszyńska-Nowak, M.: Words avoiding tangrams. Ann. Comb. (2024)
8. Fortuna, W.: O hipotezie ABBA *(in Polish)*. In: Presentation at The VIII Students' Conference *Żmirlacz*, AGH Cracow (2024)
9. Grytczuk, J., Pawlik, B., Rucinski, A.: Graphic images of shuffle squares. arXiv:2503.22043
10. Grytczuk, J., Pawlik, B., Pleszczyński, M.: Variations on shuffle squares. arxiv:2308.13882
11. He, X., Huang, E., Nam, I., Thaper, R.: Shuffle squares and reverse shuffle squares. Europ. J. Combin. **116**, 103883 (2024)
12. Henshall, D., Rampersad, N., Shallit, J.: Shuffling and unshuffling. Bullet. EATCS **107**, 131–142 (2012)
13. Komisarski, A.: Personal communication
14. Ochem, P., Pierron, T.: 4-tangrams are 4-avoidable. arxiv:2502.20774

A Succinct Study of Positionality for Dumont–Thomas Numeration Systems

Savinien Kreczman[1(✉)], Sébastien Labbé[2], and Manon Stipulanti[1]

[1] Department of Mathematics, ULiège, Liège, Belgium
{savinien.kreczman,m.stipulanti}@uliege.be
[2] LaBRI, Université de Bordeaux, Talence, France
sebastien.labbe@labri.fr

Abstract. Numeration systems are maps between a set of numbers and a set of words that act as representations of these numbers. One desirable property is positionality: the ability to relate positions in the words to values of the numbers. In general, positionality is hard to decide. In this article, we obtain a criterion to decide the positionality of so-called Dumont–Thomas numeration systems, arising from substitutions. Then, we particularize this criterion to some well-behaved classes of substitutions, allowing us to link the related systems to existing literature.

Keywords: Morphism · Substitution · Periodic points · Numeration system · Positionality · Bertrand property · Fabre substitution

1 Introduction

The need of representing natural numbers (or expressing them in writing) has occupied humans for centuries. In its most general form, this leads to so-called *abstract numeration systems*, introduced by Lecomte and Rigo in 2001 [10] (see also [1, Chapter 3]). Such a numeration system is defined by a triple $S = (L, A, \prec)$ where A is an alphabet ordered by the total order $\prec$ and L is an infinite *regular* language over A (i.e., accepted by a deterministic finite automaton). We say that L is the *numeration language* of S. When we order the words of L with the *genealogical* order (i.e., first by length, then using the dictionary order) induced by $\prec$, we obtain a one-to-one correspondence rep_S between $\mathbb{N}$ and L. The *(S-) representation* of the non-negative integer n is then the $(n+1)$st word of L, and the inverse map, called the *(S-)evaluation map*, is denoted by val_S. A simple example is given by the abstract numeration system S built on the language $L = \mathtt{1}^*\mathtt{2}^*$ over the ordered alphabet $\{\mathtt{1}, \mathtt{2}\}$. The first few words in the language are $\varepsilon, \mathtt{1}, \mathtt{2}, \mathtt{11}, \mathtt{12}, \mathtt{22}, \mathtt{111}$. For example, $\text{rep}_S(5) = \mathtt{22}$ and $\text{val}_S(\mathtt{111}) = 6$.

In general, a numeration system $S = (L, A, \prec)$ is *positional* if the underlying alphabet A is a set of consecutive integers $\{\mathtt{0}, \mathtt{1}, \dots, \mathtt{c}\}$ for some $\mathtt{c} \in \mathbb{N}$ and if there exists a sequence $(U_i)_{i\geq 0}$ of non-negative integers such that the evaluation map is of the form $\text{val}_S\colon A^* \to \mathbb{N}, w_{k-1}\cdots w_0 \mapsto \sum_{i=0}^{k-1} w_i U_i$. We use the term *positional*

G. Gamard and J. Leroy (Eds.): WORDS 2025, LNCS 15729, pp. 179–191, 2025.
https://doi.org/10.1007/978-3-031-97548-6_16

because the *positions* of letters in representations are used to generate numbers. Observe that the numeration system built on $L = 1^*2^*$ cannot be positional: indeed, we have $\mathrm{rep}_S(3) = 11$ and $\mathrm{rep}_S(5) = 22$, so there is no hope to find an integer sequence $(U_i)_{i\geq 0}$ such that $3 = 1 \cdot U_1 + 1 \cdot U_0$ and $5 = 2 \cdot U_1 + 2 \cdot U_0$ (see also [1, Example 3.1.12]). We thus raise the following question (see also [1, Exercise 3.13]):

Question 1. What are the conditions for an abstract numeration system to be positional?

As this question seems difficult to answer in its full generality, we consider a particular case with Question 2 below. Roughly, we consider numeration systems that are derived from *substitutions*, i.e., maps sending sequences to sequences and satisfying some mild properties. These numeration systems are due to Dumont and Thomas [4] in 1989 and are quite classical. See [6,15] for some examples of applications. In this paper, we exhibit conditions on the underlying substitution so that the corresponding Dumont–Thomas numeration for $\mathbb{Z}$ is positional.

The outline of the paper is as follows. Section 2 gives the necessary background and preliminary results, including the extension of Dumont–Thomas numerations to all integers and all words lengths. In Sect. 3, we study which Dumont–Thomas numeration systems are positional to answer Question 2. We start with a sketch of the argument in Sect. 3.1, then we state our main result in Sect. 3.2. We turn to particular cases in Sect. 3.3 and we finish by discussing the properties of our Dumont–Thomas numeration systems related to existing literature, e.g., the property of a numeration system to be Bertrand [2,3]. All proofs omitted due to space constraints are available online in [7].

2 Preliminaries

General Combinatorics on Words. We assume that the reader is familiar with basic notions of combinatorics on words –alphabet, word, length, factor, prefix, left- and right-infinite words, lexicographic order– and we refer the unfamiliar reader to [1, Section 1.2] for an introduction to these terms. We set some notation: we let A denote an alphabet, $|w|$ the length of a finite word w, w_i the i-th letter of w (indexed from 0), $w_{[i,j]}$ the factor of w going from positions i to j included, and $\prec_{\mathrm{lex}}$ is the lexicographic order.

We let $A^{\mathbb{D}}$ be the set of words indexed by $\mathbb{D}$ for $\mathbb{D} \in \{\mathbb{N}, \mathbb{Z}_{<0}, \mathbb{Z}\}$. We speak respectively of right-infinite, left-infinite and two-sided words. For convenience, we separate by a vertical bar the -1-th and 0-th elements of a two-sided word to indicate the origin, i.e., $u = \cdots u_{-3}u_{-2}u_{-1}|u_0u_1u_2\cdots$ when $u \in A^{\mathbb{Z}}$.

Morphisms and Substitutions. Given alphabets A, B, a *morphism* is a map $\mu\colon A^* \to B^*$ such that $\mu(uv) = \mu(u)\mu(v)$ for all words $u, v \in A^*$. A morphism is entirely determined by the images of the letters of A. A *substitution* is a morphism $\mu\colon A^* \to A^*$ such that the image $\mu(a)$ is non-empty for every letter $a \in A$ and there exists a *growing* letter $a \in A$, i.e., $\lim_{n\to+\infty} |\mu^n(a)| = +\infty$. A morphism $\mu\colon A^* \to A^*$ is *primitive* if there exists an integer $k \in \mathbb{N}$ such that for all $a, b \in A$, the letter a appears in $\mu^k(b)$. In this case, we also have that a appears in $\mu^\ell(b)$ for all $a, b \in A$ and $\ell \geq k$.

With a morphism $\mu\colon A^* \to A^*$ and a letter $a \in A$ we can associate a directed ordered tree $\mathcal{T}_{\mu,a}$ as follows. The root is labeled by a, and if a node of the tree is labeled by x and $\mu(x) = y_0 \cdots y_\ell$ then that node has $\ell + 1$ children labeled $y_0, \ldots, y_\ell$, with the edge from x to y_i labeled by i. Note that the k-th level of $\mathcal{T}_{\mu,a}$ stores the k-th iteration of μ on the letter a. We say that a node is *in column* n if it is the n-th node of its level in the order of the tree (indexed from 0).

Substitutions can naturally be applied to two-sided words by setting

$$\mu(\cdots u_{-3}u_{-2}u_{-1}|u_0u_1u_2\cdots) = \cdots\mu(u_{-3})\mu(u_{-2})\mu(u_{-1})|\mu(u_0)\mu(u_1)\mu(u_2)\cdots.$$

Let $\mathbb{D} \in \{\mathbb{N}, \mathbb{Z}, \mathbb{Z}_{<0}\}$ and consider a substitution μ over A. A word $u \in A^{\mathbb{D}}$ is a *periodic point of* μ if there exists an integer $p \geq 1$ such that $\mu^p(u) = u$. In this case, p is called a *period* of the periodic point u. The smallest such integer is called *the period* of u. A periodic point of μ with period $p = 1$ is called a *fixed point* of μ. We let $\mathrm{Per}_{\mathbb{D}}(\mu) = \{u \in A^{\mathbb{D}} \mid \mu^p(u) = u \text{ for some } p \geq 1\}$ denote the set of periodic points of μ. If $u \in \mathrm{Per}_{\mathbb{Z}}(\mu)$, then the *seed* of u is the pair of letters $u_{-1}|u_0$. If both letters of the seed of a two-sided periodic point are growing, then the periodic point is defined entirely by its seed. More precisely, we have $u = \lim_{n\to+\infty}\mu^{np}(u_{-1})|\lim_{n\to+\infty}\mu^{np}(u_0)$, where p is a period of u.

By extension of the tree associated with a substitution and a letter as in the previous paragraph, we consider two-sided trees as follow. For a substitution μ over A and two letters $a, b \in A$, the tree $\mathcal{T}_{\mu,b|a}$ is obtained by setting a start root having two children: the left (resp., right) one is reached with an edge of label $\mathtt{1}$ (resp., $\mathtt{0}$) and is the root of the tree $\mathcal{T}_{\mu,b}$ (resp., $\mathcal{T}_{\mu,a}$). We count depth in the tree such that the root is at depth -1. There is exactly one increasing bijection between depth k in this tree and an interval of $\mathbb{Z}$ that maps the rightmost node to $|\mu^k(a)| - 1$. We say that a node is *in column* n if this bijection maps it to n. Examples of such trees are given later in the paper.

Numeration Systems. A numeration system over the *domain* $\mathbb{D} \in \{\mathbb{N}, \mathbb{Z}\}$ is a pair of maps between $\mathbb{D}$ and a set of words, i.e., the *representation map* $\mathrm{rep}\colon \mathbb{D} \to A^*$ for some alphabet A, and the *evaluation map* $\mathrm{val}\colon L \to \mathbb{D}$, where $\mathrm{rep}(\mathbb{D}) \subset L \subset A^*$, such that $\mathrm{val} \circ \mathrm{rep} = id_{\mathbb{D}}$. The set $\mathrm{rep}(\mathbb{D})$ is the *language* of the numeration system, while L contains some additional, non-canonical representations.

In general, a numeration system over $\mathbb{D} = \mathbb{N}$ is *positional* if the underlying alphabet A is a set of consecutive integers $\{\mathtt{0}, \mathtt{1}, \ldots, \mathtt{c}\}$ for some $\mathtt{c} \in \mathbb{N}$ and the evaluation map is of the form $\mathrm{val}\colon A^* \to \mathbb{N}, w_{k-1}\cdots w_0 \mapsto \sum_{i=0}^{k-1} w_i U_i$ for some sequence $U = (U_i)_{i\geq 0} \in \mathbb{N}^{\mathbb{N}}$. Over $\mathbb{D} = \mathbb{Z}$, a numeration system is *positional* if the underlying alphabet A is a set of consecutive integers $\{\mathtt{0}, \mathtt{1}, \ldots, \mathtt{c}\}$ for some $\mathtt{c} \in \mathbb{N}$ and the evaluation map is of the form $\mathrm{val}\colon A^* \to \mathbb{Z}, w_{k-1}\cdots w_0 \mapsto \sum_{i=0}^{k-2} w_i U_i - w_{k-1}V_{k-1}$ for some sequences $U = (U_i)_{i\geq 0}, V = (V_i)_{i\geq 0} \in \mathbb{N}^{\mathbb{N}}$. The sequences U and V are the sequences of *weights* of the numeration system. Every position has a given weight, while the presence of an additional sequence V helps deal with the representation of negative numbers, in a fashion similar to the usual two's complement numeration system. Such a numeration system (see [8, Section 1]) can be defined by the evaluation map $w_{k-1}\cdots w_0 \mapsto -w_{k-1}2^{k-1} + \sum_{i=0}^{k-2} w_i 2^i$ together with the language $L = A^+ \setminus (00A^* \cup 11A^*)$ as the evaluation map is a bijection between L and $\mathbb{Z}$. The system is positional with weights $(2^i)_{i\geq 0}$.

Dumont–Thomas Numeration Systems. Dumont and Thomas [4] originally defined numeration systems based on substitutions and their right-infinite fixed points for the natural domain $\mathbb{D} = \mathbb{N}$. The intuition behind these systems is that the number n can be represented by the label of the shortest path from the root to a node in column n. See [7, Section 2.1] for a gentle introduction. Recently, the authors of [9] proposed an extension of these numeration systems to all integers (so $\mathbb{D} = \mathbb{Z}$) and based instead on some periodic points of substitutions. By relaxing the condition of [9] on the lengths of representations, we go even beyond and use all periodic points of substitutions to define Dumont–Thomas numeration systems.

To formally introduce our object of study, we need some additional definitions and notation. Let $\mu\colon A^* \to A^*$ be a substitution. Fix a letter $a \in A$ and an integer k. For every $i \in \{0,\ldots,k\}$, we consider a pair $(m_i, a_i) \in A^* \times A$. The sequence $((m_i, a_i))_{i=0,\ldots,k}$ is *admissible (with respect to* μ*)* if for every $i \in \{1,\ldots,k\}$, $m_{i-1}a_{i-1}$ is a prefix of $\mu(a_i)$. This sequence is *a-admissible (with respect to* μ*)* if it is admissible with respect to μ and $m_k a_k$ is a prefix of $\mu(a)$.

Theorem 1 (Extension of [9, Theorem 4.1]). *Let* $\mu : A^* \to A^*$ *be a substitution with growing letter* $a \in A$*. Consider a right-infinite periodic point* $u \in \mathrm{Per}_{\mathbb{N}}(\mu)$ *with* $u_0 = a$ *and period* $p \geq 1$*. Fix a residue* $r \in \{0,1,\ldots,p-1\}$ *modulo* p *and define* $v_r = \mu^r(u)$*. For every integer* $n \geq 0$*, there exist a unique integer* $k = k(n)$ *with* $k \equiv r \bmod p$ *and a unique sequence* $((m_i, a_i))_{i=0,\ldots,k-1}$ *such that the sequence is a-admissible,*

$$m_{k-1}m_{k-2}\cdots m_{k-p} \neq \varepsilon \text{ if } k \geq p,$$

and $(v_r)_{[0,n-1]} = \mu^{k-1}(m_{k-1})\mu^{k-2}(m_{k-2})\cdots\mu^0(m_0)$.

Theorem 2 (Extension of [9, Theorem 4.2]). *Let* $\mu : A^* \to A^*$ *be a substitution with growing letter* $b \in A$*. Consider a left-infinite periodic point* $u \in \mathrm{Per}_{\mathbb{Z}_{<0}}(\mu)$ *with* $u_{-1} = b$ *and period* $p \geq 1$*. Fix a residue* $r \in \{0,1,\ldots,p-1\}$ *modulo* p *and define* $v_r = \mu^r(u)$*. For every integer* $n \leq -1$*, there exist a unique integer* $k = k(n)$ *with* $k \equiv r \bmod p$ *and a unique sequence* $((m_i, a_i))_{i=0,\ldots,k-1}$ *such that the sequence is b-admissible,*

$$\mu^{p-1}(m_{k-1})\mu^{p-2}(m_{k-2})\cdots\mu^0(m_{k-p})a_{k-p} \neq \mu^p(b) \text{ if } k \geq p, \tag{1}$$

and $(v_r)_{[-|\mu^k(b)|,n-1]} = \mu^{k-1}(m_{k-1})\mu^{k-2}(m_{k-2})\cdots\mu^0(m_0)$.

From these results, we generalize the definition of Dumont–Thomas numeration systems [4,9].

Definition 1. *Let* $\mu\colon A^* \to A^*$ *be a substitution and let* $u \in \mathrm{Per}_{\mathbb{Z}}(\mu)$ *be a two-sided periodic point with growing seed* $u_{-1}|u_0$ *and period* $p \geq 1$*. Let* $r \in \{0, 1, \ldots, p-1\}$*. Define* $\mathtt{c} = \max_{a\in A} |\mu(a)| - 1$ *and the set* $D = \{\mathtt{0}, \mathtt{1}, \ldots, \mathtt{c}\}$*. The* Dumont–Thomas complement numeration system associated with μ, u and r is defined by the map $\mathrm{rep}_{u,r}\colon \mathbb{Z} \to \{\mathtt{0},\mathtt{1}\}D^*, n \mapsto \mathrm{rep}_{u,r}(n)$ where

$$\mathrm{rep}_{u,r}(n) = \begin{cases} \mathtt{0} \cdot |m_{k-1}| \cdot |m_{k-2}| \cdots |m_0|, & \text{if} n \geq 0; \\ \mathtt{1} \cdot |m_{k-1}| \cdot |m_{k-2}| \cdots |m_0|, & \text{if} n \leq -1; \end{cases}$$

such that $k = k(n)$ is the unique integer with $k \equiv r \bmod p$ and $((m_i, a_i))_{i=0,\ldots,k-1}$ is the unique sequence obtained from Theorem 1 (resp., Theorem 2) applied on the right-infinite periodic point $u|_{\mathbb{N}} = u_0u_1\cdots$ (resp., the left-infinite periodic point $u|_{\mathbb{Z}_{<0}} = \cdots u_{-2}u_{-1}$) with period p. Note that when the context is clear, we drop the dependence on μ, u and r.

The intuition behind this system is to use the tree $\mathcal{T}_{\mu,b|a}$ and represent n with the label of a shortest path of length congruent to $r+1$ modulo p between the root and a node in column n. Note that representations in this system have length congruent to $r+1$ modulo p, and their first digits depend only on the sign of the represented number. Note that we use a special font for the digits representing numbers to distinguish them from integers.

Example 1. Consider the substitution $\mu\colon a \mapsto ccd, b \mapsto cd, c \mapsto ab, d \mapsto a$ and its periodic point $u \in \mathrm{Per}_{\mathbb{N}}(\mu)$ with growing seed $a|a$ and period $p = 2$. The tree $\mathcal{T}_{\mu,a|a}$ is depicted in Fig. 1. Depending on the choice of even or odd length for representations, we obtain different numeration systems as illustrated in the table of Fig. 1.

The numeration systems with $r = 0$ were studied in Lepšová's PhD thesis [11]. Notably, [11, Example 6.5.6] presents two substitutions associated with the silver mean $1+\sqrt{2}$ such that one gives rise to a positional numeration system and the other does not. Lepšová raised the following natural question [11].

Question 2. [11, Question 6.5.7] What are the conditions for a complement Dumont–Thomas numeration system to be positional?

3 Positional Dumont–Thomas Numeration Systems

In this section, we study when a substitution generates a Dumont–Thomas numeration system that is positional to answer Question 2. We first sketch our argument, and give some intuition for the statement of the main result (Theorem 3). After stating the result in the most general form, we present some corollaries in Sect. 3.3. We also link our Dumont–Thomas numeration systems to previous literature.

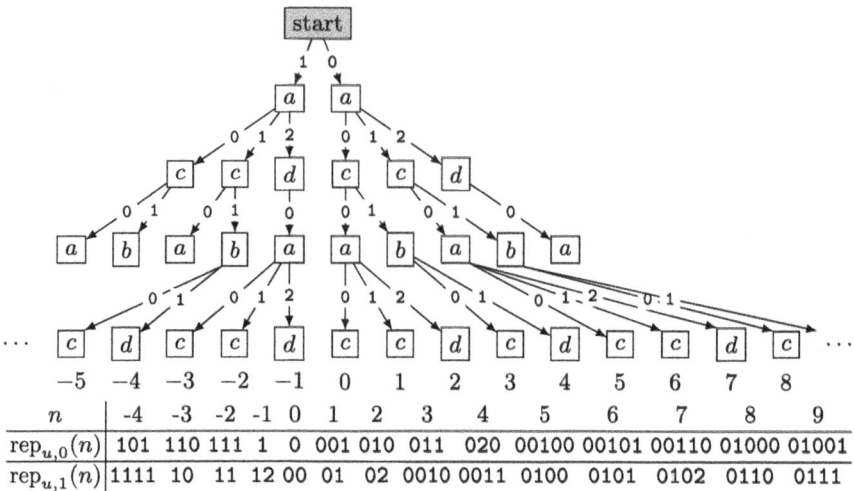

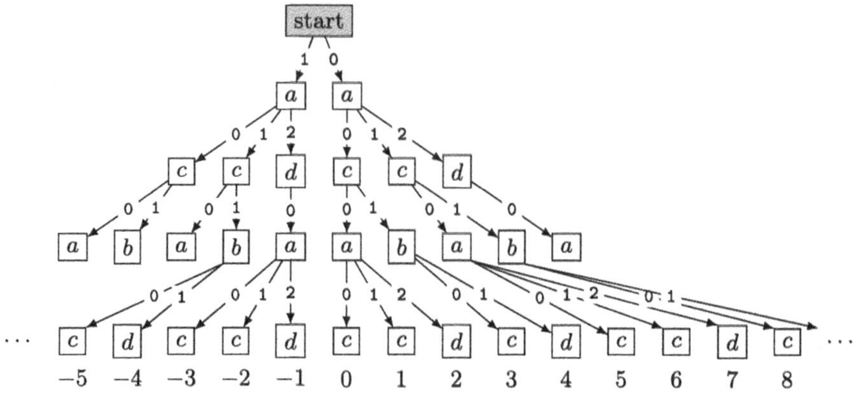

n	-4	-3	-2	-1	0	1	2	3	4	5	6	7	8	9
$\mathrm{rep}_{u,0}(n)$	101	110	111	1	0	001	010	011	020	00100	00101	00110	01000	01001
$\mathrm{rep}_{u,1}(n)$	1111	10	11	12	00	01	02	0010	0011	0100	0101	0102	0110	0111

Fig. 1. The tree $\mathcal{T}_{\mu,a|a}$ for the substitution $\mu\colon a \mapsto ccd, b \mapsto cd, c \mapsto ab, d \mapsto a$ and the periodic point u of period $p = 2$ and seed $a|a$. Below, depending on the residue $r \in \{0,1\}$, we obtain a Dumont–Thomas numeration system and we give $(\mathrm{rep}_{u,r}(n))_{-4\leq n\leq 9}$ whose lengths are congruent to $r+1 \bmod p$.

3.1 Sketch of the Argument

We informally sketch the argument that we used to solve Question 2. We then present examples where this argument fails, which allows us to motivate the technicalities that are introduced in Sect. 3.2 and to explain the reasoning without them getting in the way of the explanation.

Sketch 1. We let $\mu\colon A^* \to A^*$ be a substitution, $u \in \mathrm{Per}_{\mathbb{Z}}(\mu)$ be a two-sided periodic point of μ with growing seed $b|a$ and period $p \geq 1$, and $r \in \{0,1,\ldots,p-1\}$ be a residue. We also consider the corresponding Dumont–Thomas complement numeration system associated with μ, u and r.

Now consider two words $wt0^\ell$ and $w(t+1)0^\ell$ that label paths in the tree $\mathcal{T}_{\mu,b|a}$ (see Fig. 2, where we have arbitrarily chosen to represent the situation with $\mathcal{T}_{\mu,a}$ instead) and assume these words are the representations of the integers n_1 and n_2 respectively, i.e., the paths end in columns n_1 and n_2, in addition to which we have some conditions for admissibility. We also let c and d be the letters attained after reading wt and $w(t+1)$ respectively.

If the numeration system is positional, then $n_2 - n_1$ must equal U_ℓ, the weight in position ℓ, as the representations of n_1 and n_2 differ only by one unit in position ℓ. However, $n_2 - n_1$ is the number of columns between those two nodes in the tree, which is $|\mu^\ell(c)|$, i.e., the number of level-ℓ descendants of c in the tree (see again Fig. 2). Since this reasoning could be applied to any letter c that has a sibling to its right in $\mathcal{T}_{\mu,b|a}$, this points towards the following implication: *If the Dumont–Thomas numeration system associated with μ, u, and r is positional, then $|\mu^\ell(c)| = U_\ell$ for every letter c that has a younger sibling in $\mathcal{T}_{\mu,b|a}$.*

The converse implication is also justified, almost by definition of a Dumont–Thomas numeration system. If the positive integer $n \geq 1$ is represented by the word $0 \cdot w_{k-1} \cdots w_0$, this means that there is an a-admissible sequence $((m_i, a_i))_{i=0,\dots,k-1}$ such that

$$\mu^r(u)_{[0,n-1]} = \mu^{k-1}(m_{k-1}) \dots \mu^0(m_0) \tag{2}$$

and $|m_i| = w_i$ for every $i \in \{0, \dots, k-1\}$. Because all letters in m_i have a_i as a younger sibling, their image by μ^ℓ has length U_ℓ by assumption. Taking the length in Eq. (2) yields $n = \sum_{i=0}^{k-1} U_i w_i$, which corresponds to the numeration system being positional with weights $(U_i)_{i\geq 0}$. The case of negative numbers is similar, with one correcting term corresponding to the value of V_{k-1}.

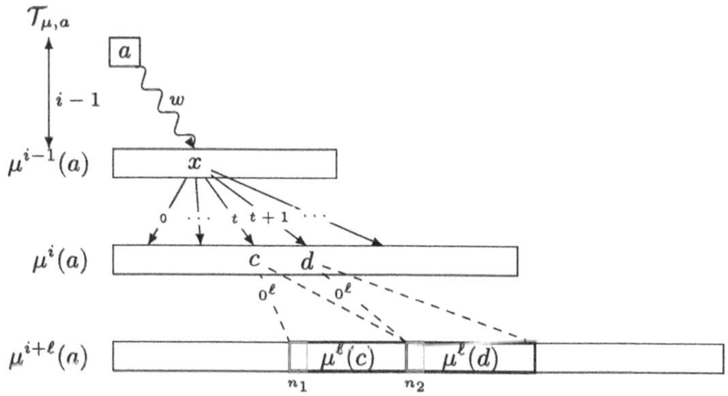

Fig. 2. Comparing the values of $wt0^\ell$ and $w(t+1)0^\ell$ in the right part of $\mathcal{T}_{\mu,b|a}$.

The above sketch leads us to formulate the next conjecture: *The Dumont–Thomas numeration system associated with μ, u, and r is positional if and only if $c \mapsto |\mu^\ell(c)|$ is constant on all letters c that have a younger sibling in $\mathcal{T}_{\mu,b|a}$, in which case this constant is the weight U_ℓ.* However, trying to prove this conjecture reveals two issues with Sketch 1, which the following examples highlight.

Example 2. Recall the substitution μ from Example 1. The letters having a younger sibling in $\mathcal{T}_{\mu,a}$ are a and c, but the sequences of the lengths of their consecutive images under μ are $(|\mu^j(a)|)_{j\geq 0} = 1, 3, 5, 13, 21, 55, 89, 233, 377, \dots$ and $(|\mu^j(c)|)_{j\geq 0} = 1, 2, 5, 8, 21, 34, 89, 144, 377, \dots$ respectively. Despite this, the corresponding Dumont–Thomas numeration system is positional for both values of r, with weights $1, 2, 5, 8, 21, 34, \dots$ for $r = 0$ and $1, 3, 5, 13, 21, 55, \dots$ for $r = 1$.

This example illustrates that, in the case where μ is not primitive, we may only control some image lengths among all different letters. Since a and c are never both present on some level of the tree, they may have different behaviors. Fixing this is the purpose of the sets E_j in Definition 2.

Example 3. Consider the substitution $\mu\colon a \mapsto bca$, $b \mapsto bb$, $c \mapsto b$ and the associated numeration system with the seed $a|b$. The letters b and c both appear in the tree with a younger sibling, and they have images of different lengths. However, the numeration system is still positional, with weights $U_i = 2^i$ and $V_0 = 1$, $V_i = 3 \cdot 2^{i-1}$ for every $i \geq 1$.

Our sketched argument fails, because if wt is a path to a node labeled by c in the tree, this node is in column -2. Then, $w(t+1)$ leads to a node in column -1 and $w(t+1)0^\ell$ is never the representation of any number, due to Eq. (1) from Theorem 2. Thus we cannot constrain the lengths of the images of c. This case and some similar ones may occur when a letter appears only in column -2, which motivates the separate treatment of column -2 found in Definition 3.

3.2 Main Result

To state the main result, we need to define some particular sets of letters. We again consider a substitution $\mu\colon A^* \to A^*$ and a two-sided periodic point $u \in \mathrm{Per}_{\mathbb{Z}}(\mu)$ of μ with growing seed $b|a$ and period $p \geq 1$. We draw the tree $\mathcal{T}_{\mu,b|a}$. For a fixed residue $r \in \{0,\dots,p-1\}$, we also consider the corresponding Dumont–Thomas numeration system $\mathrm{rep}_{u,r}\colon \mathbb{Z} \to \{0,1\}D^*$ from Definition 1.

Definition 2. *Let $j \in \{0,\dots,p-1\}$. We let E_j be the set of letters $c \in A$ such that there exist some integer $k \geq 1$ and some sequence $((m_i, a_i))_{i=0,\dots,k-1} \in (A^* \times A)^k$ that verifies the following:*

- *the sequence $((m_i, a_i))_{i=0,\dots,k-1}$ is a- or b-admissible;*
- *we have $k \equiv j \bmod p$;*
- *the letter c appears at the end of the word m_0;*
- *if $((m_i, a_i))_{i=0,\dots,k-1}$ is b-admissible, then $\mu^{k-1}(m_{k-1})\cdots\mu^0(m_0)a_0 \neq \mu^k(b)$.*

Thinking in terms of the tree $\mathcal{T}_{\mu,b|a}$, the first three conditions simply describe letters that appear in the tree at some level congruent to $j \bmod p$ and have a younger sibling on that level. The last condition excludes letters c that only appear in column -2 in the tree $\mathcal{T}_{\mu,b|a}$. To deal with these letters, a dedicated condition is required, as follows.

Definition 3. *Let $j \in \{0,\dots,p-1\}$. On level j in the tree $\mathcal{T}_{\mu,b|a}$, we let c be the letter in column -2 and d be the letter in column -1 (i.e., d is immediately to the right of c). If c and d share the same parent, we consider two cases to modify E_j. If $\left|\mu^{p-j}(d)\right| > 1$, then we add c to E_j if it was not already present. If $\left|\mu^{p-j}(d)\right| = 1$ and $j \leq r < p$, then we add the following condition:*

$$\left|\mu^{r-j}(c)\right| \textit{ must be equal to } \left|\mu^{r-j}(e)\right| \textit{ for every letter } e \in E_j. \tag{3}$$

Example 4. For the substitution μ from Example 1, consider its two-sided periodic point with seed $a|a$ and period $p = 2$. In this case, we obtain $E_0 = \{a\}$ and

$E_1 = \{c\}$. Condition (3) must only be checked for $r = 1$ and the letter c, and it is trivially satisfied.

Consider the substitution $\mu\colon a \mapsto bcd,\ d \mapsto ba,\ b \mapsto b^2,\ c \mapsto b$ and its two-sided periodic point with seed $a|b$. If we go only by Definition 2, we will find $E_0 = E_1 = \{b\}$, but if we add Definition 3, c is added to E_1, and we now correctly find that the system is not positional (which we can also see from the representations of -5, -3 and -2).

Theorem 3. *Let $\mu\colon A^* \to A^*$ be a substitution and let $u \in \mathrm{Per}_{\mathbb{Z}}(\mu)$ be a two-sided periodic point with growing seed $b|a$ and period $p \geq 1$. The Dumont–Thomas complement numeration system associated with μ, u, and r is positional if and only if for every $j \in \{0, \dots, p-1\}$, the map $c \mapsto |\mu^\ell(c)|$ is constant over E_j for every ℓ such that $\ell + j \equiv r \bmod p$, and Condition (3) is satisfied for the letters where it was added.*

In this case, the sequences U, V of weights of the numeration system are given as follows: for every $\ell \geq 0$, we define $U_\ell = |\mu^\ell(c)|$ for a letter $c \in E_j$ where $j \in \{0, \dots, p-1\}$ and $\ell + j \equiv r \bmod p$; and $V_\ell = |\mu^\ell(b)|$ for every $\ell \in \mathbb{N}$.

Remark 1. It may be, although rarely, that E_j is empty. (For example, consider the case of $\mu\colon a \mapsto b,\ b \mapsto aa$ and the set E_1.) In this case, if Condition (3) applies to some letter c, we may give the weight $|\mu^{r-j}(c)|$ to position $r - j$. Otherwise, this means that only the digit 0 appears at positions congruent to $j \bmod p$ in this numeration system, and as such the weight given to these positions is arbitrary.

Remark 2. Replacing the substitution by one of its powers may lead to the loss of positionality. This is the case of the substitution μ from Example 1 and its square, for which the numeration systems associated with seed $a|a$ are once positional, once not positional (consider e.g. the representations of 5 and 8).

3.3 Particular Cases

We now highlight some general cases where the technicalities of Sect. 3.2 do not occur, leading to results that are more concise and legible. We also discuss possible simplifications of the substitution at play, as well as a parallel to Bertrand numeration systems.

From now on, we assume that the alphabet A of the substitution μ is *minimal*, i.e., all the letters in A are present in $\mu^n(b|a)$ for some n. Given a substitution μ, we say that a letter c is *non-final* if there exist $d \in A$, $x \in A^*$, and $y \in A^+$ such that $\mu(d) = xcy$. We let E_μ denote the set of non-final letters of μ. For example, with the Fibonacci substitution $\mu\colon a \mapsto ab, b \mapsto a$, we have $E_\mu = \{a\}$.

When $\mathbb{D} = \mathbb{N}$ and the substitution μ has a fixed point or when μ is primitive (even over $\mathbb{D} = \mathbb{Z}$), the reasoning in Sketch 1 holds, as stated in the next results.

Corollary 1. *Let $\mu\colon A^* \to A^*$ be a substitution and $u = u_0u_1 \cdots$ be a right-infinite fixed point of μ with growing seed a. Then the Dumont–Thomas numeration system (for $\mathbb{N}$) associated with μ, u and $r = 0$ is positional if and only if the map $c \mapsto |\mu^\ell(c)|$ is constant over E_μ for every ℓ, in which case the sequence of weights is equal to $U_\ell = |\mu^\ell(a)|$ for every $\ell \geq 0$.*

Corollary 2. *Let $\mu\colon A^* \to A^*$ be a primitive substitution and let $u \in \mathrm{Per}_{\mathbb{Z}}(\mu)$ be a two-sided periodic point with growing seed $b|a$ and period $p \geq 1$. The Dumont–Thomas complement numeration system associated with μ, u, and r is positional if and only if the map $c \mapsto |\mu^\ell(c)|$ is constant over E_μ for every $\ell \geq 0$. In this case, for every $\ell \geq 0$, U_ℓ is the constant value of $|\mu^\ell(\cdot)|$ over E_μ and $V_\ell = |\mu^\ell(b)|$.*

In both of these cases, we may use the following lemma to simplify the substitution at play by reducing the number of non-final letters to one.

Lemma 1. *Let $\mu\colon A^* \to A^*$ be a substitution such that the map $c \mapsto |\mu^\ell(c)|$ is constant over E_μ for every integer $\ell \geq 0$. Then there exist $B \subseteq A$, a substitution $\nu\colon B^* \to B^*$ with $|E_\nu| = 1$, and $a', b' \in B$ such that the trees $\mathcal{T}_{\mu,b|a}$ and $\mathcal{T}_{\nu,b'|a'}$ differ only by their labeling. These objects can be computed effectively.*

Example 5. Consider the primitive substitution $\mu\colon a \mapsto ab, b \mapsto ba$ (often referred to as the Thue–Morse substitution). We note that both a, b are non-final letters and $|\mu^\ell(c)| = 2^\ell$ for $c \in \{a, b\}$ and for every $\ell \geq 0$. The substitution given by Lemma 1 is $\nu\colon a \mapsto a^2$. The corresponding Dumont–Thomas numeration system (over $\mathbb{N}$) is the usual binary system.

We now link some of our numeration systems to existing literature. Our target numeration systems represent natural numbers and use words of every length, so they correspond to Corollary 1. Due to Lemma 1, it is enough to consider substitutions with only one non-final letter e, which must be the seed of the fixed point. The image of any letter is defined only by the number of repetitions of e and the choice of final letter. As a result, if we operate on the minimal alphabet, the substitution is equivalent to one of the form

$$\mu : a_1 \mapsto a_1^{d_1} a_2,\ a_2 \mapsto a_1^{d_2} a_3,\ \ldots, a_n \mapsto a_1^{d_n} a_k, \tag{4}$$

where $n \geq 1$ is an integer, $\{a_1, \ldots, a_n\}$ is the alphabet, $k \in \{1, \ldots, n\}$, $d_1 > 0$ and d_i is a non-negative integer for every $i \in \{1, \ldots, n\}$. An example is the Tribonacci substitution $\tau\colon a \mapsto ab, b \mapsto ac, c \mapsto a$, for which $n = 3$, $a_1 = a$, $a_2 = b$, $a_3 = c$, $d_1 = d_2 = 1$, $d_3 = 0$, and $k = 1$. The similarity with the substitutions studied by Fabre in [5] is striking, so we call *Fabre-like* the substitutions of the form (4). We now study this particular class of substitutions in additional detail.

We quickly recall some other milestones of numeration systems. For the reader not familiar with the classical theory, see, e.g., [1, Chapter 2] and [13, Chapter 2]. In 1957, Rényi [14] introduced β-numeration systems to represent positive real numbers with a real base $\beta > 1$. We let $d_\beta(1)$ (resp., $d^*_\beta(1)$) denote the β-representation (resp., quasi-greedy β-representation) of 1. Of particular interest are the so-called *Parry numbers*, which are real numbers β such that $d_\beta(1)$ is either finite (β is then a *simple Parry* number) or ultimately periodic [12]. An article of Bertrand-Mathis [2], later corrected by Charlier, Cisternino and Stipulanti [3], studies the greedy positional numeration systems such that the associated language L verifies $w \in L \Leftrightarrow w0 \in L$ if $w \neq \varepsilon$. As it turns out, these systems are exactly those that verify the relation $U_i = d_1 U_{i-1} + d_2 U_{i-2} + \ldots + d_i U_0 + 1$ for all $i \geq 1$, where $d_1 d_2 \cdots$ is either

$d^*_\beta(1)$ for some $\beta > 1$ (called *canonical* Bertrand numeration systems), or $d_\beta(1)$ for some simple Parry number β (called *non-canonical* Bertrand numeration systems and introduced in [3]), or $d_1 d_2 \cdots = 10^\omega$ (called the *trivial* Bertrand numeration system, also from [3]). In 1995, Fabre [5] introduced another way to approach canonical Bertrand numeration systems. If β is a Parry number with $d^*_\beta(1) = d_1 \cdots d_n (d_{n+1} \cdots d_{n+m})^\omega$ for some m, n, we introduce the substitution μ_β mapping $1 \mapsto 1^{d_1}2$, ..., $(n+m-1) \mapsto 1^{d_{n+m-1}}(n+m)$, and $(n+m) \mapsto 1^{d_{n+m}}(n+1)$. Note that simple Parry numbers correspond to $n = 0$. Fabre then shows that $|\mu^\ell_\beta(1)|$ is the integer U_ℓ defined for the canonical Bertrand numeration system, and his [5, Theorem 2] establishes the equality between the Dumont–Thomas numeration system associated with μ_β and the canonical Bertrand numeration system associated with β.

Let us now go back to the study of our Dumont–Thomas numeration systems based on Fabre-like substitutions. We first note that Bertrand numeration systems based on a Parry number are a particular case of Dumont–Thomas numeration systems (and not just the canonical ones as proven by Fabre).

Proposition 1. *Every Bertrand numeration system associated with a Parry number is equal to some Dumont–Thomas numeration system associated with a Fabre-like substitution.*

Although our Dumont–Thomas numeration systems verify the property $w \in L \Leftrightarrow w0 \in L$ that characterizes Bertrand numeration systems, the converse of Proposition 1 is not true, as we see in the following example.

Example 6. Consider the Fabre-like substitution μ: $a \mapsto aab$, $b \mapsto aaaa$ with fixed point $u = aabaabaaaa \cdots$. The corresponding Dumont–Thomas numeration system is positional, with the sequence of weights starting by $1, 3, 10, 32, \ldots$. We note that $\mathrm{rep}_{u,0}(9) = 23$, but this cannot happen in a Bertrand numeration system as the representation of 9 would be 30 with the given weights.

This is because our systems are not greedy in the usual sense. This phenomenon can be understood with an adaptation of the *Parry condition.* To recall, this condition (first seen in [12, Corollary 1]) states that a word $d_1 d_2 \cdots$ is equal to $d_\beta(1)$ for some $\beta > 1$ if and only if $d_1 d_2 \cdots \succ 10^\omega$ and $d_1 d_2 \cdots \succ d_i d_{i+1} \cdots$ for every $i \geq 1$. In the case of Example 6, if the substitution μ were the Fabre substitution associated with some Parry number β, we would have $d^*_\beta(1) = (23)^\omega$ and $d_\beta(1) = 240^\omega$, which contradicts the Parry condition. Thus the system cannot be equal to a Bertrand numeration system. In fact, the Parry condition (adapted for use with $d^*_\beta(1)$ instead of $d_\beta(1)$) is all that is necessary to guarantee that the system is greedy and thus equal to a Bertrand numeration system.

Proposition 2. *Let μ be a Fabre-like substitution as in* (4)*. Construct the right-infinite word $d_1 d_2 \cdots = d_1 \cdots d_{k-1}(d_k \cdots d_n)^\omega$. The Dumont–Thomas numeration system associated with μ and the seed a_1 is equal to a Bertrand numeration system if and only if we have $d_i d_{i+1} \cdots \preccurlyeq_{lex} d_1 d_2 \cdots$ for each $i \geq 1$.*

Disclosure of Interests. The authors have no competing interests to declare that are relevant to the content of this article.

Acknowledgments.. Savinien Kreczman is supported by the FNRS Research Fellow grant 1.A.789.23F. Sébastien Labbé is supported by France's Agence Nationale de la Recherche (ANR) project IZES (ANR-22-CE40-0011), and by the *Symbolic Dynamics and Arithmetic Expansions* (SymDynAr) Project, co-funded by ANR (ANR-23-CE40-0024) and FWF (I 6750), the Austrian Science Fund. Manon Stipulanti is supported by the FNRS Research grant 1.C.104.24F. The authors want to thank Émilie Charlier for useful discussions.

References

1. Berthé, V., Rigo, M. (eds.): Combinatorics, Automata, and Number Theory, Encyclopedia of Mathematics and its Applications, vol. 135. Cambridge University Press, Cambridge (2010). https://doi.org/10.1017/CBO9780511777653
2. Bertrand-Mathis, A.: Comment écrire les nombres entiers dans une base qui n'est pas entière. Acta Math. Hungar. **54**(3–4), 237–241 (1989). https://doi.org/10.1007/BF01952053
3. Charlier, E., Cisternino, C., Stipulanti, M.: A full characterization of Bertrand numeration systems. In: Diekert, V., Volkov, M. (eds.) Developments in Language Theory, LNCS, vol. 13257, pp. 102–114. Springer, Cham (2022). https://doi.org/10.1007/978-3-031-05578-2_8
4. Dumont, J.M., Thomas, A.: Systèmes de numération et fonctions fractales relatifs aux substitutions. Theoret. Comput. Sci. **65**(2), 153–169 (1989). https://doi.org/10.1016/0304-3975(89)90041-8
5. Fabre, S.: Substitutions et β-systèmes de numération. Theoret. Comput. Sci. **137**(2), 219–236 (1995). https://doi.org/10.1016/0304-3975(95)91132-A
6. Gheeraert, F., Romana, G., Stipulanti, M.: String attractors of fixed points of k-Bonacci-like morphisms. In: Frid, A., Mercaş, R. (eds.) Combinatorics on Words, LNCS, vol. 13899, pp. 192–205. Springer, Cham (2023). https://doi.org/10.1007/978-3-031-33180-0_15
7. Kreczman, S., Labbé, S., Stipulanti, M.: Positionality of Dumont–Thomas numeration systems for integers (2025). http://arxiv.org/abs/2503.04487
8. Labbé, S., Lepšová, J.: A Fibonacci analogue of the two's complement numeration system. RAIRO Theor. Inform. Appl. (RAIRO:ITA) **57**, 12, 23 (2023). https://doi.org/10.1051/ita/2023007
9. Labbé, S., Lepšová, J.: Dumont-Thomas complement numeration systems for $\mathbb{Z}$. Integers **24**, A112, 27 (2024). https://doi.org/10.5281/zenodo.14340125
10. Lecomte, P., Rigo, M.: Numeration systems on a regular language. Theory Comput. Syst. **34**(1), 27–44 (2001). https://doi.org/10.1007/s002240010014
11. Lepšová, J.: Substitutive structures in combinatorics, number theory, and discrete geometry. Ph.D. Thesis, Université de Bordeaux and Czech Technical University in Prague (2024). https://theses.hal.science/tel-04679032
12. Parry, W.: On the β-expansions of real numbers. Acta Math. Acad. Sci. Hung. **11**, 401–416 (1960). https://doi.org/10.1007/BF02020954
13. Rigo, M.: Formal Languages, Automata and Numeration Systems, vol. 2. Networks and Telecommunications Series, ISTE, London; John Wiley & Sons, Inc., Hoboken, NJ (2014). https://doi.org/10.1002/9781119042853

14. Rényi, A.: Representations for real numbers and their ergodic properties. Acta Math. Acad. Sci. Hung. **8**(3–4), 477–493 (1957). https://doi.org/10.1007/BF02020331
15. Surer, P.: Substitutive number systems. Int. J. Number Theory **16**(8), 1709–1751 (2020). https://doi.org/10.1142/S1793042120500906

About Δ-Numeration

Bastien Laboureix[1(✉)] and Eric Domenjoud[2]

[1] Centre Borelli, ENS Paris-Saclay, Université Paris-Saclay, Gif-sur-Yvette, France
bastien.laboureix@ens-paris-saclay.fr
[2] Université de Lorraine, CNRS, LORIA, 54000 Nancy, France
eric.domenjoud@loria.fr

Abstract. In this article we study a numeration system previously used to prove combinatorial properties in discrete geometry: the Δ-numeration. Since this system, introduced via the fully subtractive algorithm, has been seen mainly as a tool, we propose here to study it from the point of view of numeration systems. In particular, we make the link with β-numeration and Cantor real bases. We reintroduce the rewriting system introduced to calculate in Δ-numeration. This systems is based on the properties of the fully subtractive algorithm and is normalising. Finally, we study the ultimately periodic case, a special case of alternate bases, and show that the ultimately periodic words represent exactly the elements of $\mathbb{Q}[\beta]$ where β is the inverse of a Pisot number.

Keywords: numeration systems · normalising rewriting system · fully subtractive algorithm · ultimately periodic words · real Cantor bases

1 Introduction

The first real base numeration systems were introduced in the middle of the 20th century by A. Rényi in [12]. The idea was to generalise the well-known notions of whole-base numeration, for any real number $\beta > 1$. A number $x \in [0, 1]$ is represented by $\sum_{n=1}^{\infty} \frac{w_n}{\beta^n}$ where w is an infinite word over the alphabet $[\![0, \lceil\beta\rceil - 1]\!]$. Since then, many numeration systems have been studied, from the point of view of word combinatorics, symbolic dynamics or rewriting. We refer the reader to [13] for an overview of the various definitions and usual properties in the field.

In 2021 and 2023, E. Charlier and C. Cisternino introduced Cantor real bases and alternate bases in [4,5]. Given a sequence $(\beta_n)_{n\in\mathbb{N}^*}$, the word $w \in \mathbb{N}^\omega$ represents, under the hypothesis of convergence of the series, the real number $x = \sum_{n=1}^{\infty} (w_n / \prod_{k=1}^{n} \beta_k)$.

In parallel and in a completely different field, E. Domenjoud, X. Provençal and L. Vuillon introduced in 2015 the foundations of a system that they called Δ-numeration [8]. This system was initially created with the aim of studying the structure of certain arithmetic hyperplanes in discrete geometry, by expressing them as a palindromic closure linked to this system. In 2019, in [6], a further study of this numeration system made it possible to define rewriting operations

G. Gamard and J. Leroy (Eds.): WORDS 2025, LNCS 15729, pp. 192–204, 2025.
https://doi.org/10.1007/978-3-031-97548-6_17

on it with the aim of obtaining a convergent rewriting system and to deduce properties on the facet connectedness of arithmetic hyperplanes. However, the main aim of the paper in question was to prove properties of discrete geometry, using numeration bases as a simple tool to achieve this goal.

In this article, we propose to study the properties of this numeration system for itself. After recalling the general principle of Δ-numeration, we show in particular that it is a special case of Cantor real bases studied a few years later by E. Charlier and characterise the differences between these two numeration systems. In particular, we look at the expansion/reduction rewriting system in Δ-numeration and recall that it converges. Finally, we study the ultimately periodic case, which is close to E. Charlier's alternate bases. Most of the results obtained are presented in the last chapter of B. Laboureix's PhD thesis [11].

2 Fully Subtractive Algorithm

We begin by presenting the fully subtractive algorithm introduced by F. Schweiger in [15]. This is a subtractive Euclidean algorithm in dimension $d \geqslant 2$ very strongly related to continued fractions. Since we are primarily interested in the dynamics of this algorithm, the version we present has no output and does not terminate. To simplify matters, we present a version that loops as soon as the vector under consideration contains a zero coordinate, which differs from Euclid's algorithm. The algorithm constructs an infinite word $\Delta = (\delta_n)_{n\in\mathbb{N}^*}$ over the alphabet $[\![1, d]\!]$, a decreasing real sequence $(\theta_n)_{n\in\mathbb{N}^*}$ and a sequence $(v^{(n)})_{n\in\mathbb{N}^*}$ of vectors of $\mathbb{R}^d$. The sequence θ allows us to define the Δ-numeration and we show that, in the framework we are interested in, the knowledge of Δ and θ are equivalent.

Algorithm: Fully subtractive algorithm [15]

Input: v a non negative vector of $\mathbb{R}^d$

Output: constructs an infinite word Δ and an infinite sequence θ

$n \leftarrow 0$;

while *TRUE* **do**

  $n \leftarrow n + 1$;

  $i \leftarrow$ index of the minimum coordinate of v (*in the event of non-uniqueness, the smallest possible value of* i *is taken*);

  $\delta_n \leftarrow i$;

  $\theta_n \leftarrow v_i$;

  $v^{(n)} \leftarrow v$;

  $v \leftarrow v - v_i\,(\mathbb{1} - e_i)$ where $\mathbb{1} \overset{\text{def}}{=} (1, \ldots, 1)$;

end

The dynamics of the fully subtractive algorithm were studied in a completely different context by C. Kraaikamp and R. Meester in [10] who were interested in a percolation model. Since the fully subtractive algorithm had not yet been defined at the time of their research, we take the liberty of describing their results using the notations and definitions previously introduced. Their study

Table 1. Fully subtractive algorithm for the example (17, 7, 19).

$v^{(n)}$	δ_n	θ_n
(17,7,19)	2	7
(10,7,12)	2	7
(3,7,5)	1	3
(3,4,2)	3	2
(1,2,2)	1	1
(1,1,1)	1	1
(1,0,0)	2	0
(1,0,0)	2	0
(1,0,0)	2	0
$\vdots$	$\vdots$	$\vdots$

focused in particular on the case where the sequence $(v^{(n)})_{n\in\mathbb{N}^*}$ converges to 0. They characterise this by looking at whether the vector is balanced or not, in the sense given by the Theorem 2.2 below (Table 1).

Definition 2.1 (Totally irrational vector). *A vector v is said to be totally irrational iff its coordinates form a free family on $\mathbb{Q}$.*

Theorem 2.2 (Kraaikamp-Meester [10, Theorem 2], [7, Lemma 11]). *Let v be a totally irrational vector. The following propositions are equivalent:*

1. $\lim_{n\to+\infty} v^{(n)} = 0$;
2. $\forall n \in \mathbb{N}^*, \left\|v^{(n)}\right\|_1 \geqslant (d-1)\left\|v^{(n)}\right\|_\infty$;
3. *each letter in* $[\![1,d]\!]$ *appears infinitely many times in* Δ.

Proof. The equivalence between (1) and (2) is Theorem 2 in [10]. The implication (1) $\Rightarrow$ (3) is Lemma 11 in [7]. For (3) $\Rightarrow$ (1), consider a vector v such that each letter in $[\![1,d]\!]$ appears infinitely many times in Δ. Since $(v^{(n)})_{n\in\mathbb{N}}$ is a positive decreasing sequence, it converges to a vector $u \geqslant 0$. Let us assume by contradiction that there exists $i \in [\![1,d]\!]$ such that $u_i > 0$ and let $A \stackrel{\text{def}}{=} \{n \in \mathbb{N}^* \mid \delta_n = i\}$. Since A is infinite, $A = \{f(m)\}_{m\in\mathbb{N}}$ where f is a strictly increasing function. Let $j \neq i$. We have

$$v_j^{f(m+1)} \leqslant v_j^{f(m)+1} = v_j^{f(m)} - v_i^{f(m)} \leqslant v_j^{f(m)} - u_i.$$

By an immediate induction, we deduce $\forall m \in \mathbb{N}, v_j^{f(m)} \leqslant v_j^{(f(0))} - m\,u_i$, which contradicts the positivity of $(v^{(n)})_{n\in\mathbb{N}}$. Hence $u = 0$, which is exactly (1). $\square$

Definition 2.3 (Kraaikamp-Meester set [10]). *A vector $v \in \mathbb{R}^d$ is said to be* Kraaikamp-Meester *iff v is totally irrational and the fully subtractive algorithm on v generates a sequence $(v^{(n)})_{n\in\mathbb{N}^*}$ of vectors of $\mathbb{R}^d$ such that $\lim_{n\to+\infty} v^{(n)} = 0$. Let K_d be the set of d-dimensional Kraaikamp-Meester vectors.*

The set $\mathcal{K}_2$ consists exactly of the vectors $(a, b) \in \mathbb{R}^{*2}$ where $\frac{a}{b}$ is irrational and C. Kraaikamp and R. Meester have shown that $\mathcal{K}_d$ is Lebesgue-negligible for all $d \geqslant 3$. We give in Fig. 1 a normalised and projected representation of $\mathcal{K}_3$ which is isometric to the Rauzy gasket studied in [1].

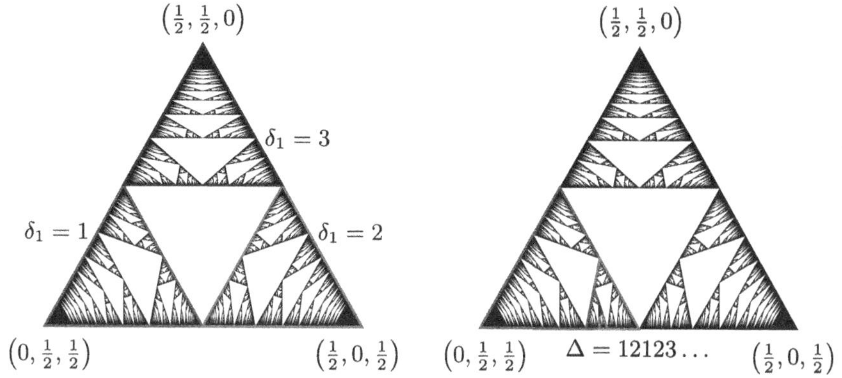

Fig. 1. A normalised representation of the set $\mathcal{K}_3$. On the left, the coloured triangles represent the 3 zones corresponding to the 3 possible values of δ_1. The figure on the right shows the first triangles T_k used to build a vector v associated with a word $\Delta = 12123\ldots$.

The characterisation of the vectors in $\mathcal{K}_d$ makes it easy to establish a link between the vector v and the sequence Δ. Given $v \in \mathcal{K}_d$, by Theorem 2.2, the infinite word Δ associated with v contains infinitely many occurrences of each letter in $[\![1, d]\!]$. Conversely, any Δ with infinitely many occurrences of each letter comes from a single vector $v \in \mathcal{K}_d$, up to a multiplicative factor. Moreover, this single vector v can be calculated from Δ. To do this, let us call T the set $\{v \in \mathbb{R}^d \mid v \geqslant 0, \|v\|_1 = 1\}$ which is compact. Since we are seeking a vector v of $\mathcal{K}_d$, it satisfies $\|v\|_1 \geqslant (d-1)\|v\|_\infty$. We are therefore interested in $T_0 \overset{\text{def}}{=} \{v \in T \mid \|v\|_1 \geqslant (d-1)\|v\|_\infty\}$. The simplex T_0 is represented in Fig. 1 in the case where $d = 3$. The first letter δ_1 of Δ is used to divide T_0 into 3 parts according to the value of δ_1. Consider the set T_1 of vectors v of T_0 the minimum coordinate of which is the one of index δ_1 and satisfying $\|v^{(1)}\|_1 \geqslant (d-1)\|v^{(1)}\|_\infty$ where $v^{(1)} \overset{\text{def}}{=} \gamma_{\delta_1}(v)$ and $\gamma_i \overset{\text{def}}{=} x \mapsto x - (\mathbb{1} - e_i)x_i$. In Fig. 1 (left), the triangle T_1 is thus one of the 3 sub-triangles of T_0 (in red if $\delta_1 = 1$, in green if $\delta_1 = 2$ and in blue if $\delta_1 = 3$). Repeat the construction to build a sequence $(T_k)_{k \in \mathbb{N}}$ where T_{k+1} is the set of vectors v of T_k such that the smallest coordinate of $v^{(k)}$ is that of index δ_{k+1} and satisfying $\|v^{(k+1)}\|_1 \geqslant (d-1)\|v^{(k+1)}\|_\infty$ with $v^{(k+1)} \overset{\text{def}}{=} \gamma_{\delta_{k+1}}(v^{(k)})$. Figure 1 (right) shows the first triangles T_k in the case $\Delta = 12123\ldots$ Thus, $(T_k)_{k \in \mathbb{N}}$ is a decreasing sequence of compacts the diameter of which tends towards 0. So this sequence converges to a single point v. The constraint that each letter must appear infinitely many times in Δ ensures that $v \in \mathcal{K}_d$. We have thus constructed a unique $v \in \mathcal{K}_d$ of norm 1 associated with Δ. All vectors in $\mathcal{K}_d$ associated with Δ are given by λv for $\lambda \in \mathbb{R}_+^*$.

3 Definition of the Δ-Numeration

The sequences generated by the fully subtractive algorithm allow us to define Δ-numeration in the Kraaikamp-Meester case, which is essential for establishing the properties of our system. We therefore fix $v \in \mathcal{K}_d$ for the rest of this article and consider the Δ and θ sequences associated with v in the fully subtractive algorithm.

Definition 3.1. *Let $w \in \mathbb{Z}^\omega$. We say that w represents the real number $\varphi(w) \stackrel{def}{=} \sum_{n=1}^{\infty} w_n \theta_n$ provided that the sum converges.*

Remark 3.2. *We are often interested in particular words w, usually bounded or with values in $\{0, 1\}$. In this case, the sum converges, a result of [10] ensuring that $\Omega \stackrel{def}{=} \varphi(1^\omega) = \sum_{n=1}^{\infty} \theta_n = \frac{\|v\|_1}{d-1}$. Moreover, we are also interested in finite words. We then artificially extend the finite word w into $w0^\omega$. In the following we will freely confuse w with $w0^\omega$.*

A first simple example of Δ-numeration was studied in 2014 in [3] for the specific Tribonacci case. The authors consider the single real root α of the polynomial X^3+X^2+X-1 and the vector $v \stackrel{\text{def}}{=} (\alpha, \alpha+\alpha^2, 1)$. The flow of the algorithm on v is given in Table 2. We then obtain $\Delta = (123)^\omega$ and $\forall n \in \mathbb{N}^*, \theta_n = \alpha^n$. This is a simple case of β-numeration in base $\frac{1}{\alpha}$.

Table 2. Fully subtractive algorithm on the vector $v = (\alpha, \alpha + \alpha^2, 1)$ where α satisfies $\alpha + \alpha^2 + \alpha^3 = 1$.

$v^{(n)}$	δ_n	θ_n
$v = (\alpha, \alpha + \alpha^2, 1)$	1	α
$(\alpha, \alpha^2, 1 - \alpha)$	2	α^2
$(\alpha - \alpha^2, \alpha^2, \alpha^3)$	3	α^3
$(\alpha^4, \alpha^4 + \alpha^5, \alpha^3) = \alpha^3 v$	1	α^4
⋮	⋮	⋮

4 Expansions and Reductions

In this section, we define a convergent rewriting system that makes it easy to calculate in Δ-numeration. This system comes from the linear recurrence relations satisfied by θ which we obtain via the fully subtractive algorithm. We call this property, derived from [9], the *liaison* Theorem.

Theorem 4.1 (Liaison [9, Lemmas 5 and 6]).

1) If the letter k has its first occurrence in Δ at position n then $v_k = \sum_{i=1}^{n} \theta_i$.
2) If the letter k has 2 successive occurrences in Δ at positions m and n with $m < n$ then $\theta_m = \sum_{i=m+1}^{n} \theta_i$.

The liaison theorem makes it possible to establish, by immediate recurrence, that the θ_n belong to $\text{Vect}_{\mathbb{Z}}(v)$, the subgroup of $\mathbb{R}$ generated by the coordinates of v. Moreover, the linking relations between the θ_n allow us to expose expansion and reduction relations on w which preserve $\varphi(w)$.

Definition 4.2 (Reduction/Expansion [7]). *Let $w \in \mathbb{Z}^\omega$. We assume that the letter $k \in [\![1, d]\!]$ has 2 successive occurrences at positions m and n in Δ, meaning that $\delta_m = \delta_n = k$ and $\delta_i \neq k$ for all $i \in [\![m+1, n-1]\!]$. The transformation of w consisting in incrementing w_m and decrementing w_i for $i \in [\![m+1, n]\!]$ to obtain a new word w' is called reduction. The reverse operation is called expansion.*

	Δ	$\cdots$ k	$(\neq k)$	$(\neq k)$	$(\neq k)$	k	$\cdots$
***Reduction*:**	w	$\cdots$ w_m	$w_{m+1}+1$	$w_{m+2}+1$	$\cdots$	w_n+1	$\cdots$
	w'	$\cdots$ w_m+1	w_{m+1}	w_{m+2}	$\cdots$	w_n	$\cdots$
***Expansion*:**	w	$\cdots$ w_m+1	w_{m+1}	w_{m+2}	$\cdots$	w_n	$\cdots$
	w'	$\cdots$ w_m	$w_{m+1}+1$	$w_{m+2}+1$	$\cdots$	w_n+1	$\cdots$

For example, consider the word $w = 004$ and Δ of the form $12132313\ldots$ By means of reductions and expansions (see Fig. 2 (left)), we get a word w' over the alphabet $\{0, 1\}$ equivalent to w, in the sense that $\varphi(w) = \varphi(w')$. Most often, working with words over the alphabet $\mathbb{N}$, we restrict the reduction and expansion operations as follows:

- reductions are only performed towards a w_m such that $w_m = 0$;
- expansions are only performed from a w_m such that $w_m + 1 \geqslant 2$.

These restrictions ensure that the word w' in the example is in normal form and, as we show below, give a canonical word in $\{0, 1\}^\omega$ to represent each real number in $[0, \Omega]$.

In order to demonstrate the normalisation property of our system, we still have to deal with the case of the improper writing of a number, equivalent in Δ-numeration to the identity $0.9999\cdots = 1$ in base ten numeration. Let us take a finite word w and apply successive expansions from its last 1. Figure 2 (right) shows a construction of the improper writing of $w = 1$ for $\Delta = 12131231\ldots$ At the limit, we obtain an infinite word $w' = (\alpha_n)_{n\in\mathbb{N}} \in \{0, 1\}^\omega$ such that $\forall n \in \mathbb{N}, \alpha_n = 1$ iff $\delta_n \neq 1$. The words w and w' are equivalent because $\varphi(w) = \varphi(w')$ but both in normal form for the expansion/reduction relations. We therefore add an infinite carry rule to avoid this problem.

1	**2**	**1**	**3**	**2**	**3**	**1**	**3**	···
0	0	4	0	0	0	0	0	0
0	0	3	1	1	1	1	0	0
0	1	2	0	0	1	1	0	0
0	1	1	1	1	2	2	0	0
1	0	0	1	1	2	2	0	0
1	0	1	0	0	1	1	0	0

1	**2**	**1**	**3**	**1**	**2**	**3**	**1**	···
1	0	0	0	0	0	0	0	···
0	1	1	0	0	0	0	0	···
0	1	0	1	1	0	0	0	···
0	1	0	1	0	1	1	1	···

Fig. 2. Left: reduction of $w \stackrel{\text{def}}{=} 004$ in the Δ-numeration system associated with $\Delta \stackrel{\text{def}}{=} 12132313\ldots$ Expansion positions are marked in red, reduction positions in blue. We obtain a word $w' = 101001100$ which represents the same value as w. Right: constructing the improper writing of a word from a finite word. (Color figure online)

Definition 4.3 (Improper writing). *Let $w \in \{0,1\}^\omega$. We say that w is in improper writing iff there exist $r \in \mathbb{N}$, $u \in \{0,1\}^r$, $v \in 0\{0,1\}^\omega$ and $k \in [\![1,d]\!]$ such that $w = uv$ and $\forall n \in \mathbb{N}^*, v_n = 0$ iff $\delta_{n+r} = k$. For such an w, we add the infinite carry rewrite rule $w \to u1$.*

Definition 4.4 (Greedy writing). *Let $x \in [0,\Omega]$. The greedy writing of x is the word $w \in \{0,1\}^\omega$ obtained from the greedy algorithm, defined by recurrence as follows:*

- $x_1 = x$
- *for all $n \in \mathbb{N}^*$, let* $w_n \stackrel{def}{=} \begin{cases} 1 \text{ if } x_n \geqslant \theta_n \\ 0 \text{ if } x_n < \theta_n \end{cases}$ *and* $x_{n+1} = x_n - w_n\theta_n$.

We define a normalisation property expressing the behaviour of the Δ-numeration. For finite words, the system is convergent. For infinite words, it is necessary to allow a potentially infinite number of reductions.

Definition 4.5 (Normalisation). *We say that a rewriting system $\to$ is normalising on A^ω iff:*

1. *$\to$ is strongly normalising for finite words, i.e. the restriction of $\to$ to A^* is well founded.*
2. *$\to$ is weakly normalising for infinite words, i.e. for all $u_0 \in A^\omega$, there exists a sequence $(u_n)_{n\in\mathbb{N}}$ decreasing for $\to$ and converging in the Cantor set A^ω to a limit in normal form.*

Theorem 4.6 (normalisation [7, Theorem 16], [6, Section 2]).

1. *the rewriting system associated with the expansion/reduction/infinite carry rules is normalising and confluent;*
2. *$\varphi(\{0,1\}^\omega) = [0,\Omega]$ and any real number $x \in [0,\Omega]$ is uniquely written as $\varphi(w)$ for a $w \in \{0,1\}^\omega$ in normal form for the expansion/reduction/infinite carry system;*
3. *with the previous notations, w is the greedy representation of x.*

Note that this normalisation property is very strong and is found in very few numeration systems. However, its presence can be explained by the very strong constraints on the θ sequence. In particular, it allows comparisons, additions and subtractions to be carried out easily in our numeration system. For comparisons, all we have to do is put the words w, w' to be compared in normal form and then compare the normal forms lexicographically. For additions, we sum w and w' term by term and then normalise. Note that if $\varphi(w) + \varphi(w')$ falls outside the interval $[0, \Omega]$, it is still possible to calculate the sum by extending Δ to the left and θ accordingly. Finally, for subtractions, if $\varphi(w) \geqslant \varphi(w')$, we can subtract the 2 words term by term to obtain a word over the alphabet $\{-1, 0, 1\}$. By allowing expansions from 1 and reductions towards -1, we then get a word over the alphabet $\{0, 1\}$ in normal form representing $\varphi(w) - \varphi(w')$.

5 Link with β-Numeration and Cantor Bases

We have already shown an example, for $\Delta = (123)^\omega$ and the Tribonacci number, where Δ-numeration meets β-numeration. In fact, this is the only case where these two systems coincide, up to a generalisation to d-dimensions. Let us start with a technical lemma before proving this property.

Lemma 5.1. *Let $k, r \in \mathbb{N}^*$ such that $\theta_k = \sum_{i=1}^{r} \theta_{k+i}$. Then $\delta_k = \delta_{k+r}$ and δ_{k+r} is the next occurrence of δ_k in Δ.*

Proof. Should we modify the initial vector v, we can assume without loss of generality that $k = 1$ and $\delta_k = 1$. Let us call $q \overset{\text{def}}{=} \min\{i \geqslant 2 \mid \delta_i = 1\}$ the next occurrence of 1 in Δ. By the liaison Theorem 4.1, we have $\sum_{i=2}^{q} \theta_i = \theta_1 = \sum_{i=2}^{r} \theta_i$. By subtracting, we obtain that $\sum_{i=\min(q,r)+1}^{\max(q,r)} \theta_i = 0$. Since $v \in \mathcal{K}_d$, θ_i's are positive so that $q = r$. □

Proposition 5.2. *Let $v \in \mathcal{K}_d$ and $\beta > 1$ be such that the Δ-numeration system associated with v is base β numeration. Then β is the Pisot number root of $X^d - \sum_{k=0}^{d-1} X^k$ and, up to a permutation of the coordinates, $\Delta = (123\ldots d)^\omega$ and v is proportional to $\left(\sum_{i=1}^{k} \frac{1}{\beta^i}\right)_{k \in [\![1,d]\!]}$.*

Proof. Let $v \in \mathcal{K}_d$ and Δ, θ be the sequences associated by the fully subtractive algorithm. It is assumed that there exists $\beta > 1$ such that $\forall n \in \mathbb{N}^*, \theta_n = \beta^{-n}$. Should we permute the coordinates of v, we can assume without loss of generality that v is increasing and therefore that $\delta_1 = 1$. Since Δ admits infinitely many occurrences of each letter, 1 appears infinitely many times in Δ. In particular, let $r \overset{\text{def}}{=} \min\{n \geqslant 2 \mid \delta_n = 1\} - 1$ be the second occurrence of 1 in Δ. Then, by the liaison theorem, $\frac{1}{\beta} = \theta_1 = \sum_{i=2}^{r+1} \theta_i = \sum_{i=2}^{r+1} \frac{1}{\beta^i}$. Multiplying by β^{r+1}, we find that β is a root of $P \overset{\text{def}}{=} X^r - \sum_{i=0}^{r-1} X^i$.

Let $k \in \mathbb{N}^*$. Since β is a root of P, $\beta^{-k} = \sum_{i=1}^{r} \beta^{-(i+k)}$ and hence $\theta_k = \sum_{i=1}^{r} \theta_{i+k}$. By Lemma 5.1, we deduce that the next occurrence of δ_k in Δ is δ_{k+r}. Since Δ admits infinitely many occurrences of each letter, this imposes $r = d$

and $\delta = (123\dots d)^\omega$ because v is increasing. By posing $v_0 \stackrel{\text{def}}{=} \left(\sum_{i=1}^{k} \frac{1}{\beta^i}\right)_{k\in[\![1,d]\!]}$, we can easily check, like in [3], that $v_0 \in \mathcal{K}_d$ has Δ as associated word. It suffices to note that a step in the fully subtractive algorithm applied to v_0 gives the vector $\frac{1}{\beta}v_0$, up to the permutation of the coordinates by the cycle $(d\dots 321)$. By uniqueness, we deduce that v is proportional to v_0. □

Definition 5.3 (Real Cantor bases [5]). *Let $(\beta_n)_{n\in\mathbb{N}^*}$ be a sequence of real numbers in $[1,+\infty[$ such that $\prod_{k=1}^{\infty}\beta_k = +\infty$. The word $w \in \mathbb{N}^\omega$ represents for real Cantor bases the real number $\sum_{n=1}^{\infty}(w_n/\prod_{k=1}^{n}\beta_k)$, provided that the sum exists.*

We can easily create a link between real Cantor bases and Δ-numeration, simply by looking for a sequence β such that $\forall n \in \mathbb{N}^*, \theta_n = 1/\prod_{k=1}^{n}\beta_k$. If we have such a relationship, we can state $\frac{\theta_n}{\theta_{n+1}} = \left(\prod_{k=1}^{n+1}\beta_k\right)/\left(\prod_{k=1}^{n}\beta_k\right) = \beta_{n+1}$. Thus, given a sequence θ given by the fully subtractive algorithm, it suffices to put $\beta_1 \stackrel{\text{def}}{=} \frac{1}{\theta_1}$ and $\forall n \in \mathbb{N}^*, \beta_{n+1} \stackrel{\text{def}}{=} \frac{\theta_n}{\theta_{n+1}}$ to see Δ-numeration as a special case of Cantor real bases. The reals β_n (for $n \geqslant 2$) are indeed in $[1,+\infty[$ because θ is weakly decreasing, and $\forall n \in \mathbb{N}^*, \prod_{k=1}^{n}\beta_k = \frac{1}{\theta_n} \xrightarrow{n\to+\infty} +\infty$ because θ converges to 0. As for $\beta_1 = \frac{1}{\theta_1}$, we cannot be sure that it belongs to $[1,+\infty[$ but, since Δ-numeration is defined up to a multiplicative constant, we can always return to the case where $\beta \in [1,+\infty[$.

While Δ-numeration can be seen as a special case of Cantor real bases, the converse is false. First of all, finite words in Δ-numeration all represent reals in $\mathrm{Vect}_{\mathbb{Z}}(v)$ which is a module of finite type. Cantor real bases do not have this type of constraint. Moreover, Cantor real bases generalise numeration in base $\beta > 1$, whereas Δ-numeration is limited to very particular β, as explained in Proposition 5.2. Cantor real bases therefore offer a much more general framework than Δ-numeration. However, this generality prevents these bases from being equipped with an efficient normalisation system, such as expansion/reduction. The linking relations on the θ_n are therefore what restricts our system the most, but also what makes the calculations in it so simple.

6 The (Ultimately) Periodic Case

We are interested here in the case where the word Δ is periodic. Let us first note that the reasoning in this section is stable by any finite left extension of Δ, so that these results generalise without difficulty to the ultimately periodic case. For the sake of simplicity, however, we restrict ourselves here to the periodic case, where $\Delta = \pi^\omega$ where $\pi \in [\![1,d]\!]^*$ is of length p. The properties of Δ-numeration in the periodic case were first set out in [6], as a tool for solving a discrete geometry problem. In parallel, they were studied by E. Charlier [4] in the more general context of alternate bases, the periodic case of real Cantor bases. Both papers independently involve a Pisot number β and the number field $\mathbb{Q}[\beta]$.

Theorem 6.1 (Periodicity [6, Section 6]). *Let us note, for* $i \in [\![1, d]\!]$, $\gamma_i \overset{def}{=} x \mapsto x - (\mathbb{1} - e_i)x_i$ *the transformations of the fully subtractive algorithm, where* $\mathbb{1} \overset{def}{=} (1, \ldots, 1)$. *Let* $\Gamma \overset{def}{=} \gamma_{\pi_p} \circ \cdots \circ \gamma_{\pi_2} \circ \gamma_{\pi_1}$. *Then*

- *there exists a real number* β *such that* $\forall n \in \mathbb{N}^*, \theta_{n+p} = \beta\theta_n$;
- *the vector* v *is an eigenvector of* Γ *associated with the eigenvalue* β;
- Γ^{-1} *is a cyclic endomorphism (in a* $\mathbb{K}$*-vectorial space* E, $u \in \mathcal{L}(E)$ *is called cyclic iff* $\exists x \in E, E = \mathbb{K}[u]x$*) product of Pisot matrices [2];*
- $\frac{1}{\beta}$ *is a Pisot number of degree* d*, with minimal polynomial equal to the minimal (and characteristic) polynomial of* Γ.

The previous theorem thus allows us to easily find the vector v from the period π of Δ. In the case where $\Delta = (123)^\omega$, the vector v can be found as the eigenvector of $\gamma_3 \circ \gamma_2 \circ \gamma_1$ the matrix of which is $\begin{pmatrix} 2 & 0 & -1 \\ -1 & 2 & -1 \\ 0 & -1 & 1 \end{pmatrix}$. This matrix has a single real eigenvalue α^3 (which is the inverse of a Pisot number), where α is the single real root of $X^3 + X^2 + X - 1$, and an associated eigenvector is $v = (\alpha, \alpha + \alpha^2, 1)$. This gives us the Tribonacci vector studied in [3].

In the case where $\Delta = (1213)^\omega$, the example we will use later, the vector v can be found as the eigenvector of $\gamma_3 \circ \gamma_1 \circ \gamma_2 \circ \gamma_1$ the matrix of which is $\begin{pmatrix} 4 & -1 & -1 \\ -1 & 2 & -1 \\ -2 & 0 & 1 \end{pmatrix}$. The characteristic polynomial of this matrix is $X^3 - 7X^2 + 11X - 1$ which has 3 real roots. Only one of them has modulus < 1, which is called β. In particular, $\beta \approx 0.09678807$ is the inverse of a Pisot number. An eigenvector v associated with the eigenvalue β is $(2, -\beta^2 + 4\beta + 3, \beta^2 - 6\beta + 5)$.

Note that, in the ultimately periodic case where $\Delta = u\pi^\omega$, it is also possible to calculate v easily. We start by calculating v', the vector associated with π^ω. Then, by posing $A \overset{\text{def}}{=} \gamma_{u_k} \circ \cdots \circ \gamma_{u_1}$, we obtain $Av = v'$. Since A is invertible, we deduce $v = A^{-1}v'$.

The periodicity theorem can also be used to demonstrate the link with alternate Cantor bases. The property $\forall n \in \mathbb{N}^*, \theta_{n+p} = \beta\theta_n$ shows the periodicity of the associated sequence $(\beta_n)_{n \in \mathbb{N}^*}$: $\forall n \in \mathbb{N}^*, \beta_{n+1+p} = \frac{\theta_{n+p}}{\theta_{n+1+p}} = \frac{\beta\theta_{n+1}}{\beta\theta_n} = \beta_{n+1}$. To prove $\beta_{p+1} = \beta_1$, however, it is necessary to set $\theta_p = \beta$, which is the same as setting v. Finally, the periodic character of Δ allows us to characterise the image by φ of the set of ultimately periodic words.

Corollary 6.2. $\forall n \in \mathbb{N}^*, \theta_n \in v_1\mathbb{Q}[\beta]$.

Theorem 6.3. *If we set* $U \overset{def}{=} \{\sigma \in \mathbb{Z}^\omega \mid \sigma \text{ is ultimately periodic}\}$, *we have* $\varphi(U) = v_1\mathbb{Q}[\beta]$.

Before proving the result, let us note that analogous proofs of this result are proposed in [4] for alternate bases and in [14] for Rényi bases. However, our proof has an effective side which makes it easier to perform calculations in our numeration system.

Proof. Let us first show that $\varphi(U) \subset v_1\mathbb{Q}[\beta]$. Recall that $v_1\mathbb{Q}[\beta]$ is stable by sum and by product by an element of $\mathbb{Q}[\beta]$. Since β is algebraic, we also know that $\mathbb{Q}[\beta]$ is a field. Since all θ_n live in $v_1\mathbb{Q}[\beta]$ which is closed by sum, every finite word has its image by φ in $v_1\mathbb{Q}[\beta]$. Let us now look at ultimately periodic words in general. Let us call p the period of Δ and take $\sigma \in U$. Since σ is ultimately periodic, there are $u, w \in \{0,1\}^*$ such that $\sigma = u\,w^\omega$. Should we increase the sizes of u and w, we can assume without loss of generality that $|u|$ and $|w|$ are multiples of p. Let k and n be such that $|u| = kp$ and $|w| = np$, where $|\cdot|$ denotes the word length function. Note finally that σ is ultimately periodic and therefore bounded, and that $(\theta_n)_{n\in\mathbb{N}}$ is summable so that $\varphi(\sigma)$ exists. Thus, since $\varphi(u)$ and $\varphi(w)$ are in $v_1\mathbb{Q}[\beta]$,

$$\varphi(\sigma) = \varphi(u) + \beta^k\varphi(w^\omega) = \varphi(u) + \beta^k\left(\sum_{j=0}^{\infty}\varphi(w)\beta^{jn}\right) = \varphi(u) + \frac{\beta^k}{1-\beta^n}\varphi(w) \in v_1\mathbb{Q}[\beta]$$

Let us now show that every element of $v_1\mathbb{Q}[\beta]$ is the image of an element of U. Note that U is stable by sum so that $\varphi(U)$ also is. Since U is stable by shift, $\varphi(U)$ is stable by multiplication by β. It therefore remains to show that $(\frac{v_1}{b})$ for $b \in \mathbb{N}^*$ are representable by words in U. Let us reason by analysis-synthesis. We are looking for $\lambda \in \mathbb{Z}^\omega$ periodic of period np such that (note that the family $(\lambda_k\beta^k)_{k\in\mathbb{N}}$ is summable because λ is periodic and therefore bounded):

$$\begin{aligned}\frac{1}{b} &= \sum_{k=0}^{+\infty}\lambda_k\beta^k = \sum_{m=0}^{+\infty}\sum_{q=0}^{np-1}\lambda_{mnp+q}\beta^{mnp+q} = \sum_{m=0}^{+\infty}\beta^{mnp}\sum_{q=0}^{np-1}\lambda_q\beta^q\\ &= \frac{P(\beta)}{1-\beta^{np}} \quad \text{where } P(\beta) \stackrel{\text{def}}{=} \sum_{q=0}^{np-1}\lambda_q\beta^q \in \mathbb{Z}[\beta]\end{aligned}$$

From now on, we wish to find $P \in \mathbb{Z}[X]$ and $n \in \mathbb{N}^*$ such that $1 - \beta^{np} = bP(\beta)$. Let us consider $A = \mathbb{Z}[X]/(\mu_\beta, b)$, the quotient ring of $\mathbb{Z}[X]$ by the ideal generated by μ_β and b. A is a finite ring. Since μ_β is unitary, by the isomorphism theorem, we have the ring isomorphisms $\mathbb{Z}[X]/(\mu_\beta, b) \simeq \mathbb{Z}_{d-1}[X]/(b) \simeq (\mathbb{Z}/b\mathbb{Z})_{d-1}[X]$ which is finite. Recall then that μ_β is the characteristic polynomial of a product matrix of $\gamma_i \stackrel{\text{def}}{=} x \mapsto x - (\mathbf{1} - e_i)\,x_i$. Since $\forall i \in [\![1,d]\!]\,, \det(\gamma_i) = 1$, the constant coefficient of μ_β is then $(-1)^d$. Thus, there exists $Q \in \mathbb{Z}[X]$ such that $\mu_\beta = XQ + (-1)^d$. Passing to the quotient in A, we obtain $0 = XQ + (-1)^d$ from which $X((-1)^{d+1}Q) = 1$. X is therefore invertible in A. Since the multiplicative group $A^\times$ is finite, by Lagrange's theorem there exists $n \in \mathbb{N}^*$ such that $X^n = 1$ in A, hence $X^{np} = 1$ in A. Thus, there exists $P, R \in \mathbb{Z}[X]$ such that $X^{np} = 1 + R\mu_\beta - bP$. Evaluating in β, we then find $\beta^{np} = 1 - bP(\beta)$ so $1 - \beta^{np} = bP(\beta)$ which was the equality we were looking for.

The fractions $\frac{v_1}{b}$ are thus representable by words of U, and since $\varphi(U)$ is closed by sum and by product by β, we get the reciprocal inclusion $v_1\mathbb{Q}[\beta] \subset \varphi(U)$. □

The effective nature of the proof of Theorem 6.3 means that, when $v_1 \in \mathbb{Q}[\beta]$, we can effectively perform new operations in Δ-numeration, in particular multiplication and division. In fact, we can move from U to $\mathbb{Q}[\beta]$ effectively and vice versa. As it is simple to perform the usual operations in $\mathbb{Q}[\beta]$, we can perform them without difficulty in U.

Let us illustrate the effectiveness of the proof using the example of $\Delta = (1213)^\omega$. The word Δ has the associated Pisot inverse $\beta \approx 0,09678807$ which is the root of the polynomial $\mu_\beta = X^3 - 7X^2 + 11X - 1$. We obtain $v \stackrel{\text{def}}{=} (2, -\beta^2 + 4\beta + 3, \beta^2 - 6\beta + 5)$ as the vector associated with Δ and we can calculate the sequence θ by the liaison Theorem 4.1,

$$\begin{cases} \theta_1 = v_1 & = 2 \\ \theta_2 = v_2 - \theta_1 & = -\beta^2 + 4\beta + 1 \\ \theta_3 = \theta_1 - \theta_2 & = \beta^2 - 4\beta + 1 \\ \theta_4 = v_3 - \theta_1 - \theta_2 - \theta_3 & = \beta^2 - 6\beta + 1 \end{cases} \quad \textit{and } \forall n \in \mathbb{N}^*, \theta_{n+4} = \beta\theta_n.$$

Let us take the example $\sigma \stackrel{\text{def}}{=} 10(1201)^\omega \in U$. We begin by writing σ in blocks of period 4 as in the proof of the previous theorem: $\sigma = u\,w^\omega$ with $u \stackrel{\text{def}}{=} 1012$ and $w \stackrel{\text{def}}{=} 0112$. Thus,

$$\begin{cases} \varphi^v(u) = \theta_1 + \theta_3 + 2\theta_4 = 3\beta^2 - 16\beta + 5 \\ \varphi^v(w) = \theta_1 + 2\theta_2 + \theta_4 = -\beta^2 + 2\beta + 5 \end{cases}$$

hence

$$\varphi^v(\sigma) = \varphi^v(u) + \frac{\beta}{1-\beta}\varphi^v(w) = \frac{13\beta^2 - 102\beta + 21}{4} \in \mathbb{Q}[\beta].$$

To obtain all the polynomials in β as elements of $\varphi^v(U)$, let us start by noting that $v_1 = 2$. So $\varphi^v(1) = 2$ and $\varphi^v(0^{4n}1) = 2\beta^n$. So we already have $2\mathbb{Z}[\beta] \subset \varphi^v(U)$. To obtain $\mathbb{Q}[\beta] \subset \varphi^v(U)$, it remains to obtain codes for $\frac{2}{b}$ with $b \in \mathbb{N}^*$. For example, let us take $b = 2$. We place ourselves in $A \stackrel{\text{def}}{=} \mathbb{Z}[X]/(\mu_\beta, 2)$ and look at the order of X in $A^\times$. By definition of A, $0 = X^3 - 7X^2 + 11X - 1 = X^3 + X^2 + X + 1$. Hence $X^3 = X^2 + X + 1$ and $X^4 = X^3 + X^2 + X = 1$ so X is of order 4 in $A^\times$. We then divide $X^4 - 1$ by μ_β: $X^4 - 1 = Q\mu_\beta + 2(19X^2 - 38X + 3)$ so $\beta^4 - 1 = 2(19\beta^2 - 38\beta + 3)$ and, posing $P \stackrel{\text{def}}{=} -19X^2 + 38X - 3$, we obtain:

$$\frac{1}{2} = \frac{P(\beta)}{1-\beta^4} = P(\beta)\sum_{k=0}^{\infty} \beta^{4k}$$

We can therefore encode $\frac{2}{2}$ by the word $((-3)000(38)000(-19)0000000)^\omega$ which we can then normalise to $(0011001000001101)^\omega$. So the normalised code for $\frac{2}{2}$ is $(0011001000001101)^\omega$.

7 Conclusion

This article reviews the known properties of Δ-numeration from the point of view of numeration systems. We make the link with β-numeration and with

Cantor real bases. While Δ-numeration is a special case of Cantor bases, the linear constraints induced by the fully subtractive algorithm make it possible to define expansion/reduction operations and to obtain normalisation properties for this system. Finally, we look at the ultimately periodic case, analogous in Δ-numeration to alternating bases, where the inverse β of a Pisot number appears. We then characterise the set of numbers represented by ultimately periodic words. After multiplication by a scalar, this is the number field $\mathbb{Q}[\beta]$, which makes it possible to perform additions, subtractions, multiplications and divisions on the ultimately periodic words. Beyond the ultimately periodic case, it would be interesting to look at more general cases of Δ words, in the case of words resulting from substitution processes for example, or even for general uniformly recurrent words. Although the linking relations between θ_n are of bounded size, they nevertheless seem much more difficult to exploit without the periodicity hypothesis.

References

1. Arnoux, P., Starosta, S.: The rauzy gasket. In: Further Developments in Fractals and Related Fields: Mathematical Foundations and Connections. Springer (2013)
2. Avila, A., Delecroix, V.: Some monoids of pisot matrices. In: New Trends in One-Dimensional Dynamics: IMPA 2016, Rio de Janeiro. Springer (2019)
3. Berthé, V., Domenjoud, E., Jamet, D., Provençal, X.: Fully subtractive algorithm, tribonacci numeration and connectedness of discrete planes. RIMS Lecture notes 'Kokyuroku Bessatu' **46**, 159–174 (2014)
4. Charlier, E.: Alternate base numeration systems. In: International Conference on Combinatorics on Words, pp. 14–34. Springer (2023)
5. Charlier, E., Cisternino, C.: Expansions in cantor real bases. Monatshefte für Mathematik **195**(4), 585–610 (2021)
6. Domenjoud, E., Laboureix, B., Vuillon, L.: Facet connectedness of arithmetic discrete hyperplanes with non-zero shift. In: DGCI. Springer (2019)
7. Domenjoud, E., Provençal, X., Vuillon, L.: Facet connectedness of discrete hyperplanes with zero intercept: the general case. In: DGCI. Springer (2014)
8. Domenjoud, E., Provençal, X., Vuillon, L.: Palindromic language of thin discrete planes. Theoret. Comput. Sci. **624**, 101–108 (2016)
9. Domenjoud, E., Vuillon, L.: Geometric palindromic closure. Uniform Distrib. Theory **7**(2), 109–140 (2012)
10. Kraaikamp, C., Meester, R.: Ergodic properties of a dynamical system arising from percolation theory. Ergodic Theory Dyn. Syst. (1995)
11. Laboureix, B.: Hyperplans arithmétiques: connexité, reconnaissance & transformations. Ph. D. thesis, Université de Lorraine (2024)
12. Rényi, A.: Representations for real numbers and their ergodic properties. Acta Math. Acad. Sci. Hungar **8**(3–4), 477–493 (1957)
13. Rigo, M.: Numeration systems: a link between number theory and formal language theory. In: International Conference on Developments in Language Theory, pp. 33–53. Springer (2010)
14. Schmidt, K.: On periodic expansions of pisot numbers and salem numbers. Bull. Lond. Math. Soc. **12**(4), 269–278 (1980)
15. Schweiger, F.: Multidimensional Continued Fractions. Oxford University (2000)

Purely Automatic Sequences with the Uniform Distribution Property

Shuo Li(✉) and Narad Rampersad

Department of Mathematics, University of Winnipeg, 515 Portage Avenue, Winnipeg, Manitoba R3B 2E9, Canada
{sh.li,n.rampersad}@uwinnipeg.ca

Abstract. An infinite sequence with finitely many distinct letters is said to have the *uniform distribution property* if all letters in the alphabet of the sequence have the same density in all arithmetical progressions. This property was first studied by Gelfond for *sum of digits* functions. In this note, we characterize a larger class of purely automatic sequences with the same property.

1 Introduction

Let $W = (W[n])_{n\in\mathbb{N}}$ be an infinite sequence over the finite alphabet $\{1,\ldots,M\}$ for some positive integer M. We say that W has the *uniform distribution property* if and only if for all pairs of integers $a, b \in \mathbb{N}$ such that $a \geq 1$, $0 \leq b < a$ and all $m \in \{1,\ldots,M\}$, one has

$$d_{W,a,b}(m) = \lim_{n\to\infty} \frac{\#\{k|W[ak+b] = m, 0 \leq ak+b < n\}}{n} = \frac{1}{aM}.$$

The uniform distribution property was first studied by Gelfond in [8] for *sum of digits* functions. More precisely, he proved that for integers $q, M, d, b, a \in \mathbb{N}$ such that $q, M, a \geq 2$ and $\gcd(M, q-1) = 1$,

$$\lim_{n\to\infty} \frac{\#\{k|s_q(ak+b) \equiv d \bmod M, 0 \leq ak+b < n\}}{n} = \frac{1}{aM},$$

where $s_q(n)$ is the sum of all digits in the q-expansion of n for all $n \in \mathbb{N}$. In particular, the sequence $(s_2(n) \bmod 2)_{n\in\mathbb{N}}$ is the well-known Thue-Morse sequence and the sequences of the form $(s_q(n) \bmod q)_{n\in\mathbb{N}}$ are recognized as a type of *generalized Thue-Morse sequences*. It is "folklore" that all of these sequences are *purely automatic*; Séébold has written down a proof [18, Proposition 5.2]. For the precise definition of purely automatic sequences, see Sect. 2. Gelfond's result implies immediately that all the sequences $(s_q(n) \bmod q)_{n\in\mathbb{N}}$ have the uniform distribution property. In the same paper, Gelfond posed several influential questions on the distribution of the sum of digits function along different subsequences, which motivated much subsequent research [5,10,13–17,19]. Other works related to Gowers norms can be found in [2,7,11]. Note that the density of letters or

G. Gamard and J. Leroy (Eds.): WORDS 2025, LNCS 15729, pp. 205–216, 2025.
https://doi.org/10.1007/978-3-031-97548-6_18

factors are also considered in [12] to study the arithmetical complexity, which was introduced in [1], of infinite sequences. Also, some similar methods as the one in this article have been introduced in studying the substitutive dynamical systems based on uniform substitutions, see, for example, [4].

The motivation of this work is to characterize all automatic sequences satisfying the uniform distribution property and to generalize Gelfond's work to all automatic sequences. In this note, we focus on purely automatic sequences. We first interpret the uniform distribution property in terms of matrices defined by the generating morphism of the sequence. This result is stated in Theorem 10. Then we give a criterion for the uniform distribution property from the combinatorial structure of the generating morphism. This is given in Theorem 20.

2 Preliminaries

Let A be a finite *alphabet* and let A^* and A^∞ denote respectively the set of finite and infinite strings with elements in A. For a finite or infinite string w, let $|w|$ be its length. In particular, $|w| = \infty$ if w is an infinite word. The word of length 0 is the *empty word*, which is also an element in A^* and denoted by ε. Let us write $w = w[0]w[1]\cdots w[|w|-1]$, where $w[i]$ are *letters* in A. For two finite words u, v, let $uv = u[0]u[1]\cdots u[|u|-1]v[0]v[1]\cdots v[|v|-1]$ be the *concatenation* of u and v. A morphism $\phi : A^* \to A^*$ is a function satisfying that for any $a, b \in A^*$, $\phi(ab) = \phi(a)\phi(b)$.

In the rest of this paper: let $M, L \in \mathbb{N}^+$; let $\phi : \{1,\ldots,M\}^* \to \{1,\ldots,M\}^*$ be a morphism such that for all integers $i \in \{1,\ldots,M\}$, the length of $\phi(i)$, denoted by $|\phi(i)|$, equals L, and that $\phi(1) = 1w$ for some nonempty word $w \in \{1,\ldots,M\}^*$; let W be the fixed point of ϕ beginning with 1, i.e. $W = \lim_{i\to\infty}\phi^i(1)$. The sequence W is a *purely automatic sequence* and ϕ is a *generating morphism* of W. We first give a necessary condition for W to satisfy the uniform distribution property.

Lemma 1. *If W satisfies the uniform distribution property, then, necessarily, for each $0 \le l < L$, $\phi(1)[l], \phi(2)[l], \cdots, \phi(M)[l]$ are pairwise distinct.*

Proof. If there exists a l such that $\phi(1)[l], \phi(2)[l], \cdots, \phi(M)[l]$ are not pairwise distinct, then there exists a letter m that does not appear in the arithmetic progression $Ln + l$. Thus, $d_{W,L,l}(m) = 0$. Consequently, W does not satisfy the uniform distribution property for the pair L, l. □

From now, let us suppose that for each $0 \le l < L$, $\phi(1)[l], \phi(2)[l], \cdots, \phi(M)[l]$ are pairwise distinct. Let $P_l : \{1,\ldots,M\} \to \{1,\ldots,M\}$ be a permutation such that $P_l(m) = \phi(m)[l]$ for all $m \in \{1,\ldots,M\}$. Note that P_l is a bijection; let us write $m = P_l^{-1}(\phi(m)[l])$. In particular, since $\phi(1) = 1w$ for some finite string w, we have $P_0(1) = 1$.

Now, let us recall some basic definitions and notation concerning matrices. Let A, B be two $n\times n$ matrices. We write $A > B$ (resp. $A \ge B$) if $A(i,j) > B(i,j)$ (resp. $A(i,j) \ge B(i,j)$) for all $1 \le i,j \le n$. For any real number r, we write

$A > r$ (resp. $A \geq r$) if $A(i,j) > r$ (resp. $A(i,j) \geq r$) for all $1 \leq i,j \leq n$. A matrix A is called nonnegative (resp. positive) if $A \geq 0$ (resp. $A > 0$). A is called *primitive* if $A^k > 0$ for some positive integer k, and is called *irreducible* if for any pair of integers $1 \leq i,j \leq n$, $A^k(i,j) > 0$ for some positive integer k.

The primitivity and the irreducibility of an nonnegative matrix A can also be defined in terms of the *associated digraph.* The associated digraph of A, denoted by $G(A) = \{V, E\}$, is the digraph with vertex set $V = \{1, 2, \ldots, n\}$ such that there is an edge from i to j in the edge set E if and only if $a(i,j) > 0$.

A digraph G is *strongly connected* if for each ordered pair of vertices $i, j \in V$, there exists a directed path from i to j in G. A digraph G is *primitive* if G is strongly connected and $\gcd(r_1, \ldots, r_s) = 1$, where $\{r_1, \ldots, r_s\}$ is the set of distinct lengths of the elementary circuits of G. From [6, Chapter 6, Sect. 1, Remarks 6, 8], one has:

I) A is irreducible if and only if $G(A)$ is strongly connected.
II) A is primitive if and only if $G(A)$ is primitive.
By *an elementary circuit*, we mean a path going through the vertices $v_1, v_2, \ldots, v_n, v_1$ in order, such that $v_i \neq v_j$ for all $i \neq j$.

3 Matrix Representation

Let W be the fixed point of the morphism ϕ over the alphabet $\{1, \ldots, M\}$ such that $W[0] = 1$ as defined in Sect. 2 and let $a \geq 2$ be a positive integer. For $i, j, b \in \mathbb{N}$, such that $0 \leq b \leq a-1$ and $m \in \{1, \ldots, M\}$, let

$$v_{a,b,m,i,j} = \#\{k | W[ak+b] = m, iL^j \leq ak+b < (i+1)L^j\}.$$

Let $V_a(i,j) \in \mathcal{M}^{aM \times 1}$ be a vector such that the r-th component in $V_a(i,j)$ is $v_{a,b,m,i,j}$ when $r = bM + m$ with $b \geq 0$ and $1 \leq m \leq M$.

Lemma 2. *For any a, b, i, j defined as above, one has*

$$v_{a,b,m,i,j+1} = \sum_{0 \leq t < L} v_{a,b_t,P_{l_t}^{-1}(m),i,j},$$

where (b_t, l_t) is the unique pair of integers satisfying $at + b = b_t L + l_t$ such that $0 \leq b_t < a$ and $0 \leq l_t < L$.

Proof.

$$\begin{aligned}
v_{a,b,m,i,j+1} &= \#\{k | w[ak+b] = m, iL^{j+1} \leq ak+b < (i+1)L^{j+1}\} \\
&= \sum_{0 \leq t < L} \#\{k | w[aLk + at + b] = m, iL^{j+1} \leq aLk + at + b < (i+1)L^{j+1}\} \\
&= \sum_{0 \leq t < L} \#\{k | w[L(ak + b_t) + l_t] = m, iL^{j+1} \leq aLk + b_t L + l_t < (i+1)L^{j+1}\} \\
&= \sum_{0 \leq t < L} \#\{k | w[ak + b_t] = P_{l_t}^{-1}(m), iL^j \leq ak + b_t < (i+1)L^j\} \\
&= \sum_{0 \leq t < L} v_{a,b_t,P_{l_t}^{-1}(m),i,j}.
\end{aligned}$$

Example 3. Let $M = 2$, let $\psi : \{1,2\}^* \to \{1,2\}^*$ such that $\psi(1) = 1121$ and $\psi(2) = 2212$ and let $U = \lim_{n\to\infty} \psi^n(1)$. In this case, $L = 4$. Let $a = 3$, $b = 0$, $i = 1$ and $j = 5$ then one has:

$$\begin{aligned}
v_{3,0,1,1,6} = &\#\{k|U[3k] = 1, 1 \times 4^6 \leq 3k < 2 \times 4^6\} \\
=&\#\{k|U[12k] = 1; 4 \times 4^5 \leq 12k < 8 \times 4^5\} \\
+&\#\{k|U[12k+3] = 1, 4 \times 4^5 \leq 12k+3 < 8 \times 4^5\} \\
+&\#\{k|U[12k+6] = 1; 4 \times 4^5 \leq 12k+6 < 8 \times 4^5\} \\
+&\#\{k|U[12k+9] = 1, 4 \times 4^5 \leq 12k+9 < 8 \times 4^5\} \\
=&\#\{k|U[3k] = P_0^{-1}(1), 1 \times 4^5 \leq 3k < 2 \times 4^5\} \\
+&\#\{k|U[3k] = P_3^{-1}(1), 1 \times 4^5 \leq 3k < 2 \times 4^5\} \\
+&\#\{k|U[3k+1] = P_2^{-1}(1), 1 \times 4^5 \leq 3k+1 < 2 \times 4^5\} \\
+&\#\{k|U[3k+2] = P_1^{-1}(1), 1 \times 4^5 \leq 3k+2 < 2 \times 4^5\} \\
=&v_{3,0,1,1,5} + v_{3,0,1,1,5} + v_{3,1,2,1,5} + v_{3,2,1,1,5}.
\end{aligned}$$

Similarly, one has

$$\begin{aligned}
v_{3,0,2,1,6} &= v_{3,0,2,1,5} + v_{3,0,2,1,5} + v_{3,1,1,1,5} + v_{3,2,2,1,5}; \\
v_{3,1,1,1,6} &= v_{3,0,1,1,5} + v_{3,1,1,1,5} + v_{3,1,1,1,5} + v_{3,2,2,1,5}; \\
v_{3,1,2,1,6} &= v_{3,0,2,1,5} + v_{3,1,2,1,5} + v_{3,1,2,1,5} + v_{3,2,1,1,5}; \\
v_{3,2,1,1,6} &= v_{3,0,2,1,5} + v_{3,1,1,1,5} + v_{3,2,1,1,5} + v_{3,2,1,1,5}; \\
v_{3,2,2,1,6} &= v_{3,0,1,1,5} + v_{3,1,2,1,5} + v_{3,2,2,1,5} + v_{3,2,2,1,5}.
\end{aligned}$$

Let $M_a = (M_a(p,q))_{1\leq p,q\leq aM} \in \mathcal{M}^{aM\times aM}$ be a matrix such that

$$M_a(p,q) = \#\{(t,l_t)|t,l_t \in \mathbb{N},\ 0 \leq t,\ l_t < L,\ at + b_p = b_qL + l_t,\ P_{l_t}(m_q) = m_p\},$$

where b_p, b_q, m_p, m_q are integers satisfying:

$$\begin{cases} 0 \leq b_p, b_q < a;\ 1 \leq m_p, m_q \leq M; \\ p = b_pM + m_p, q = b_qM + m_q. \end{cases}$$

Lemma 4. *For any $i, j \in \mathbb{N}$, one has*

$$V_a(i, j+1) = M_aV_a(i,j) = M_a^{j+1}V_a(i,0).$$

Proof. With the notation as above, for any $1 \leq p \leq aM$, the p-th component of $V_a(i,j+1)$ is $v_{a,b_p,m_p,i,j+1}$. From Lemma 2, one has

$$v_{a,b_p,m_p,i,j+1} = \sum_{0\leq t<L} v_{a,b_t,P_{l_t}^{-1}(m_p),i,j},$$

where (b_t, l_t) is the unique pair of integers satisfying $at + b_p = b_tL + l_t$ for the given t such that $0 \leq b_t < a$ and $0 \leq l_t < L$. From the definition of the matrix M_a, the number of the occurrences of the q-th component of $V_a(i,j)$ in $\sum_{0\le t<L} v_{a,b_t,P_{l_t}^{-1}(m_p),i,j}$ is exactly $M_a(p,q)$. Indeed, write $q = b_qM + m_q$ as above; then $v_{a,b_t,P_{l_t}^{-1}(m_p),i,j}$ coincides with $v_{a,b_q,m_q,i,j}$ only if there exist t and l_t satisfying $at + b_p = b_qL + l_t$ and $P_{l_t}(m_q) = m_p$. Thus, the number of occurrences of $v_{a,b_q,m_q,i,j}$ in $\sum_{0\le t<L} v_{a,b_t,P_{l_t}^{-1}(m_p),i,j}$ equals $M_a(p,q)$. □

Example 5. From Example 3, the associated matrix M_3 for U is the following

$$M_3 = \begin{bmatrix} 2&0&0&1&1&0\\ 0&2&1&0&0&1\\ 1&0&2&0&0&1\\ 0&1&0&2&1&0\\ 0&1&1&0&2&0\\ 1&0&0&1&0&2 \end{bmatrix}.$$

Using the same method, one can prove

$$M_6 = \begin{bmatrix} 1&0&0&1&0&0&1&0&0&1&0&0\\ 0&1&1&0&0&0&0&1&1&0&0&0\\ 1&0&1&0&0&0&1&0&1&0&0&0\\ 0&1&0&1&0&0&0&1&0&1&0&0\\ 0&1&0&0&1&0&0&1&0&0&1&0\\ 1&0&0&0&0&1&1&0&0&0&0&1\\ 1&0&0&0&1&0&1&0&0&0&1&0\\ 0&1&0&0&0&1&0&1&0&0&0&1\\ 0&0&1&0&0&1&0&0&1&0&0&1\\ 0&0&0&1&1&0&0&0&0&1&1&0\\ 0&0&1&0&1&0&0&0&1&0&1&0\\ 0&0&0&1&0&1&0&0&0&1&0&1 \end{bmatrix}.$$

Notice that, although matrices like M_3, which have many distinct entries, rarely appear in the study of the automatic sequences, the binary matrices like M_6 have already appeared in the literature when studying the behavior of the partial sums of the *binary digital sequences*, see, for example, [3].

Now let us introduce some basic properties of the matrix M_a.

Lemma 6. *For all integers $a > 0$,*
I) $M_a(1,1) > 0$;
II) for all integers $1 \le p \le aM$, $\sum_{q=1}^{am} M_a(p,q) = L$.

Proof. For the first part, set $p = q = 1$. Then, from the definition of M_a, $b_p = b_q = 0$ and $m_p = m_q = 1$. Set $t = 0$, then $l_t = l_0 = 1$ and $P_{l_0}(1) = 1$. Thus, $M_a(1,1) \ge 1$.

For the second part, it is enough to figure out that, for any $0 \le t < L$, there exists a unique l_t such that $0 \le l_t < L$, $at + b_p = b_qL + l_t$ and $P_{l_t}(m_q) = m_p$. □

Proposition 7. *Let $a \in \mathbb{N}$. If M_a is primitive, then $d_{W,a,b}(m) = \frac{1}{aM}$ for all $0 \le b < a$ and all $1 \le m \le M$.*

Proof. First, M_a is nonnegative from its definition. If it is primitive, then, from [9, Definition 8.5.0, Theorem 8.5.2], there exists a positive number λ, such that λ is an eigenvalue of M_a and all other eigenvalues are strictly smaller than λ in norm. Moreover, from [9, Theorem 8.5.1],

$$\lim_{n\to\infty} \left(\frac{M_a}{\lambda}\right)^n = U,$$

where $U = xy^T$, $M_a x = \lambda x$, $M_a^T y = \lambda y$, $x > 0$, $y > 0$, $x^T y = 1$.

For the matrix M_a, set $x^T = y^T = \frac{1}{\sqrt{aM}}[1, 1, \ldots, 1] \in \mathcal{M}^{1\times aM}$. From Lemma 6(II), $M_a^T y = Ly$ and, from Lemma 2, $M_a x = Lx$. Thus, U is a matrix of size $aM \times aM$ such that all its entries are $\frac{1}{aM}$. Consequently,

$$\lim_{n\to\infty} \left(\frac{M_a(p,q)}{L}\right)^n = \frac{1}{aM}$$

for all $1 \le p, q \le aM$. For any integers $b, m, i, j \in \mathbb{N}$ such that $0 \le b < a$ and $m \in \{1, \ldots, M\}$, $v_{a,b,m,i,j}$ is the $(bM + m)$-th component of $V_a(i, j)$. Thus, from Lemma 4,

$$v_{a,b,m,i,j} = \sum_{k=1}^{aM} (M_a)^j (bM + m, k) V_a(i, 0)[k],$$

where $V_a(i, 0)[k]$ is the k-th component of the vector $V_a(i, 0)$. Since $(M_a)^j(p, q) = \frac{L^j}{aM} + o(L^j)$ for all $1 \le p, q \le aM$ and all $j \ge 0$,

$$v_{a,b,m,i,j} = \left(\sum_{k=1}^{aM} V_a(i,1)[k]\right)\left(\frac{L^j}{aM} + o(L^j)\right) = \frac{L^j}{aM} + o(L^j).$$

The last equation holds because $\sum_{k=1}^{aM} V_a(i,1)[k] = \#\{j | i \le j < (i+1)\} = 1$.

Let $R \in \mathbb{N}$ and write $R = \sum_{i=0}^{k} r_i L^i$ where $k \ge 0$, $0 \le r_i \le L-1$ and $r_k > 0$. Let $s_j = \sum_{i=j}^{k} r_i L^{i-j}$ for all $0 \le j \le k$ and let $s_{k+1} = 0$, then $r_j = s_j - s_{j+1}L$ for all $0 \le j \le k$. Thus,

$$R = \sum_{j=0}^{k} \sum_{t=s_{j+1}L}^{s_j - 1} L^j + 1.$$

For any $0 \leq b \leq a-1$ and $m \in \{1, \dots, M\}$, let $N(a,b,m,R) = \#\{k | W(ak+b) = m, 0 \leq ak+b < R\}$, then

$$\begin{aligned} N(a,b,m,R) &= \sum_{j=0}^{k} \sum_{t=s_{j+1}L}^{s_j-1} v_{a,b,m,t,j} \\ &= \sum_{j=0}^{k} (r_j \frac{L^j}{aM} + o(L^j)) \\ &= \frac{R}{aM} + o(R). \end{aligned}$$

□

Example 8. Let U be the sequence defined in Example 3. Let us compute $N(3,0,1,945)$. Notice that the expansion of 945 in base 4 is 32301. With the notation as above, we have $s_5 = [0]_4$, $s_4 = [3]_4$, $s_3 = [32]_4$, $s_2 = [323]_4$, $s_1 = [3230]_4$, $s_0 = [32301]_4$. Thus,

$$\begin{aligned} N(3,0,1,945) &= \sum_{j=0}^{4} \sum_{t=s_{j+1}L}^{s_j-1} v_{3,0,1,t,j} \\ &= v_{3,0,1,0,4} + v_{3,0,1,1,4} + v_{3,0,1,2,4} + v_{3,0,1,30,3} + v_{3,0,1,31,3} \\ &+ v_{3,0,1,320,2} + v_{3,0,1,321,2} + v_{3,0,1,322,2} + v_{3,0,1,32300,0}. \end{aligned}$$

Proposition 9. *Let $a \geq 1$ be an integer, if $d_{W,a,b}(m) = \frac{1}{aM}$ for all $0 \leq b \leq a-1$ and all $m \in \{1, \dots, M\}$, then the associated matrix M_a is primitive.*

Proof. We first prove that, for any ordered pair of integers $1 \leq p, q \leq aM$, there is a $K \in \mathbb{N}$ such that $(M_a)^K(p,q) > 0$, which implies that M_a is irreducible. Then we conclude by using the fact that $M_a(1,1) > 0$, which is from Lemma 6(I), and [9, Theorem 8.5.10], that M_a is primitive.

Let us write $p = b_p M + m_p$ and $q = b_q M + m_q$ such that $b_p, b_q \geq 0$ and $1 \leq m_p, m_q \leq M$. Let I be an integer in $(an + b_p)_{n \in \mathbb{N}}$ such that $W[I] = m_p$. Since, for all $0 \leq b \leq a-1$ and all $m \in \{1, \dots, M\}$,

$$\lim_{j \to \infty} \frac{\#\{k | W[ak+b] = m, 0 \leq ak+b < (I+1)L^j\}}{(I+1)L^j} = d_{W,a,b}(m) > 0,$$

and

$$\lim_{j \to \infty} \frac{\#\{k | W[ak+b] = m, IL^j \leq ak+b < (I+1)L^j\}}{\#\{k | W[ak+b] = m, 0 \leq ak+b < (I+1)L^j\}} = \frac{1}{I+1} > 0,$$

then

$$\lim_{j \to \infty} \frac{\#\{k | W[ak+b] = m, IL^j \leq ak+b < (I+1)L^j\}}{L^j} = d_{W,a,b}(m) > 0.$$

In particular, there exists a $K \in \mathbb{N}$ such that $C = \#\{k | W[ak + b_q] = m_q, IL^K \leq ak + b < (I+1)L^K\} > 0$. Let us prove $C = (M_a)^K(p,q) > 0$. First, from the choices of I and C, one has $v_{a,b_p,m_p,I,0} = 1$ and $v_{a,b_q,m_q,I,K} = C$. Set $e_r \in \mathcal{M}^{aM\times 1}$ such that $e_r(i) = 1$ if $i = r$ and $e_r(i) = 0$ if not for $r = p, q$. One has $V_a(I,0) = e_p$ and $V_a(I,K)^T e_q = C$. Second, from Lemma 4, $V_a(I,K) = (M_a)^K V_a(I,0)$. Thus, $C = e_q^T (M_a)^K e_p = (M_a)^K(p,q) > 0$. □

Theorem 10. *The sequence W satisfies the uniform distribution property if and only if M_a is primitive for all $a \in \mathbb{N}^*$.*

Proof. It is a direct consequence of Proposition 9 and Proposition 7. □

Example 11. Let $M = 2$, $\kappa : \{1,2\}^* \to \{1,2\}^*$ such that $\kappa(1) = 121$ and $\kappa(2) = 212$ and let $V = \lim_{n\to\infty} \kappa^n(1)$. Then the associated matrix for $a = 4$ is

$$M_4 = \begin{bmatrix} 1&0&0&1&1&0&0&0 \\ 0&1&1&0&0&1&0&0 \\ 0&1&1&0&0&0&1&0 \\ 1&0&0&1&0&0&0&1 \\ 1&0&0&0&1&0&0&1 \\ 0&1&0&0&0&1&1&0 \\ 0&0&1&0&0&1&1&0 \\ 0&0&0&1&1&0&0&1 \end{bmatrix}.$$

One can check that this matrix is not primitive because $(M_4)^k(1,2) = 0$ for all $k \in \mathbb{N}$. From Theorem 10, this sequence does not satisfy the uniform distribution property. In fact, V is a periodic sequence with the period 1, 2. Hence, V trivially does not satisfy the uniform distribution property.

Lemma 12. *For any integers $0 \leq r, s < a$, the square matrix*

$$M_{a,r,s} = (M_a(p,q))_{\substack{rM+1 \leq p < r(M+1) \\ sM+1 \leq q < s(M+1)}}$$

satisfies the equation

$$M_{a,r,s} = \sum_{\substack{0 \leq t, l_t < L \\ at+r=sL+l_t}} M_{P_{l_t}},$$

where M_{P_l} is the matrix representation of the permutation P_l over the basis $1, 2, \ldots, M$.

Proof. Let $T = \{(t, l_t) | 0 \leq t, l_t < L, at + r = sL + l_t\}$. For any pair $(t, l_t) \in T$ and any p such that $rM + 1 \leq p = rM + m_p < r(M+1)$, there exists a unique $1 \leq m_q < M$ such that $P_{l_t}(m_q) = m_p$. Let $M_{P_l} = (M_{P_l}(p,q))_{1\leq p,q\leq M} \in \mathcal{M}^{M\times M}$ satisfying $M_{P_l}(p,q) = 1$ if $P_{l_t}(q) = p$ and $M_{P_l}(p,q) = 0$ otherwise. Then M_{P_l} is the matrix representation of the permutation P_{l_t} over the basis $1, 2, \ldots, M$ and $M_{a,r,s} = \sum_{\substack{0 \leq t, l_t < L \\ at+r=sL+l_t}} M_{P_{l_t}}$. □

Example 13. Let us consider the sequence $U = \lim_{n\to\infty} \psi^n(1)$ studied in Example 3 and Example 5, where $\psi : \{1,2\}^* \to \{1,2\}^*$ satisfies $\psi(1) = 1121$ and $\psi(2) = 2212$. The permutation matrices are

$$P_0 = P_1 = P_3 = \begin{bmatrix} 1 & 0 \\ 0 & 1 \end{bmatrix} \text{ and } P_2 = \begin{bmatrix} 0 & 1 \\ 1 & 0 \end{bmatrix}.$$

Then the block matrices in M_3 satisfy

$$\begin{aligned} M_{3,0,0} &= M_{3,1,1} = M_{3,2,2} = P_0 + P_3; \\ M_{3,0,1} &= M_{3,1,2} = M_{3,2,0} = P_2; \\ M_{3,0,2} &= M_{3,1,0} = M_{3,2,1} = P_1. \end{aligned}$$

Lemma 14. *For any integer $a > 0$, the square matrix $M_{a,a-1,a-1}$ satisfies*

$$M_{a,a-1,a-1} = M_{P_{L-1}} + M',$$

where $M' \geq 0$.

Proof. From Lemma 12, $M_{a,a-1,a-1} = \sum_{\substack{0\le t,l_t<L \\ at+(a-1)=(a-1)L+l_t}} M_{P_{l_t}}$. Let $t = L-1$, then $l_t = L-1$. Thus, setting $M' = \sum_{\substack{0\le t,l_t<L-1 \\ at+(a-1)=(a-1)L+l_t}} M_{P_{l_t}}$, one has $M' \geq 0$ and $M_{a,a-1,a-1} = M_{P_{L-1}} + M'$. □

Now let us introduce the reduced matrix $M'_a = (M'_a(p,q))_{1\le p,q\le a} \in \mathcal{M}^{a\times a}$ such that, for $0 \le r, s \le a-1$, $M'_a(r+1, s+1) = 1$ if $M_{a,r,s}$ is a non-zero matrix and $M'_a(r+1, s+1) = 0$ if not.

Example 15. The reduced matrices of the two matrices in Example 5 are the following:

$$M'_3 = \begin{bmatrix} 1 & 1 & 1 \\ 1 & 1 & 1 \\ 1 & 1 & 1 \end{bmatrix}.$$

$$M'_6 = \begin{bmatrix} 1&1&0&1&1&0 \\ 1&1&0&1&1&0 \\ 1&0&1&1&0&1 \\ 1&0&1&1&0&1 \\ 0&1&1&0&1&1 \\ 0&1&1&0&1&1 \end{bmatrix}.$$

Lemma 16. *For any integers $0 \le r, s \le a-1$, $M'_a(r+1, s+1) = 1$ if and only if $r - sL \mod a \in \{0, 1, \ldots, L-1\}$.*

Proof. For any integer $0 \le s \le a-1$, $M_{a,r,s}$ is non-zero if and only if there exists a t such that $sL \le at + r < (s+1)L$. In this case, $r - sL \mod a \in \{0, 1, \ldots, L-1\}$. □

Lemma 17. *For the matrix M'_a, one has $M'_a(1,1) = M'_a(a,a) = 1$.*

Proof. Obviously, setting $r = s = 0$ in Lemma 16, one has $0 \mod a \in \{0,1,\ldots,L-1\}$. Similarly, setting $r = s = a-1$, one has $a-1-(a-1)L \mod a \in \{0,1,\ldots,L-1\}$. □

Lemma 18. *The matrix M'_a is primitive.*

Proof. Let us prove by induction that for all positive integers k and all integers s such that $0 \le s \le a-1$, $(M'_a)^k(r+1,s+1) > 0$ in the matrix $(M'_a)^k = ((M'_a)^k(p,q))_{1\le p,q\le a} \in \mathcal{M}^{a\times a}$ if and only if $r - sL^k \mod a \in \{0,1,\ldots,L^k-1\}$. This statement is true for $k=1$ from Lemma 16 and let us suppose that is true for $(M'_a)^k$. Then, for $(M'_a)^{k+1}$, one has for any $0 \le r,s \le a-1$

$$(M'_a)^{k+1}(r+1,s+1) = \sum_{i=0}^{a-1} M'_a(r+1,i+1)(M'_a)^k(i+1,s+1).$$

Hence, $(M'_a)^{k+1}(r+1,s+1) > 0$ if and only if there exists i such that

$$\begin{cases} M'_a(r+1,i+1) > 0, \\ (M'_a)^k(i+1,s+1) > 0. \end{cases}$$

From the hypothesis of the induction,

$$\begin{cases} r - iL \mod a \in \{0,1,\ldots,L-1\}; \\ i - sL^k \mod a \in \{0,1,\ldots,L^k-1\}. \end{cases}$$

Thus, $(M'_a)^{k+1}(r+1,s+1) > 0$ if and only if $r - sL^{k+1} \mod a \in \{0,1,\ldots, L^{k+1}-1\}$. Let $K \in \mathbb{N}$ such that $L^K > a$, then the matrix $(M'_a)^K > 0$. □

Lemma 19. *For all $a \in \mathbb{N}^*$, the matrix M_a is irreducible if P_{L-1} is a cyclic permutation.*

Proof. Let us prove that for any pair of integers (p,q) such that $1 \le p,q \le aM$, there exists a directed path from p to q in the associated digraph of M_a.

From Lemma 18, the matrix M'_a is primitive and, from Lemma 12, the matrices $M_{a,r,s}$ are all sums of permutation matrices for $0 \le r,s \le a-1$. Then there exists $k \in \mathbb{N}$ such that, for all $0 \le r,s \le a-1$, the matrix

$$(M_a)^k_{r,s} = \left((M_a)^k(p,q)\right)_{\substack{rM+1\le p<r(M+1)\\ sM+1\le q<s(M+1)}}$$

is a sum of permutation matrices and is nonzero. Consequently, for all integers r,s,p' such that $0 \le r,s \le a-1$ and $rM+1 \le p' < r(M+1)$, there is a q' such that $sM+1 \le q' < s(M+1)$ and $(M_a)^k(p',q') = 1$. Thus, for all $0 \le r,s \le a-1$ and all $1 \le i \le M$, there is $1 \le j \le m$ such that there is a direct path from

$ra+i$ to $sa+j$. Similarly, there is $1 \leq j' \leq m$ such that there is a direct path from $sa+j'$ to $ra+i$.

Now, let p, q be two arbitrary numbers such that $1 \leq p, q \leq aM$. There exist $1 \leq i, j \leq M$ such that there are a direct path from p to $(a-1)M+i$ and a direct path from $(a-1)M+j$ to q. Since P_{L-1} is a cyclic permutation, from Lemma 14, there is a direct path from $(a-1)M+i$ to $(a-1)M+j$. Thus, there is a direct path from p to q. □

Theorem 20. *With the notation as above, the infinite word W has the uniform distribution property if the following conditions hold:*
I) The maps $P_l : i \rightarrow \phi(i)[l]$ are all permutations over $\{1, \ldots, M\}$ for $0 \leq l < L$;
II) The map P_{L-1} is cyclic.

Proof. From condition II) and Lemma 19, all matrices M_a are irreducible, and primitive by [9, Theorem 8.5.10]. Thus, we conclude using Theorem 10. □

Remark 21. The conditions given in Theorem 20 are sufficient but not necessary. Indeed, all digital sequences of the form $(s_q(n) \bmod q)_{n\in\mathbb{N}}$ can be generated by morphisms satisfying this criterion, but can also be generated by some other morphisms not satisfying these conditions. Let us consider the Thue-Morse sequence as an example. This sequence can be generated by $\chi : \{1,2\}^* \rightarrow \{1,2\}^*$ such that $\chi(1) = 12$ and $\chi(2) = 21$. So χ satisfies the two conditions in Theorem 20. However, the Thue-Morse sequence can also be generated by $\chi^2(1) = 1221$ and $\chi^2(2) = 2112$, which does not satisfy the condition II in Theorem 20.

Remark 22. The condition II in Theorem 20 cannot be replaced by other P_k such that $0 \leq k < L-1$. The reason is that Lemma 14 only works for the last block-matrix on the diagonal of M_a and P_{L-1} "appears" in this matrix. This phenomena cannot be generalized to other P_k because we cannot guarantee that P_k appears in a block matrix on the diagonal of M_a if $k \neq 0$ or $L-1$. Indeed, P_0 appears always in the block matrix $M_{a,0,0}$ of M_a. However, P_0 is not cyclic because $P_0(1) = 1$. Let us consider Example 11. Although P_1 of the morphism satisfies the condition II in Theorem 20, it does not appear in a matrix $M_{4,i,i}$ of M_4 for some $i \in \{0,1,2,3\}$. Thus, M_4 may not be primitive and the sequence V may not satisfy the uniform distribution property, which is the case in Example 11.

4 Conclusion and Remarks

We studied the uniform distribution property of purely automatic sequences. We first gave a necessary and sufficient condition for a sequence to have the uniform distribution property in terms of matrices, then we gave a sufficient condition from the structure of the generating morphisms. The last condition can be applied to all generalized Thue-Morse sequences of the form $(s_q(n) \bmod q)_{n\in\mathbb{N}}$, which have been studied by Gelfond. We finally close this note by asking two questions:

Question 23. Is there a characterization of the uniform distribution property in terms of the combinatorial properties of the generating morphism?

Question 24. Is it true that a purely automatic sequence has the uniform distribution property if and only if it is aperiodic and satisfies the condition I in Theorem 20?

Acknowledgments. This work was funded by NSERC grant number RGPIN-2019-04111.

References

1. Avgustinovich, S.V., Fon-Der-Flaass, D.G., Frid, A.E.: Arithmetical Complexity of Infinite Words, pp. 51–62 (2003)
2. Byszewski, J., Konieczny, J., Müllner, C.: Gowers norms for automatic sequences. Discrete Anal. (2023)
3. Boyd, D.W., Janice Cook, P.M.: On sequences of ±1's defined by binary patterns. Instytut Matematyczny Polskiej Akademi Nauk (1989). http://eudml.org/doc/268620
4. Dekking, F.M., Keane, M.S.: On the conjugacy class of the fibonacci dynamical system. Theor. Comput. Sci. **668**, 59–69 (2017)
5. Drmota, M., Mauduit, C., Rivat, J.: The sum-of-digits function of polynomial sequences. J. Lond. Math. Soc. **84**(1), 81–102 (2011)
6. Dulmage, A.L., Mendelsohn, N.S.: Graphs and matrices. In: Harary, F. (ed.) Graph Theory and Theoretical Physics, pp. 167–227. Academic, London (1967)
7. Fan, A., Konieczny, J.: On uniformity of q-multiplicative sequences. Bull. London Math. Soc. (2019)
8. Gelfond, A.: Sur les nombres qui ont des propriétés additives et multiplicatives données. Acta Arithmetica **13**(3), 259–265 (1968). http://eudml.org/doc/204828
9. Horn, R.A., Johnson, C.R.: Matrix Analysis. Cambridge University Press, 2 edn. (2012)
10. Kim, D.H.: On the joint distribution of q-additive functions in residue classes. J. Number Theory **74**(2), 307–336 (1999)
11. Konieczny, J.: Gowers norms for the Thue-Morse and Rudin-Shapiro sequences. Annales de l'Institut Fourier **69**(4), 1897–1913 (2019)
12. Konieczny, J., Müllner, C.: Arithmetical subword complexity of automatic sequences (2023). https://hal.science/hal-04201450
13. Mauduit, C., Rivat, J.: La somme des chiffres des carrés. Acta Math. **203**(1), 107–148 (2009)
14. Mauduit, C., Rivat, J.: Sur un problème de Gelfond: la somme des chiffres des nombres premiers. Ann. Math. (2) **171**(3), 1591–1646 (2010)
15. Mauduit, C., Rivat, J.: Prime numbers along Rudin-Shapiro sequences. J. Eur. Math. Soc. (JEMS) **17**(10), 2595–2642 (2015)
16. Mauduit, C., Rivat, J.: Rudin-Shapiro sequences along squares. Trans. Amer. Math. Soc. **370**(11), 7899–7921 (2018)
17. Müllner, C.: The Rudin-Shapiro sequence and similar sequences are normal along squares. Canad. J. Math. **70**(5), 1096–1129 (2018)
18. Séébold, P.: On some generalizations of the Thue-Morse morphism. Theor. Comput. Sci. **292**(1), 283–298 (2003)
19. Spiegelhofer, L.: The level of distribution of the Thue-Morse sequence. Compos. Math. **156**(12), 2560–2587 (2020)

On the Closed-Rich Constant of Infinite Words

Anuran Maity[1,2] and Svetlana Puzynina[2](✉)

[1] Indian Institute of Technology Guwahati, Department of Mathematics, Guwahati, India
[2] Saint Petersburg State University, 7/9 Universitetskaya nab., 199034 St. Petersburg, Russia
s.puzynina@gmail.com

Abstract. A finite word w is called *closed* if it has length at most 1 or it contains a proper factor that occurs both as a prefix and as a suffix but does not have internal occurrences. An infinite word u is called *closed-rich* if the infimum of all possible ratios between the number of closed factors within any factor w of u and square of the length of w exists and is positive. We define this infimum as the closed-rich constant C_u of the infinite closed-rich word u. Puzynina and Parshina (2024) proved that infinite closed-rich words exist. In this paper, we estimate possible values of C_u for an infinite closed-rich word u, and apply these results to estimate the supremum C_{sup} of the closed-rich constants of infinite closed-rich words. We show that $0.0952 < C_{sup} \leq 0.165964$, where the lower bound comes from the Fibonacci word.

Keywords: Closed word · closed-rich word · Fibonacci word

1 Introduction

We investigate a problem related to the distribution of closed factors in infinite closed-rich words. A finite word is called *closed* if it has length at most 1 or it contains a proper factor that occurs both as a prefix and as a suffix but does not have internal occurrences. The notion of a closed word is closely related to the notion of a return word. The concept of first return to a factor, often called "return word", can be seen as a discrete counterpart to the first return map in dynamical systems. It serves as a powerful tool for analyzing symbolic dynamical systems and related fields (see [7,16]). A lot of study has been done on closed words from various perspectives, such as the occurrence of closed factors in prefixes of Sturmian words [6], the relationship between closed factors and palindromic factors in a word [2], closed factorizations of a word [1], reconstruction of a string from its longest closed

The results from Sect. 4 were performed at the Saint Petersburg Leonhard Euler International Mathematical Institute and supported by the Ministry of Science and Higher Education of the Russian Federation (agreement no. 075 – 15 – 2022 – 287). The results from Sect. 3 were supported by the Russian Science Foundation, project 23-11-00133.

G. Gamard and J. Leroy (Eds.): WORDS 2025, LNCS 15729, pp. 217–229, 2025.
https://doi.org/10.1007/978-3-031-97548-6_19

factor array [4], expressing the property of being closed in first-order logic for automatic sequences [15], closed complexities for the family of Arnoux-Rauzy words [14], a refinement of the Morse-Hedlund theorem using closed complexities [12] and other (see, e.g., a survey [8]). Possible numbers of closed factors in a finite word have been studied in [2]. Most recently, Parshina and Puzynina [13] provided the asymptotics for the maximum number of distinct closed factors a finite word can contain, and introduced the concept of closed-rich words for infinite words.

The study of words rich in specific types of factors is an important topic in modern combinatorics on words. Glen et al. [10] investigated finite and infinite words rich in palindromes. Bannai et al. [3] resolved the renowned 'runs conjecture', which concerns finite words rich in maximal repetitions. Fici et al. [9] explored finite and infinite words rich in abelian squares. Recently, Brlek and Li [5] solved a long standing conjecture which focuses on finite words rich in squares. Similarly to the concept of these rich words, Parshina and Puzynina [13] defined a word as closed-rich if it is rich in distinct closed factors, i.e., it contains the maximal number of distinct closed factors among words of the same length. They showed that a finite closed-rich word of length n contains asymptotically $\sim \frac{n^2}{6}$ distinct closed factors. Let $\texttt{Cl}(w)$ denote the number of disinct closed factors in w. An infinite word u is called closed-rich if the constant $C_u = \inf\left\{\frac{\texttt{Cl}(w)}{|w|^2} : w \in \mathrm{Fac}(u)\right\}$ is positive. Here C_u is called the closed-rich constant of u. In [13], the authors asked a question about the supremum C_{sup} of the closed-rich constants of infinite closed-rich words. Since a finite closed-rich word of length n contains $\sim \frac{n^2}{6}$ distinct closed factors, one has $C_{sup} \leq \frac{1}{6}$.

In this paper, we have improved this bound proving that $C_{sup} \leq 0.165964$. Besides that, we found a lower bound for C_{sup} using the Fibonacci word. Namely, we proved that for the Fibonacci word f, we have $\frac{\phi^3+3}{\phi^9} < C_f \leq 0.11022$, where $\phi = \frac{1+\sqrt{5}}{2}$ is the golden ratio. Thus, we have $0.0952 < C_{sup} \leq 0.165964$.

The infinite Fibonacci word, which is known as the simplest Sturmian word providing essential insights to the properties of all Sturmian words, has been shown to be an infinite closed-rich word ([13]). Parshina and Puzynina [13] conjectured that finite closed-rich words are cubes or words of exponent close to 3. Since every sufficiently long finite factor of f always contains sufficiently long factors with an exponent close to 3 and f often serves as an extremal case and an example in combinatorics on words, f is one of the best possible candidates for estimating C_{sup}. In addition, the techniques and observations developed for the Fibonacci word can be useful to study other Sturmian words by utilizing the standard sequences associated with them.

The paper is organized as follows. In the next section, we give necessary definitions and notation. In Sect. 3, we prove the upper bound for C_{sup}. In Sect. 4, we discuss the Fibonacci word. In particular, we find an explicit formula for the number of closed words in its prefixes, and find lower and upper bounds for its closed-rich constant mentioned above. In Sect. 5, we discuss the bounds of C_{sup}, propose a conjecture about the closed-rich constant of the Fibonacci word, and discuss some open questions.

2 Preliminaries

For the basic notation of combinatorics on words, we follow Lothaire book [11]. Throughout the paper, we denote the alphabet by Σ. If a finite word w has a proper prefix u which is also its suffix, then u is called a *border* of w and w is called a *bordered word.* For a finite word w of length n, we denote its prefix of length $n-1$ by w^- and its suffix of length $n-1$ by ^-w. If $x, u, y \in \Sigma^+$ and $w = xuy$, then we say u occurs internally in w. Let $v = v_0v_1v_2\cdots$ be an infinite word. Then the *critical exponent* $E(v)$ of v is defined as $E(v) = \sup\{e \in \mathbb{Q} : u^e$ is a factor of v for a non empty word $u\}$. The *asymptotic critical exponent* $E^*(v)$ of v is

$$E^*(v) = \limsup_{n\to\infty}\{e \in \mathbb{Q} \ : \ u^e \text{ is a factor of } v \text{ for some } u \text{ of length } n\}.$$

A finite word is called *closed* if it has length at most 1 or it contains a proper factor that occurs as a prefix and as a suffix but does not have internal occurrences. In other words, a closed word of length at least 2 has a border which has exactly two occurrences in the word. For a finite word w, we let $\mathtt{Cl}(w)$ denote the number of distinct closed factors of w. The following upper bound for $\mathtt{Cl}(w)$ is introduced in [13]:

Theorem 1. *[13] For a finite word w of length n, $\mathtt{Cl}(w) \leq \frac{n^2}{6} + \frac{7}{6}n + 1$.*

Besides that, [13] provides examples of words containing $\sim \frac{n^2}{6}$ distinct closed factors, thus proving the asymptotics $\sim \frac{n^2}{6}$ for the maximum number of distinct closed factors in a word. The definition of a closed-rich word can be extended to infinite words as follows.

Definition 1. *For an infinite word u, we define the following real number C_u:*

$$C_u = \inf\left\{\frac{\mathtt{Cl}(w)}{|w|^2} \ \mid \ w \in \mathrm{Fac}(u), |w| \geq 1\right\}.$$

If C_u is positive, then the word u is called closed-rich, and C_u is called the closed-rich constant of u.

Remark 1. In [13], infinite closed-rich words have been defined in a slightly different although equivalent way. According to [13], an infinite word u is closed-rich if there exists a positive constant C' such that for each n, each factor of length n of u contains at least $C'n^2$ distinct closed factors. It is not hard to show that the two definitions are equivalent, and in fact the constant C_u from Definition 1 can be defined as the supremum of the constants C':

$$C_u = \sup\left\{C' : \mathtt{Cl}(w) \geq C'|w|^2 \text{ for each } w \in \mathrm{Fac}(u)\right\}.$$

3 Upper Bound for the Closed-Rich Constant of Any Infinite Closed-Rich Word

In this section, we show that the closed rich constant of any infinite closed-rich word is less than or equal to 0.165964. We will make use of the following auxiliary results. Let $\texttt{Cl}'(w)$ denote the number of closed factors of w of length at least 2 ("long" closed factors), and $\texttt{Cl}^0(w)$ denote the number of closed factors of w of length ≤ 1 ("short" closed factors). So, $\texttt{Cl}(w) = \texttt{Cl}'(w) + \texttt{Cl}^0(w)$. If w is closed and v is its longest border, then we write $\texttt{LB}(\texttt{w}) = v$. Now, we have the following.

Lemma 1. *Let $w, u \in \Sigma^+$ such that* $\mathrm{Alph}(u) = \mathrm{Alph}(w)$ *and $w = ux$ where x is a letter. If t is the longest repeated suffix of w and z is the longest repeated suffix of t, then* $\texttt{Cl}(w) - \texttt{Cl}(u) = |t| - |z|$.

Proof. Since $\mathrm{Alph}(u) = \mathrm{Alph}(w)$, there always exists a non-empty repeated suffix of w and $\texttt{Cl}(w) - \texttt{Cl}(u) = \texttt{Cl}'(w) - \texttt{Cl}'(u)$. Then, $\texttt{Cl}'(w) - \texttt{Cl}'(u) \leq |t| - |z|$. Now, the repetition of t implies that for each suffix p of t, there always exists a closed factor β of w such that $\texttt{LB}(\beta) = p$.

Now, we show that $\texttt{Cl}'(w) - \texttt{Cl}'(u) = |t| - |z|$. Assume the converse, i.e., that $\texttt{Cl}'(w) - \texttt{Cl}'(u) < |t| - |z|$. Then, there exists a long closed factor α of u such that α is a suffix of w and $|z| + 1 \leq |\texttt{LB}(\alpha)| \leq |t|$. Now, $|\alpha|$ can not be greater than $|t|$ as t is the longest repeated suffix of w. Thus, $|\alpha| \leq |t|$. But as $|\texttt{LB}(\alpha)| \geq |z| + 1$ and α is a suffix of w, z is not the longest repeated suffix of t, which is a contradiction. Thus, there is no long closed factor α of u such that α is a suffix of w and $|z| + 1 \leq |\texttt{LB}(\alpha)| \leq |t|$. Therefore, $\texttt{Cl}'(w) - \texttt{Cl}'(u) = |t| - |z|$.

Lemma 2. *Let w and u be two non-empty words such that* $\mathrm{Alph}(u) \neq \mathrm{Alph}(w)$. *If $w = ux$ or $w = xu$ for some $x \in \Sigma$, then* $\texttt{Cl}(w) - \texttt{Cl}(u) = 1$.

Proof. For both cases, since $\mathrm{Alph}(u) \neq \mathrm{Alph}(w)$, $x \notin \mathrm{Alph}(u)$. This implies $\texttt{Cl}(w) = \texttt{Cl}(u) + 1$, where the new closed factor in w compare to u is the letter x.

The main result of this section is the following theorem giving an upper bound for the closed-rich constant of infinite closed-rich word.

Theorem 2. *The closed-rich constant of an infinite word is less than or equal to* $\frac{4603}{27735}$.

The general strategy of the proof is as follows. We will distinguish between several cases depending on the asymptotic critical exponent of the infinite word. We will say that an infinite word v *has long* α-powers if for every $m \in \mathbb{N}$, there exists a factor u of v of length $n \geq m$ such that the exponent of u is at least α. Now, if α is the asymptotic critical exponent of an infinite word v, then two cases are possible: either v has long α-powers and does not have long α^+-powers, or v has long factors with exponents $\geq (\alpha - \epsilon)$ for any $\epsilon > 0$ and does not have long α-powers. We split possible values of the asymptotic critical exponent of infinite closed-rich words into four intervals and obtain an upper bound for the closed-rich constant for each interval, and the upper bound in Theorem 2 is given by the maximum of the bounds obtained for each interval. For each of the intervals we treat the two cases above, and the intevals are the following: between 1 and 1.4, between 1.4 and 2, between 2 and 2.5, finally greater than 2.5. Finally, the proof of Theorem 2 is given by combining the above cases and the fact that infinite words with unbounded critical exponent are not closed-rich [13]. The worst case is given by the interval between 2 and 2.5, where we get the bound $\frac{4603}{27735}$.

Due to space constraints, we provide the proof for only one interval, for asymptotic critical exponent $1.4 < \alpha < 2$, and the case when the word has long α-powers but it does not have long α^+-powers.

Proposition 1. *Let v be an infinite closed-rich word of asymptotic critical exponent α, where $1.4 < \alpha < 2$. Suppose that it has long α-powers but it does not have long α^+-powers. Then the closed-rich constant of v is at most $\frac{13}{80}$.*

Proof. Since v has long α-powers but it does not have long α^+-powers, then for each $\epsilon > 0$, there exists a positive integer M such that the length of any factor of v with exponent $\alpha + \epsilon$ is at most M. For a fixed $\delta > 0$, we set $N = \lceil \frac{M}{\delta} \rceil$. Let w be an α-power factor of v of length $n \geq N$ (such w exists in v by the condition of the proposition). Note that for each $(\alpha + \epsilon)$-power factor z of w, where $\epsilon > 0$, its length is at most $M \leq \delta N \leq \delta n$; we will use this fact several times in the proof. We set $\gamma = \alpha - 1$; so $0.4 < \gamma < 1$. Then w is of the form $w = upu$, where $u, p \in \Sigma^+$, $|u| = l$, $|p| = r$ and $\frac{2l+r}{l+r} = \alpha$. Since $|w| = n = 2l + r$, we have $l = \frac{\gamma}{1+\gamma} n$ and $r = \frac{1-\gamma}{1+\gamma} n$. Consider some occurrence of w in v and its extension to the right by a factor of length ξn for some $\xi > 0$ (we will choose the value of ξ later). We let x denote this extension, and w' the extended factor, i.e. $w' = upux \in \mathrm{Fac}(v)$, where $|x| = \xi n$. We find an upper bound for the number of distinct closed factors in w', which will give us an upper bound for the closed-rich constant of v. Let $x^{(i)}$ be a prefix of x of length i and $t^{(i)}$ be the longest repeated suffix of $upux^{(i)}$. If $t^{(i)} = \lambda$, then $\mathtt{Cl}(upux^{(i)}) - \mathtt{Cl}(upux^{(i)-}) = 1$. Now, consider $t^{(i)} \in \Sigma^+$. Let us denote the rightmost $t^{(i)}$ in $upux^{(i)}$ by $t^{(i1)}$ and the rightmost $t^{(i)}$ in $upux^{(i)-}$ by $t^{(i2)}$. We consider two cases depending on whether $t^{(i1)}$ and $t^{(i2)}$ intersect or not, and in each case we get an upper bound on the length of $t^{(i)}$. We will later use these bounds to obtain an upper bound for the number of distinct closed-rich factors of w', which immediately gives an upper bound for the closed-rich constant of v.

If $t^{(i1)}$ and $t^{(i2)}$ intersect (see Fig. 1), then $t^{(i2)} = s^{(1)}s^{(2)}$ and $t^{(i1)} = s^{(2)}s^{(3)}$ for some $s^{(1)}, s^{(2)}, s^{(3)} \in \Sigma^+$. Then, the exponent of $s^{(1)}s^{(2)}s^{(3)}$ is greater than or equal to 2. Since all factors of v of exponent greater or equal to 2 are of length at most $n\delta$, we have $|s^{(1)}s^{(2)}s^{(3)}| < n\delta$. Since $|t^{(i)}| < |s^{(1)}s^{(2)}s^{(3)}|$, we also have $|t^{(i)}| < n\delta$.

Now suppose that $t^{(i1)}$ and $t^{(i2)}$ do not intersect. Then as $t^{(i1)}, t^{(i2)} \in \mathrm{Fac}(upux^{(i)})$ and $|upux^{(i)}| = 2l + r + i$, we have $|t^{(i)}| \leq l + \frac{i+r}{2}$.

Now, we consider the following cases, depending on the length of $t^{(i)}$:

Case 1: $|t^{(i)}| \leq l + i$.

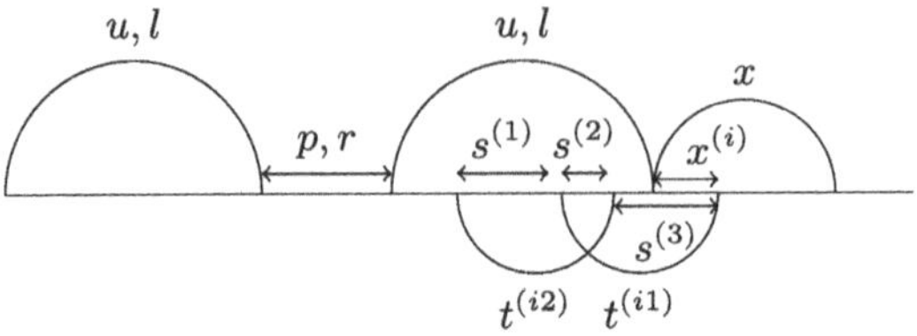

Fig. 1. Illustration to the proof of Proposition 1 in the case when $t^{(i1)}$ and $t^{(i2)}$ intersect.

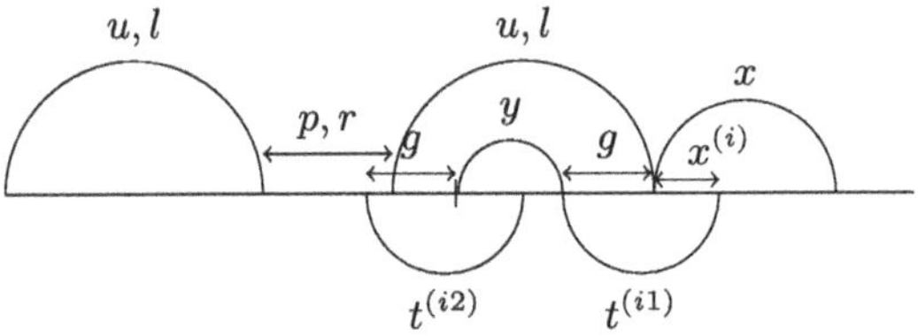

Fig. 2. Illustration to the proof of Proposition 1, Case 1.1.

In this case we have either $|t^{(i)}| \leq i$ or $t^{(i1)} = gx^{(i)}$ for some $g \in \Sigma^+$. In the first case we will use this bound for $|t^{(i)}|$, so we now consider the case $g \in \Sigma^+$. We consider two subcases depending on the occurrence of $t^{(i2)}$:

Case 1.1: $t^{(i2)} \in \text{Fac}(pu)$ (see Fig. 2).

In this case, since $g \in \text{Pref}(t^{(i2)})$, we have $gyg \in \text{Suff}(pu)$ for some $y \in \Sigma^+$. Now, $gyg = (gy)^{\frac{|g|+|y|+|g|}{|g|+|y|}}$. If $\frac{|g|+|y|+|g|}{|g|+|y|} > \alpha$, then $|gyg| < n\delta$, so we also have $|g| < n\delta$. Then $|t^{(i)}| < n\delta + i$.

If $\frac{|g|+|y|+|g|}{|g|+|y|} \leq \alpha$, then $\frac{(1-\gamma)|g|}{\gamma} \leq |y|$. Since $gyg \in \text{Fac}(pu)$, $|g|+|y|+|g| \leq l+r$. Then, as $\frac{(1-\gamma)|g|}{\gamma} \leq |y|$, we have $|g| + \frac{(1-\gamma)|g|}{\gamma} + |g| \leq |g| + |y| + |g| \leq l+r$. Since $l + r = \frac{n}{1+\gamma}$, we get $|g| \leq \frac{n\gamma}{(1+\gamma)^2}$. Thus, $|t^{(i)}| = |g| + i \leq \frac{n\gamma}{(1+\gamma)^2} + i$.

Case 1.2: $t^{(i2)} \in \text{Fac}(upu)$ and $t^{(i2)} \notin \text{Fac}(pu)$ (see Fig. 3).

In this case, we have either $t^{(i2)}$ is completely contained in the first occurrence of u in w' (which is not possible as there would be another occurrence of $t^{(i)}$ inside the rightmost occurrence of u, and this occurrence of $t^{(i)}$ would be the second right occurrence of $t^{(i)}$ in w' instead of $t^{(i2)}$), or $t^{(i2)} = \beta^{(1)}\beta^{(2)}$ for some non-empty words $\beta^{(1)}$ and $\beta^{(2)}$ such that $\beta^{(1)} \in \text{Suff}(u)$ and $\beta^{(2)} \in \text{Pref}(pu)$. Since w is an α-power and not an α^+-power, we have that $w_{l+1} \neq w_{2l+r+1}$. So, we have $|g| \neq |\beta^{(1)}|$.

Case 1.2.1: If $|\beta^{(1)}| < 2|g|$, then the occurrence of g as a prefix of $t^{(i2)}$ and the occurrence of g as a suffix of the first occurrence of u in w intersect. Thus we have a factor of exponent greater than 2, so $|g| < n\delta$. This implies $|t^{(i)}| < n\delta + i$.

Case 1.2.2: If $|\beta^{(1)}| \geq 2|g|$, then as $g \in \text{Pref}(t^{(i2)})$ and $g \in \text{Suff}(u)$, gqg is a suffix of u for some $q \in \Sigma^*$. This implies $gqg \in \text{Pref}(t^{(i2)})$, i.e., $qg \in \text{Pref}(x^{(i)})$. Then, $gqgqg \in \text{Fac}(ux^{(i)})$. Since exponent of $gqgqg$ is greater than or equal to 2, $|gqgqg| < n\delta$, i.e., $|g| < n\delta$. Thus, $|t^{(i)}| < n\delta + i$.

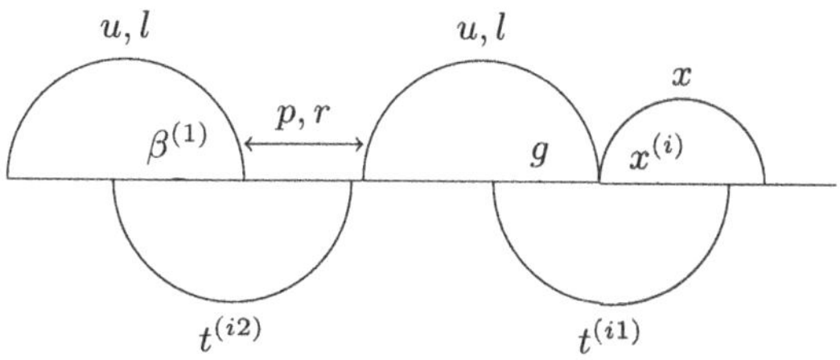

Fig. 3. Illustration to the proof of Proposition 1, Case 1.2.

Case 2: $l + i < |t^{(i)}| \leq l + \frac{i+r}{2}$.

In this case, $t^{(i1)} = y'ux^{(i)}$ for some $y' \in \Sigma^+$.

Case 2.1: $t^{(i2)} \in \text{Fac}(p)$ (see Fig. 4).

In this subcase, as $u \in \text{Fac}(t^{(i2)})$, $r > l$. Then $1.4 < \alpha < 1.5$. This implies $l < r < \frac{3l}{2}$. Now, $u \in \text{Fac}(t^{(i2)})$ and $u \in \text{Suff}(w)$ implies $uu'u \in \text{Suff}(w)$ for some $u' \in \Sigma^+$. Since $r < \frac{3l}{2}$ and $|u| = l$, we have $|u'| < l$. Then the exponent of $uu'u$ is > 1.5, i.e., the exponent of $uu'u$ is greater than α. This implies $|uu'u| < n\delta$. Since $|u'| > |y'|$, we have $|t^{(i)}| = |y'| + |u| + i \leq |u'| + |u| + i \leq n\delta + i$.

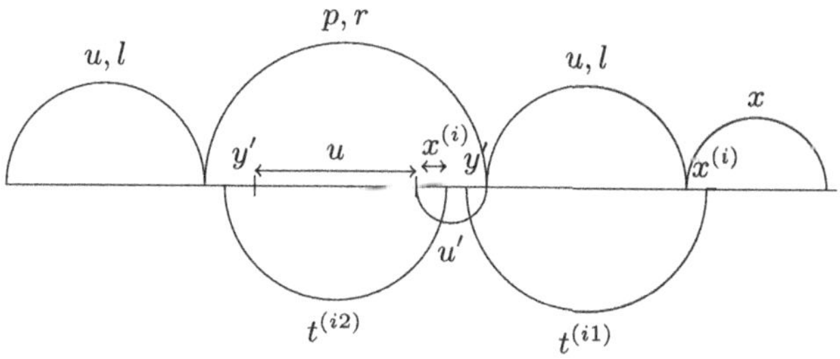

Fig. 4. Illustration to the proof of Proposition 1, Case 2.1.

Case 2.2: $t^{(i2)} \in \text{Fac}(up)$ and $t^{(i2)} \notin \text{Fac}(p)$ (see Fig. 5)

In this subcase, $t^{(i2)} = g^{(1)}g^{(2)}$ for some nonempty $g^{(1)}, g^{(2)}$, such that $g^{(1)} \in \text{Suff}(u)$ and $g^{(2)} \in \text{Pref}(p)$. Then $g^{(1)} \in \text{Pref}(t^{(i1)})$.

Now, if the occurrence of $g^{(1)}$ as a prefix of $t^{(i1)}$ and the occurrence of $g^{(1)}$ as a suffix of the second occurrence of u intersect, then the exponent of the factor given by this intersection of the two occurrences of $g^{(1)}$ (in other words, the factor $t^{(i1)}(x^{(i)})^{-1}$) is greater than or equal to 2, so its length is at most $n\delta$, and hence $|t^{(i)}| < n\delta + i$.

If $g^{(1)} \in \text{Pref}(t^{(i1)})$ and $g^{(1)} \in \text{Suff}(u)$ do not intersect, then $t^{(i1)} = g^{(1)}y''g^{(1)}x^{(i)}$ for some $y'' \in \Sigma^*$. Now, we have the following cases:

Case 2.2.1: $|g^{(1)}| \geq |y'|$ (see Fig. 6).

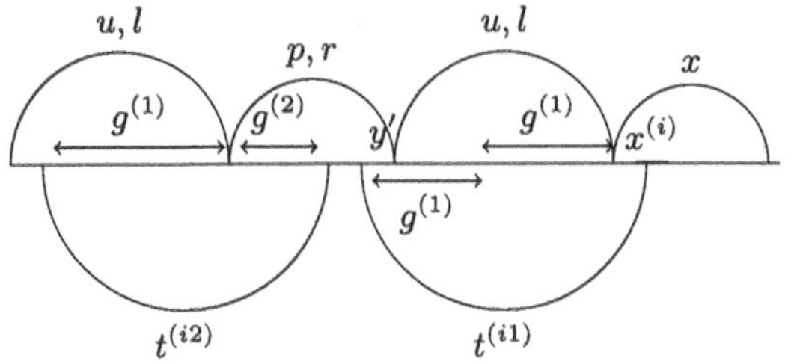

Fig. 5. Illustration to the proof of Proposition 1, Case 2.2.

Then, $y'' \in \mathrm{Fac}(u)$. So, $y''g^{(1)} \in \mathrm{Suff}(u)$. Then as $g^{(1)}y''g^{(1)} \in \mathrm{Pref}(t^{(i2)})$, $y''g^{(1)}y''g^{(1)} \in \mathrm{Fac}(up)$. Since exponent of $y''g^{(1)}y''g^{(1)}$ is 2, $|y''g^{(1)}y''g^{(1)}| < n\delta$. This implies $|t^{(i)}| < n\delta + i$.

Case 2.2.2: $|g^{(1)}| < |y'|$ (see Fig. 7).

Then, as $y' \in \mathrm{Pref}(t^{(i2)})$, we have $u \in \mathrm{Fac}(p)$. Therefore, $r > l$, and so similarly to the Case 2.1, we have $1.4 < \alpha < 1.5$. This implies $l < r < \frac{3l}{2}$. Now, $u \in \mathrm{Fac}(p)$ and $u \in \mathrm{Suff}(w)$ implies $uu''u \in \mathrm{Suff}(w)$ for some $u'' \in \Sigma^+$. Since $r < \frac{3l}{2}$ and $|u| = l$, $|u''| < l$. Then the exponent of $uu''u$ is greater than 1.5, i.e., greater than α. This implies $|uu''u| < n\delta$. Since $|u''| > |y'|$, we have $|t^{(i)}| = |y'| + |u| + i \leq |u''| + |u| + i \leq n\delta + i$.

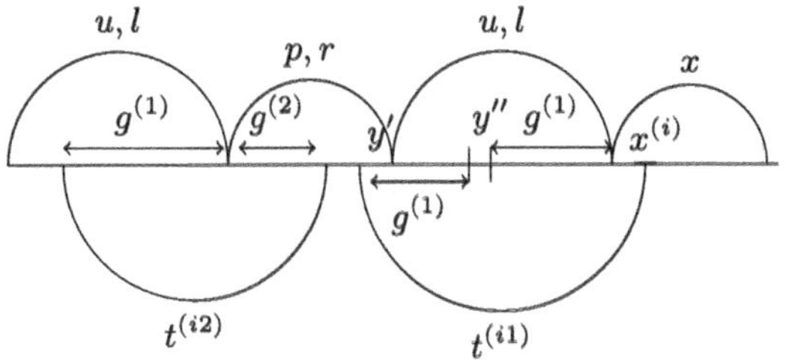

Fig. 6. Illustration to the proof of Proposition 1, Case 2.2.1

Now we are ready to estimate the number of closed factors in w'. Combining the cases above, we have

$$\mathtt{Cl}(upux^{(i)}) - \mathtt{Cl}(upux^{(i)-}) \leq \max\left\{\frac{n\gamma}{(1+\gamma)^2} + i, \delta n + i\right\}.$$

So, we can take $\delta = \frac{\gamma}{(1+\gamma)^2}$ (or in fact any number smaller than that). Then, by Lemma 1, we have $\mathtt{Cl}(upux^{(i)}) - \mathtt{Cl}(upux^{(i)-}) \leq \frac{n\gamma}{(1+\gamma)^2} + i$. We set $|x| = \xi n$ for some $\xi > 0$. Thus using Theorem 1, we have

$$\mathtt{Cl}(w') \leq \frac{n^2}{6} + \frac{7n}{6} + 1 + \sum_{i=1}^{\xi n}\left(\frac{n\gamma}{(1+\gamma)^2} + i\right) = \frac{n^2}{6} + \frac{7n}{6} + 1 + \frac{\xi n^2\gamma}{(1+\gamma)^2} + \frac{\xi^2 n^2}{2} + \frac{\xi n}{2}.$$

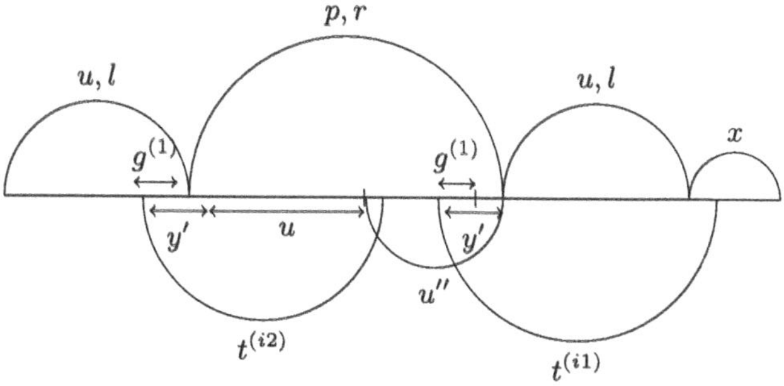

Fig. 7. Illustration to the proof of Proposition 1, Case 2.2.2

Now, since $|w'| = n + |x| = (1+\xi)n$, we have

$$\inf\left\{\frac{\mathtt{Cl}(w')}{(1+\xi)^2 n^2} : w' = wx \in \mathrm{Fac}(v), |w| = n, |x| = \xi n, exp(w) = \alpha, n \geq N\right\}$$
$$\leq \frac{1}{(1+\xi)^2}\left(\frac{1}{6} + \frac{\xi\gamma}{(1+\gamma)^2} + \frac{\xi^2}{2}\right).$$

We let denote the obtained function by $f(\xi, \gamma)$:

$$f(\xi,\gamma) = \frac{1}{(1+\xi)^2}\left(\frac{1}{6} + \frac{\xi\gamma}{(1+\gamma)^2} + \frac{\xi^2}{2}\right).$$

The previous inequality implies by Definition 1 that $f(\xi, \gamma)$ is an upper bound for the closed-rich constant C_v of v: $C_v \leq f(\xi, \gamma)$.

So, we can choose for each γ the value of $\xi \in (0, +\infty)$ giving minimum of our bound $f(\xi, \gamma)$, and then to get an upper bound for all $\gamma \in (0.4, 1)$, take maximum over γ in this interval. (Here we ignore the fact that ξn can be non-integer, since taking the previous or the next length does not affect the infimum. This follows, for example, from the fact that adding one letter to a word adds only a sublinear number of closed factors). So,

$$C_v \leq \max_{\gamma\in(0.4,1)} \min_{\xi\in(0,+\infty)} f(\xi,\gamma).$$

For each $\gamma \in (0.4, 1)$, considering f as a function of one argument ξ, we get that f attains its minimum for $\xi = \xi_0(\gamma) = \frac{\gamma^2+1-\gamma}{3(\gamma^2+1+\gamma)}$.

Now, considering $f(\xi_0(\gamma), \gamma)$ as a function of γ, with standard arguments one can see that it is growing on the interval $(0.4, 1)$ (since its derivative is positive), so its maximum is attained when $\gamma = 1$. This gives an upper bound $f(\xi_0(1), 1) = \frac{13}{80}$.

4 Closed-Rich Properties of the Fibonacci Word

In this section, we calculate the number of closed factors in any prefix of the Fibonacci word f and provide upper and lower bounds for C_f.

Let $(f_n)_{n\geq -1}$ denote the sequence of *Fibonacci words* where $f_{-1} = b$, $f_0 = a$, and $f_n = f_{n-1}f_{n-2}$ for $n \geq 1$. The word f_n is called the n-th Fibonacci word, and the limit $f = \lim_{n\to\infty} f_n$ is the *infinite Fibonacci word.* We let denote $F_n = |f_n|$; it is straightforward that the number F_n is the n-th Fibonacci number.

The following two theorems give explicit formulas for the sequence of numbers of distinct closed factors in a prefix of length n of the Fibonacci word and its first difference sequence (the proofs are omitted due to space constraints).

Theorem 3. *The first difference sequence $(s(n))_{n\geq 1}$ of the sequence $\mathtt{PCl}(n)$ of the number of closed factors in the prefix of length n of the Fibonacci word is given by the following:*

$$(s(n))_{n\geq 1} = \underbrace{F_{-1}, F_{-1},}_{2F_{-1}\,terms} \underbrace{F_0, F_0,}_{2F_0\,terms} \underbrace{F_1, \ldots, F_1,}_{2F_1\,terms} \ldots, \underbrace{F_m, F_m, \ldots, F_m,}_{2F_m\ terms} \ldots$$

In other words, if we denote repetitions in a sequence by exponents and concatenation by the symbol $\prod$, then we can rephrase the theorem as follows:

$$(s(n))_{n\geq 1} = \prod_{i=-1}^{\infty} {F_i}^{2F_i}.$$

Theorem 4. *For $n \geq 4$, let p be a prefix of f of length l where $l = F_n + k$ and $0 \leq k < F_{n-1}$. Then*

$$\mathtt{Cl}(p) = \begin{cases} 1 + F_{n-3}(F_{n-2} + F_{n-4} + k + 2), & \text{if } 0 \leq k \leq F_{n-3} - 2, \\ 1 + F_{n-2}(F_{n-3} + k + 2), & \text{if } F_{n-3} - 2 < k \leq F_{n-1} - 1; \end{cases}$$

and the initial values are $\mathtt{PCl}(1) = 2$, $\mathtt{PCl}(2) = 3$, $\mathtt{PCl}(3) = 4$, $\mathtt{PCl}(4) = 5$, $\mathtt{PCl}(5) = 7$, $\mathtt{PCl}(6) = 9$, $\mathtt{PCl}(7) = 11$.

Remark 2. The statement of this theorem can be reformulated using Zeckendorf representation. Namely, every non-negative integer can be represented as a sum of Fibonacci numbers $(F_i)_{i\geq 0}$ with no two consecutive Fibonacci numbers, and such representation is unique [17]. So, the condition $F_n \leq l < F_{n+1}$ on the length l can be reformulated as follows: in Zeckendorf representation, l has the most significant digit at position $n + 1$ (corresponding to F_n).

In [13], it was proved that f is an infinite closed-rich word. We now provide the following bounds for C_f (the proof is omitted due to space constraints):

Theorem 5. *The following bounds hold for the closed rich constant C_f of the Fibonacci word f:*

$$\frac{\phi^3 + 3}{\phi^9} < C_f \leq 0.11022,$$

where $\phi = \frac{1+\sqrt{5}}{2}$.

Here, for the lower bound, one has $\frac{\phi^3+3}{\phi^9} \approx 0.0952$. The upper bound is obtained from numerical experiments: we find that the minimum number of closed factors among factors of length 341 of f is 12816, so $C_f \le \frac{12816}{341^2}$. The proof of the lower bound makes use of Theorem 4. Note that the upper bound could be easily improved by calculating for longer factors of Fibonacci; however, it would be more interesting to prove the conjectured closed-rich constant of f (see Conjecture 2).

5 Conclusions and Open Problems

Let $C_{sup} = \sup\{C_u : C_u$ is the closed-rich constant of an infinite closed-rich word $u\}$. From Theorems 2 and 5, we have the following bounds for C_{sup}:

Corollary 1. $0.0952 < C_{sup} \le 0.165964$.

With computational experiments we observed that f provides a better lower bound for C_{sup} than several well-known infinite closed-rich words. For instance, for the Thue-Morse word we obtained $C_{TM} \le 0.09$, for ternary Thue-Morse word $C_{TTM} \le 0.092$, for the period-doubling word $C_{PD} \le 0.093$, for Dejean's word $C_D \le 0.08$, for Mephisto-Waltz word $C_{MW} \le 0.08$, for Leech's word $C_{LE} \le 0.07$, for Fibonacci-Thue-Morse word $C_{FTM} \le 0.062$, etc.

Based on numerical experiments, we propose the following conjecture regarding $M_n = \min\{\mathtt{Cl}(w) : w \in \mathrm{Fac}(f) \cap \Sigma^n\}$. The following conjecture gives a formula for its first difference sequence $R_n = M_n - M_{n-1}$ (it is not hard to derive from it an explicit formula for M_n).

Conjecture 1. For the Fibonacci word, we have

$$\begin{aligned}(R_n)_{n\ge 1} &= 1,1,1,1,F_0,F_2,F_0,F_2,F_1,F_1,F_3,F_1,F_1,F_3,F_2,F_2,F_2,F_4,F_4,F_2,F_2,\\ &\quad F_2,F_4,F_4,\cdots\\ &= 1,1,1,1,\prod_{m=0}^{\infty} {F_m}^{F_m}, {F_{m+2}}^{F_{m-1}}, {F_m}^{F_m}, {F_{m+2}}^{F_{m-1}}.\end{aligned}$$

Based on Conjecture 1, we conjecture the value of C_f:

Conjecture 2. For f, the closed-rich constant $C_f = \frac{5\phi+3}{45\phi+29} \approx 0.10893$.

Remark 3. From the conjectured sequence, the minimum value of closed factors occurs for the lengths $F_n + 2F_{n-3} - 2$ (in fact, between these and the next lengths the first difference function increases). Computer experiments show that for each of these lenghts there is only one factor giving the minimal value. Moreover, these factors are palindromes of a special kind: they have bispecial factors of length F_n in the middle, which are continued in the unique way to the left and to the right with length $F_{n-3} - 1$. However, for now we do not have a proof of the fact that these factors give the minimum.

The question about the closed-rich constant of the Fibonacci word can be naturally extended to closed-rich Sturmian words:

Question 1. Given a Sturmian word s of slope α such that its continued fraction expansion has bounded partial quotients. In [13] is has been shown that s is closed-rich. What is the closed-rich constant of s?

Question 2. Is it true that any real number in $(0, C_{sup})$ is a closed-rich constant of some infinite closed-rich word? If not, is the set of closed-rich constants dense in $(0, C_{sup})$?

References

1. Badkobeh, G., et al.: Closed factorization. Discret. Appl. Math. **212**, 23–29 (2016)
2. Badkobeh, G., Fici, G., Lipták, Z.: On the number of closed factors in a word. In: Dediu, AH., Formenti, E., Martín-Vide, C., Truthe, B. (eds) LATA 2015, LNCS,volume 8977. pp. 381–390. Springer, Cham (2015). https://doi.org/10.1007/978-3-319-15579-1_29
3. Bannai, H., Tomohiro, I., Inenaga, S., Nakashima, Y., Takeda, M., Tsuruta, K.: The "runs" theorem. SIAM J. Comput. **46**(5), 1501–1514 (2017)
4. Bannai, H., et al.: Efficient algorithms for longest closed factor array. In: Iliopoulos, C., Puglisi, S., Yilmaz, E. (eds.) String Processing and Information Retrieval: 22nd International Symposium, SPIRE 2015, London, UK, September 1-4, 2015, Proceedings, pp. 95–102. Springer International Publishing, Cham (2015). https://doi.org/10.1007/978-3-319-23826-5_10
5. Brlek, S., Li, S.: On the number of squares in a finite word. arXiv preprint arXiv:2204.10204 (2022)
6. De Luca, A., Fici, G.: Open and closed prefixes of Sturmian words. In: Karhumäki, J., Lepistö, A., Zamboni, L. (eds) WORDS 2013, LNCS 8079. pp. 132–142. Springer, Berlin, Heidelberg (2013). https://doi.org/10.1007/978-3-642-40579-2_15
7. Durand, F., Host, B., Skau, C.: Substitutional dynamical systems, Bratteli diagrams and dimension groups. Ergodic Theory Dynam. Syst. **19**(4), 953–993 (1999)
8. Fici, G.: Open and closedwords. Bull. EATCS **123** (2017). http://eatcs.org/beatcs/index.php/beatcs/article/view/508
9. Fici, G., Mignosi, F., Shallit, J.: Abelian-square-rich words. Theor. Comput. Sci. **684**, 29–42 (2017)
10. Glen, A., Justin, J., Widmer, S., Zamboni, L.Q.: Palindromic richness. Eur. J. Comb. **30**(2), 510–531 (2009)
11. Lothaire, M.: Algebraic Combinatorics on Words. Cambridge University Press, Cambridge (2001)
12. Parshina, O., Postic, M.: Open and closed complexity of infinite words. arXiv preprint arXiv:2005.06254 (2020)
13. Parshina, O., Puzynina, S.: Finite and infinite closed-rich words. Theoret. Comput. Sci. **984**, 114315 (2024)
14. Parshina, O., Zamboni, L.Q.: Open and closed factors in Arnoux-Rauzy words. Adv. Appl. Math. **107**, 22–31 (2019)

15. Schaeffer, L., Shallit, J.: Closed, palindromic, rich, privileged, trapezoidal, and balanced words in automatic sequences. Electron. J. Comb., P1–25 (2016)
16. Vuillon, L.: A characterization of Sturmian words by return words. Eur. J. Comb. **22**(2), 263–275 (2001)
17. Zeckendorf, É.: Representations des nombres naturels par une somme de nombres de Fibonacci on de nombres de lucas. Bulletin de La Society Royale des Sciences de Liege, 179–182 (1972)

Note on Dissecting Power of Regular Languages

Josef Rukavicka(✉)

Department of Mathematics, Faculty of Nuclear Sciences and Physical Engineering, Czech Technical University in Prague, Prague, Czech Republic
josef.rukavicka@seznam.cz

Abstract. Let $c > 1$ be a real constant. We say that a language L is *c-constantly growing* if for every word $u \in L$ there is a word $v \in L$ with $|u| < |v| \leq c + |u|$. We say that a language L is *c-geometrically growing* if for every word $u \in L$ there is a word $v \in L$ with $|u| < |v| \leq c|u|$. Given a language L, we say that L is REG-*dissectible* if there is a regular language R such that $|L \setminus R| = \infty$ and $|L \cap R| = \infty$. In 2013, it was shown that every c-constantly growing language L is REG-dissectible. In 2023, the following open question has been presented: "Is the family of geometrically growing languages REG-dissectible?" For every $c > 1$, we construct a c-geometrically growing language L that is not REG-dissectible. Hence we answer negatively to the open question.

1 Introduction

Let L_1 and L_2 be two infinite languages. We say that L_1 *dissects* L_2 if

$$|L_2 \setminus L_1| = \infty \quad \text{and} \quad |L_1 \cap L_2| = \infty.$$

Dissection of infinite languages have been investigated in recent years. Most notably, in [3], the reader finds some results about the dissecting power of regular languages.

Let $\mathcal{C}$ be a family of languages. We say that a language L_2 is $\mathcal{C}$-*dissectible* if there is $L_1 \in \mathcal{C}$ such that L_1 dissects L_2. Let REG denote the family of regular languages and let CFL denote the family of context free languages. In [3], several examples of language families that are REG-dissectible have been presented. In addition, it was proved that the language

$$L = \{a^{n!} \mid n \text{ is a positive integer } \}$$

is not REG-dissectible.

In 1980, the following open problem was presented in [1]: Given two context-free languages L_1, L_2 such that $L_1 \subset L_2$ and $L_2 \setminus L_1$ is an infinite language, is there a context-free language L_3 such that $L_3 \subset L_2$, $L_1 \subset L_3$, and both the languages $L_3 \setminus L_1$ and $L_2 \setminus L_3$ are infinite? This question was mentioned also in

G. Gamard and J. Leroy (Eds.): WORDS 2025, LNCS 15729, pp. 230–237, 2025.
https://doi.org/10.1007/978-3-031-97548-6_20

[3] using the so-called i-separation. For $L_2 = \Sigma^*$ the question is, if the differencc of two context free languages is CFL-dissectible.

Related results on the dissection of infinite languages may be found in [6]. In [4], the construction of minimal covers of languages is investigated.

A language $L_1 \subseteq \Sigma_n^*$ is called $\mathbb{C}$-*immune* if there is no infinite language $L_2 \subseteq L_1$ such that $L_2 \in \mathbb{C}$; some results about the immunity can be found in [5,7,8].

An infinite language L is called *constantly growing* or *c-constantly growing* if there is a real constant $c \geq 1$ such that for every word $u \in L$ there is a word $v \in L$ with $|u| < |v| \leq c + |u|$. In [3], it has been proved that every constantly growing language L is REG-dissectible.

In [2], a "natural" generalization of constantly growing languages has been introduced as follows. A language L is called *c-geometrically growing* (or just *geometrically growing*) if there is a real constant $c > 1$ such that for every $u \in L$ there exists $v \in L$ with $|u| < |v| \leq c|u|$.

In [2] it was shown how to dissect a geometrically growing language by a homomorphic image of intersection of two context-free languages. Consider two alphabets Σ and Θ such that $|\Sigma| = 1$ and $|\Theta| = 4$; that is Σ and Θ denote alphabets with one letter and four letters, respectively. The next corollary is the main result of [2].

Corollary 1 *(see [2, Corollary 1.3]). There are context-free languages* $\mathrm{M}_1, \mathrm{M}_2 \subseteq \Theta^*$, *an erasing alphabetical homomorphism* $\pi : \Theta^* \rightarrow \Sigma^*$, *and a nonerasing alphabetical homomorphism* $\varphi : \Gamma^* \rightarrow \Sigma^*$ *such that: If* $L \subseteq \Gamma^*$ *is a geometrically growing language then there is a regular language* $R \subseteq \Theta^*$ *such that* $\varphi^{-1}(\pi(R \cap \mathrm{M}_1 \cap \mathrm{M}_2))$ *dissects the language* L.

Remark 1 (see [2, Remark 1.4]). Since the intersection of a regular language and a context-free language is a context-free language we have $R \cap \mathrm{M}_1 \cap \mathrm{M}_2$ is also intersection of two context-free languages. This explains why we do not mention the regular language in the title of the article.

In the current article, for every real constant $c > 1$, we construct a c-geometrically growing language L that is not REG-dissectible. The language L is constructed over an alphabet with one letter. As such, the result gives the negative answer to the open question from [2]: "Is the family of geometrically growing languages REG-dissectible?".

Our current result implies also that L is not CFL-dissectible; just realize that every context free language over one letter alphabet is a regular language. For the same reason, L can not be dissected by an intersection of context free languages. Hence if we consider only regular languages, context free languages, their intersections, unions, and homomorphic images, then we can claim that Corollary 1 presents the "best (easiest) possible" way how to dissect geometrically growing languages.

For a more detailed overview and other open questions concerning the dissection of infinite languages, see [2,3].

2 Preliminaries

Let $\mathbb{N}_0$ denote the set of all non-negative integers, let $\mathbb{N}_1$ denote the set of all positive integers and let $\mathbb{R}^+$ denote the set of all positive real numbers.

Let $\Sigma = \{a\}$ denote the alphabet containing exactly one letter a.

3 Geometrically Growing Language

Suppose $\alpha, \beta \in \mathbb{N}_1$ with $\alpha < \beta$. Let $\gamma = \frac{1}{\ln \beta - \ln \alpha}$.

Let

$$\Delta = \{(j, n) \mid j, n \in \mathbb{N}_0 \text{ and } n > \alpha \lfloor \gamma \ln n \rfloor \text{ and } j \in \{0, 1, \ldots, \lfloor \gamma \ln n \rfloor\}\}.$$

Given $(j, n) \in \Delta$, let

$$\phi(j, n) = \left(\frac{\beta}{\alpha}\right)^j n!$$

and let

$$\omega(n) = \frac{n!}{(\lfloor \gamma \ln n \rfloor \alpha + 1)!}.$$

We will use the function $\phi(j, n)$ to define the lengths of words. Hence we need the value of $\phi(j, n)$ to be a positive integer; the next lemma shows that this requirement is satisfied. In addition we show that $\phi(j, n)$ is divisible by $\omega(n)$.

Lemma 1. *If $(j, n) \in \Delta$ then $\phi(j, n)$ is a positive integer and*

$$0 \equiv \phi(j, n) \pmod{\omega(n)}.$$

Proof. From $n > \alpha \lfloor \gamma \ln n \rfloor$ we have that

$$n! = n(n-1) \cdots (\lfloor \gamma \ln n \rfloor \alpha) \cdots ((\lfloor \gamma \ln n \rfloor - 1)\alpha) \cdots (2\alpha) \cdots \alpha \cdots 1.$$

It follows that $n! = m\alpha^{\lfloor \gamma \ln n \rfloor}$ for some $m \in \mathbb{N}_1$. Consequently $\phi(j, n)$ is a positive integer. From the definitions of $\phi(j, n)$ and $\omega(n)$ we have that

$$\frac{\phi(j, n)}{\omega(n)} = \frac{\left(\frac{\beta}{\alpha}\right)^j n!}{\frac{n!}{(\lfloor \gamma \ln n \rfloor \alpha + 1)!}} = (\lfloor \gamma \ln n \rfloor \alpha + 1)! \left(\frac{\beta}{\alpha}\right)^j.$$

Since $j \leq \lfloor \gamma \ln n \rfloor$ it is clear that

$$0 \equiv \phi(j, n) \pmod{\omega(n)}.$$

This ends the proof.

We show that for given $(j, n) \in \Delta$ there is $(\overline{j}, \overline{n}) \in \Delta$ such that $\phi(\overline{j}, \overline{n})$ is strictly bigger that $\phi(j, n)$ and that $\phi(\overline{j}, \overline{n})$ and $\phi(j, n)$ are "close" to each other.

Proposition 1. *If* $(j, n) \in \Delta$ *then there is* $(\overline{j}, \overline{n}) \in \Delta$ *such that* $\overline{n} \geq n$ *and*

$$\phi(j,n) < \phi(\overline{j},\overline{n}) \leq \frac{(n+1)\beta}{n\alpha}\phi(j,n).$$

Proof. If $j < \lfloor \gamma \ln n \rfloor$ then

$$\frac{\phi(j+1,n)}{\phi(j,n)} = \frac{\left(\frac{\beta}{\alpha}\right)^{j+1} n!}{\left(\frac{\beta}{\alpha}\right)^{j} n!} = \frac{\beta}{\alpha}. \tag{1}$$

We have that

$$\left(\frac{\beta}{\alpha}\right)^{\lfloor \gamma \ln n \rfloor} = \left(\frac{\beta}{\alpha}\right)^{\left\lfloor \frac{\ln n}{\ln \beta - \ln \alpha} \right\rfloor} \leq \left(\frac{\beta}{\alpha}\right)^{\frac{\ln n}{\ln \beta - \ln \alpha}} = e^{\frac{\ln n}{\ln \beta - \ln \alpha} \ln \frac{\beta}{\alpha}} = e^{\ln n} = n. \tag{2}$$

From (2) it follows that

$$\frac{\phi(0,n+1)}{\phi(\lfloor \gamma \ln n \rfloor, n)} = \frac{\left(\frac{\beta}{\alpha}\right)^{0} (n+1)!}{\left(\frac{\beta}{\alpha}\right)^{\left\lfloor \frac{\ln n}{\ln \beta - \ln \alpha} \right\rfloor} n!} = \frac{(n+1)!}{\left(\frac{\beta}{\alpha}\right)^{\left\lfloor \frac{\ln n}{\ln \beta - \ln \alpha} \right\rfloor} n!} \geq \frac{(n+1)}{n} > 1 \tag{3}$$

and

$$\frac{\phi(0,n+1)}{\phi(\lfloor \gamma \ln n \rfloor, n)} = \frac{\left(\frac{\beta}{\alpha}\right)^{0} (n+1)!}{\left(\frac{\beta}{\alpha}\right)^{\left\lfloor \frac{\ln n}{\ln \beta - \ln \alpha} \right\rfloor} n!} \leq \frac{(n+1)!}{\left(\frac{\beta}{\alpha}\right)^{\frac{\ln n}{\ln \beta - \ln \alpha} - 1} n!} = \frac{(n+1)!}{n\left(\frac{\beta}{\alpha}\right)^{-1} n!} = \frac{\beta(n+1)}{\alpha n}. \tag{4}$$

We distinguish the following two cases when proving the proposition:

- If $0 \leq j < \lfloor \gamma \ln n \rfloor$ then from (1) it follows that

$$\phi(j,n) < \phi(j+1,n) \leq \frac{\beta}{\alpha}\phi(j,n).$$

- If $j = \lfloor \gamma \ln n \rfloor$ then from (3) and (4) we have that

$$\phi(j,n) < \phi(0,n+1) \leq \frac{(n+1)\beta}{n\alpha}.$$

It follows that for every $(j, n) \in \Delta$ there is $(\overline{j}, \overline{n}) \in \Delta$ such that

$$\phi(j,n) < \phi(\overline{j},\overline{n}) \leq \frac{(n+1)\beta}{n\alpha}\phi(j,n).$$

This ends the proof.

Remark 2. From the proof of Proposition 1 it follows also that

$$\phi(j,n) < \phi(\bar{j},\bar{n})$$

if and only if $n \leq \bar{n}$ or ($n = \bar{n}$ and $j \leq \bar{j}$).

Given the letter $a \in \Sigma$, let

$$\Pi = \{a^{\phi(j,n)} \mid (j,n) \in \Delta\}.$$

Remark 3. Lemma 1 implies that $\phi(j,n)$ is a positive integer. Hence the definition of the language Π makes sense. Also it is clear that if $a^{\phi(j,n)} \in \Pi$ then $|a^{\phi(j,n)}| = \phi(j,n)$.

We show that for any real constant $c > \frac{\beta}{\alpha}$ there is c-geometrically growing language that we construct from the language Π by removing a finite number of its elements.

Corollary 2. *If $c \in \mathbb{R}^+$ and $c > \frac{\beta}{\alpha}$ then there is $\overline{\Pi} \subseteq \Pi$ such that the language $\overline{\Pi}$ is c-geometrically growing and $|\Pi \setminus \overline{\Pi}| < \infty$.*

Proof. Let $n_0 \in \mathbb{N}_1$ be such that

$$\frac{(n_0+1)\beta}{n_0\alpha} < c.$$

Clearly such n_0 exists and also for every $n > n_0$ we have that

$$\frac{(n+1)\beta}{n\alpha} < c.$$

Proposition 1 asserts that for every $(j,n) \in \Delta$ with $n > n_0$ there is $(\bar{j},\bar{n}) \in \Delta$ such that

$$\bar{n} \geq n \quad \text{and} \quad \phi(j,n) < \phi(\bar{j},\bar{n}) \leq \frac{(n+1)\beta}{n\alpha}\phi(j,n) < c\phi(j,n). \tag{5}$$

Let

$$\overline{\Pi} = \{a^{\pi(j,n)} \in \Pi \mid n > n_0\}.$$

From (5) it follows that for every $a^{\phi(j,n)} \in \overline{\Pi}$ there is $a^{\phi(\bar{j},\bar{n})} \in \overline{\Pi}$ such that

$$|a^{\phi(j,n)}| < |a^{\phi(\bar{j},\bar{n})}| \leq c|a^{\phi(j,n)}|.$$

We conclude that $\overline{\Pi}$ is a c-geometrically growing language. Also it is clear that

$$|\Pi \setminus \overline{\Pi}| < \infty.$$

This ends the proof.

4 Dissecting by a Regular Language

Remark 4. We suppose the reader to be familiar with the regular languages and deterministic finite automata. As such, we present the proofs of Lemma 2 and Lemma 3 in a less formal way.

Given a deterministic finite automaton D, let $\mathcal{L}(D) \subseteq \Sigma^*$ denote the set of words that D accepts.

Let

$$\begin{aligned} \mathrm{REG}_1 = \{R \subseteq \Sigma^* \mid |R| = \infty \text{ and there is a deterministic finite} \\ \text{automaton } D \text{ such that } \mathcal{L}(D) = R \\ \text{and } D \text{ has exactly one accepting state}\}. \end{aligned}$$

Less formally said, REG_1 is the set of all infinite regular languages that can be accepted by a deterministic finite automaton having exactly one accepting state. The next lemma illuminates the properties of such regular languages.

Lemma 2. *If $R \in \mathrm{REG}_1$ then there are $q \in \mathbb{N}_0$ and $r \in \mathbb{N}_1$ such that*

$$R = \{a^{q+ir} \mid i \in \mathbb{N}_1\}.$$

Proof. Consider a deterministic finite automaton D accepting R such that D has exactly one accepting state. Since $R \in \mathrm{REG}_1$ we have that such D exists. Since the alphabet Σ has only one letter, it is easy to see that D can be represented as a directed graph with $q+r$ vertices, where r vertices form a cycle. Since $|R| = \infty$, the cycle contains the single accepting state of D. The lemma follows.

This ends the proof.

We show that when investigating the dissecting power of regular languages on one letter alphabet, it suffices to consider regular languages from REG_1.

Lemma 3. *If a regular language $R \subseteq \Sigma^*$ dissects a language $L \subseteq \Sigma^*$ then there is $\overline{R} \in \mathrm{REG}_1$ such that $\overline{R}$ dissects L.*

Proof. Let D be a deterministic finite automaton accepting R and let T be the set of accepting states of D. Given an accepting state $t \in T$, we define

$$R(t) = \{u \in R \mid \text{ Given the input } u, D \text{ halts in the state } t\}.$$

Since $|R| = \infty$, it is clear that if R dissects L then obviously there is at least one accepting state $t \in T$ such that $|R(t) \cap L| = \infty$. Since $|L \setminus R| = \infty$ and $R(t) \subseteq R$, we have that $|L \setminus R(t)| = \infty$. It follows that $R(t)$ dissects L. The definition of $R(t)$ implies that $R(t) \in \mathrm{REG}_1$. The lemma follows.

This ends the proof.

Equipped with Lemma 2 and Lemma 3 we can prove that there is no regular language R such R dissects Π.

Proposition 2. *The language Π is not REG-dissectible.*

Proof. Suppose that the language Π is REG-dissectible. Then Lemma 2 and Lemma 3 imply that there are $q, r \in \mathbb{N}_0$ and $R \in \text{REG}_1$ such that

$$R = \{a^{q+ir} \mid i \in \mathbb{N}_1\}$$

and R dissects Π.

Let n_0 be such that $n_0 - \alpha\lfloor\gamma \ln n_0\rfloor - 1 > r$. Realize that

$$\lim_{n\to\infty} (n - \alpha\lfloor\gamma \ln n\rfloor) = \infty,$$

hence such n_0 exists. Obviously $n - \alpha\lfloor\gamma \ln n\rfloor - 1 > r$ for all $n > n_0$. It follows that for every $n > n_0$ there is

$$m \in \{\alpha\lfloor\gamma \ln n\rfloor + 2, \alpha\lfloor\gamma \ln n\rfloor + 3, \dots, n\}$$

such that $0 \equiv m \pmod r$. In consequence, we have that

$$0 \equiv \omega(n) \pmod r.$$

Lemma 1 implies that

$$0 \equiv \phi(j, n) \pmod{\omega(n)}.$$

It follows that if $n > n_0$ then

$$0 \equiv \phi(j, n) \pmod r. \tag{6}$$

Let

$$\widehat{\Pi} = \{a^{\phi(j,n)} \mid (j, n) \in \Delta \text{ and } n > n_0\}.$$

Obviously $|\Pi \setminus \widehat{\Pi}| < \infty$ and (6) implies that

$$\widehat{\Pi} \subseteq \{a^{ir} \mid i \in \mathbb{N}_1\}.$$

We distinguish two following cases:

- If $0 \equiv q \pmod r$ then $\widehat{\Pi} \subseteq R$.
- If $0 \not\equiv q \pmod r$ then $\widehat{\Pi} \cap R = \emptyset$.

We conclude that both $\widehat{\Pi}$ and Π are not REG-dissectible.

This ends the proof.

Now we step to the main result of the current article.

Theorem 1. *For every $c \in \mathbb{R}^+$ with $c > 1$ there is a c-geometrically growing language L that is not* REG*-dissectible.*

Proof. Let $\alpha, \beta \in \mathbb{N}_1$ be such that $c > \frac{\beta}{\alpha}$ and $\beta > \alpha$. Since $c > 1$ we have that such α, β exist. Then Corollary 2 asserts that there is $\overline{\Pi} \subseteq \Pi$ such that $\overline{\Pi}$ is c-geometrically growing and $|\Pi \setminus \overline{\Pi}| < \infty$. From Proposition 2 we have that Π is not REG-dissectible. Then it is easy to see that for every $S \subseteq \Pi$ with $|\Pi \setminus S| < \infty$ we have that S is not REG-dissectible. Hence we conclude $\overline{\Pi}$ is not REG-dissectible.

This completes the proof.

References

1. Bucher, W.: A density problem for context-free languages. Bull. Eur. Assoc. Theor. Comput. Sci. EATCS 10 (1980)
2. Rukavicka, J.: Dissecting power of intersection of two context-free languages. Discrete Math. Theoret. Comput. Sci. **25**, 2 (2023)
3. Yamakami, T., Kato, Y.: The dissecting power of regular languages. Inf. Process. Lett. **113**, 116–122 (2013)
4. Domaratzki, M., Shallit, J., Yu, S.: Minimal covers of formal languages. In: Developments in Language Theory (2001)
5. Flajolet, P., Steyaert, J.M.: On sets having only hard subsets. In: Loeckx, J. (ed.) ICALP 1974. LNCS, vol. 14, pp. 446–457. Springer, Heidelberg (1974). https://doi.org/10.1007/978-3-662-21545-6_34
6. Julie, J., Baskar Babujee, J., Masilamani, V.: Dissecting power of certain matrix languages. In: Arumugam, S., Bagga, J., Beineke, L.W., Panda, B.S. (eds.) ICTCSDM 2016. LNCS, vol. 10398, pp. 98–105. Springer, Cham (2017). https://doi.org/10.1007/978-3-319-64419-6_13
7. Post, E.L.: Recursively enumerable sets of positive integers and their decision problems. Bull. Amer. Math. Soc. **50**, 284–316 (1944)
8. Yamakami, T.: Intersection and union hierarchies of deterministic context-free languages and pumping lemmas. In: Leporati, A., Martín-Vide, C., Shapira, D., Zandron, C., (eds.) Language and Automata Theory and Applications, pp. 341–353. Springer International Publishing, Cham (2020)

Author Index

G. Gamard and J. Leroy (Eds.): WORDS 2025, LNCS 15729, p. 239, 2025.
https://doi.org/10.1007/978-3-031-97548-6

The manufacturer's authorised representative in the EU is Springer Nature Customer Service Centre GmbH, Europaplatz 3, 69115 Heidelberg, Germany. If you have any concerns regarding our products, please contact ProductSafety@springernature.com

Printed and bound by CPI Group (UK) Ltd, Croydon, CR0 4YY
24/04/2026
02096364-0002